MULTIMEDIA MATHS

Bieke Masselis and Ivo De Pauw

Multimedia Maths

The original, Dutch version of this book was first published by LannooCampus in 2009.
This publication was entitled Wiskunde voor multimedia.

Chapter 1, David Ritter; 2, John Evans; 3, Wouter Verweirder; 4, Daryl Beggs, Juan Pablo Arancibia Medina; 5, Stephanie Berghaeuser; 6, Martin Walls; 7, 14, Wouter Tansens; 8, Danie Pratt; 9, Caetano Lacerda; 10, Ken Munyard; 11, Bieke Masselis; 12, Cornelia Roessing; 13, Sofie Eeckeman; p.21, p.91, Wouter Tansens; p.42, Wouter Verweirder; p.46, Leo Storme; p.250, p.265, Owen Barrett; p.261, p.267, Mark McClure; p.263, Felipe Dimer de Oliveira; p.264, Sy M. Blinder; p.268, Bart Gardin; p.290, Yu-Sung Chang.

D/2013/45/270 – ISBN 978 94 014 1066 3 – NUR 918

Layout: Jurgen Leemans, Peter Flynn and Conny Meuris
Cover design: Jan Middendorp, credits to Ellen Deketele

LannooCampus is part of the Lannoo Publishing Group

Publisher LannooCampus
Erasme Ruelensvest 179 bus 101
B - 3001 Leuven
Belgium
www.lannoocampus.com.

Content

Credits 11

Chapter 1 · Arithmetic Refresher 13

1.1 Algebra 14
Real Numbers 14
Real Polynomials 19
1.2 Equations in one variable 21
Linear Equations 21
Quadratic Equations 22
1.3 Exercises 28

Chapter 2 · Linear systems 31

2.1 Definitions 32
2.2 Methods for solving linear systems 34
Solving by substitution 34
Solving by elimination 35
2.3 Exercises 39

Chapter 3 · Trigonometry 41

3.1 Angles 42
3.2 Triangles 44
3.3 Right Triangle 48
3.4 Unit Circle 49
3.5 Special Angles 51
Trigonometric ratios for an angle of $45° = \frac{\pi}{4}$ rad 52
Trigonometric ratios for an angle of $30° = \frac{\pi}{6}$ rad 52
Trigonometric ratios for an angle of $60° = \frac{\pi}{3}$ rad 53
Overview 53
3.6 Pairs of Angles 54
3.7 Sum Identities 54
3.8 Inverse Trigonometric Functions 57
3.9 Exercises 59

Chapter 4 · Functions 61

4.1 Basic concepts on real functions 62

4.2	*Polynomial functions*	63
	Linear functions	63
	Quadratic functions	65
4.3	*Intersecting functions*	67
4.4	*Trigonometrical functions*	69
	Elementary sine function	69
	Generalized sine function	69
4.5	*Inverse trigonometrical functions*	73
4.6	*Exercises*	76

Chapter 5 · The Golden Section 79

5.1	*The Golden Number*	80
5.2	*The Golden Section*	82
	The Golden Triangle	82
	The Golden Rectangle	83
	The Golden Spiral	84
	The Golden Pentagon	86
	The Golden Ellipse	86
5.3	*Golden arithmetics*	87
	Golden Identities	87
	The Fibonacci Numbers	88
5.4	*The Golden Section worldwide*	90
5.5	*Exercises*	93

Chapter 6 · Coordinate systems 95

6.1	*Cartesian coordinates*	96
6.2	*Parametric curves*	96
6.3	*Polar coordinates*	99
6.4	*Polar curves*	102
	A polar superformula	103
6.5	*Exercises*	105

Chapter 7 · Vectors 107

7.1	*The concept of a vector*	108
	Vectors as arrows	108
	Vectors as arrays	109
	Free Vectors	112
	Base Vectors	112
7.2	*Addition of vectors*	112
	Vectors as arrows	113

Vectors as arrays 113
Vector addition summarized 113
7.3 Scalar multiplication of vectors 114
Vectors as arrows 115
Vectors as arrays 115
Scalar multiplication summarized 116
Properties 116
Vector subtraction 117
Decomposition of a plane vector 118
Base vectors defined 119
7.4 Dot product 120
Definition 120
Angle between vectors 121
Orthogonality 123
Vector components in 3D 124
7.5 Cross product 126
Definition 126
Parallelism 128
7.6 Normal vectors 129
7.7 Exercises 131

Chapter 8 · Parameters 133
8.1 Parametric equations 134
8.2 Vector equation of a line 135
8.3 Intersecting straight lines 139
8.4 Vector equation of a plane 141
8.5 Exercises 145

Chapter 9 · Collision detection 147
9.1 Collision detection and frame rate 148
9.2 Collision detection using circles and spheres 149
Circles and spheres 149
Intersecting line and circle 151
Intersecting circles and spheres 153
9.3 Collision detection using vectors 156
Location of a point with respect to other points 156
Altitude to a straight line 157
Altitude to a plane 159
Frame rate issues 161
Location of a point with respect to a polygon 162

9.4 Exercises 165

Chapter 10 · Matrices 167

10.1 The concept of a matrix 168
10.2 Determinant of a square matrix 169
10.3 Addition of matrices 171
10.4 Scalar multiplication of matrices 173
10.5 Transpose of a matrix 174
10.6 Dot product of matrices 174
Introduction 174
Condition 176
Definition 176
Properties 177
10.7 Inverse of a matrix 179
Introduction 179
Definition 179
Conditions 180
Row reduction 180
Matrix inversion 181
Inverse of a product 184
Solving systems of linear equations 185
10.8 The Fibonacci operator 187
10.9 Exercises 189

Chapter 11 · Linear transformations 191

11.1 Translation 192
11.2 Scaling 197
11.3 Rotation 200
Rotation in 2D 200
Rotation in 3D 202
11.4 Reflection 204
11.5 Shearing 205
11.6 Composing transformations 208
2D rotation around an arbitrary center 210
3D scaling about an arbitrary center 213
2D reflection over an axis through the origin 214
2D reflection over an arbitrary axis 215
3D combined rotation 218
11.7 Conventions 219
11.8 Exercises 220

Chapter 12 · Hypercomplex numbers 223
12.1 Complex numbers 224
12.2 Complex number arithmetics 227
Complex conjugate 227
Addition and subtraction 228
Multiplication 229
Division 231
12.3 Complex numbers and transformations 233
12.4 Complex continuation of the Fibonacci numbers 235
Integer Fibonacci numbers 235
Complex Fibonacci numbers 236
12.5 Quaternions 237
12.6 Quaternion arithmetics 238
Addition and subtraction 239
Multiplication 239
Quaternion conjugate 241
Inverse quaternion 242
12.7 Quaternions and rotation 242
12.8 Exercises 247

Chapter 13 · Fractals 249
13.1 The concept of a fractal 250
The Sierpinski Gasket 251
The Koch Snowflake 251
The Minkowski Island 252
The Cantor set 253
The Pythagoras Tree 253
13.2 Self-similarity 254
13.3 Fractal dimension 258
Euclidean dimension 258
Hausdorff dimension 258
The concept of a logarithm 259
Illustrations 259
13.4 The Mandelbrot and Julia Sets 260
Dynamical systems 260
The Mandelbrot Set 262
The Julia Sets 263
13.5 Exercises 268

Chapter 14 · Bezier curves 271
14.1 Vector equation of segments 272
Linear Bezier segment 272
Quadratic Bezier segment 273
Cubic Bezier segment 274
Bezier segments of higher degree 276
14.2 De Casteljau algorithm 277
14.3 Bezier curves 278
Concatenation 278
Linear transformations 280
Illustrations 280
14.4 Matrix representation 282
Linear Bezier segment 282
Quadratic Bezier segment 283
Cubic Bezier segment 284
14.5 B-splines 286
Cubic B-splines 286
Matrix representation 287
De Boor's algorithm 289
14.6 Exercises 291

Annex A · Real numbers in computers 293
A.1 Scientific notation 293
A.2 The decimal computer 293
A.3 Special values 294

Annex B · Notations and Conventions 295
B.1 Alphabets 295
Latin alphabet 295
Greek alphabet 295
B.2 Mathematical symbols 296
Sets 296
Mathematical symbols 297
Mathematical keywords 297
Numbers 298

Annex C · Companion website 299
C.1 Interactivities 299
C.2 Solutions 299
Bibliography 300
Index 303

Credits

We hereby insist to thank a lot of people who made this book possible: Prof. Dr. Leo Storme, Wim Serras, Wouter Tansens, Wouter Verweirder, Koen Samyn, Hilde De Maesschalck, Ellen Deketele, Conny Meuris, Hans Ameel, Dr. Rolf Mertig, Dick Verkerck, ir. Gose Fischer, Prof. Dr. Fred Simons, Sofie Eeckeman, Dr. Luc Gheysens, Dr. Bavo Langerock, Wauter Leenknecht, Marijn Verspecht, Sarah Rommens, Prof. Dr. Marcus Greferath, Dr. Cornelia Roessing, Tim De Langhe, Tim Vandendriessche, Peter Saerens, Peter Flynn, Jurgen Leemans, Jan Middendorp, Hilde Vanmechelen, Jef De Langhe, Ann Deraedt, Rita Vanmeirhaeghe, Prof. Dr. Jan Van Geel, Dr. Ann Dumoulin, Bart Uyttenhove, Rik Leenknegt, Peter Verswyvelen, Roel Vandommele, ir. Lode De Geyter, Bart Leenknegt, Olivier Rysman, ir. Johan Gielis, Frederik Jacques, Kristel Balcaen, ir. Wouter Gevaert, Bart Gardin, Dieter Roobrouck, Dr. Yu-Sung Chang (*WolframDemonstrations*), Prof. Dr. Sy Blinder (*WolframDemonstrations*), Prof. Dr. Mark McClure (*WolframDemonstrations*), Dr. Felipe Dimer de Oliveira (*WolframDemonstrations*) and anyone whom we might have forgotten!

Chapter 1 · Arithmetic Refresher

As this chapter offers all necessary mathematical skills for a full mastering of all further topics explained in this book, we strongly recommend it. To serve its purpose, the successive paragraphs below refresh some required aspects of mathematical language as used on the applied level.

1.1 Algebra

Real Numbers

We typeset the set of:

- ▷ natural numbers (unsigned integers) as $\mathbb{N}$ including zero,
- ▷ integer numbers as $\mathbb{Z}$ including zero,
- ▷ rational numbers as $\mathbb{Q}$ including zero,
- ▷ real numbers (floats) as $\mathbb{R}$ including zero.

All the above make a chain of subsets: $\mathbb{N} \subset \mathbb{Z} \subset \mathbb{Q} \subset \mathbb{R}$.

To avoid possible confusion, we outline a brief glossary of mathematical terms. We recall that using the correct mathematical terms reflects a correct mathematical thinking. Putting down ideas in the correct words is of major importance for a profound insight.

Sets

- ▷ We recall writing all **subsets** in between braces, e.g. the **empty set** appears as $\{\}$.
- ▷ We define a **singleton** as any subset containing only one element, e.g. $\{5\} \subset \mathbb{N}$, as a subset of natural numbers.
- ▷ We define a **pair** as any subset containing just two elements, e.g. $\{115, -4\} \subset \mathbb{Z}$, as a subset of integers. In programming the boolean values *true* and *false* make up a pair $\{true, false\}$ called the boolean set which we typeset as $\mathbb{B}$.
- ▷ We define $\mathbb{Z}^- = \{\ldots, -3, -2, -1\}$ whenever we need negative integers only. We express symbolically that -1234 is an **element** of $\mathbb{Z}^-$ by typesetting $-1234 \in \mathbb{Z}^-$.
- ▷ We typeset the **setminus** operator to delete elements from a set by using a backslash, e.g. $\mathbb{N} \setminus \{0\}$ reading all natural numbers except zero, $\mathbb{Q} \setminus \mathbb{Z}$ meaning all pure rational numbers after all integer values left out and $\mathbb{R} \setminus \{0, 1\}$ expressing all real numbers apart from zero and one.

Calculation basics

operation	**example**	a	b	c
to add	$a+b=c$	term	term	sum
to subtract	$a-b=c$	term	term	difference
to multiply	$a \cdot b=c$	factor	factor	product
to divide	$\frac{a}{b}=c, b \neq 0$	numerator	divisor or denominator	quotient or fraction
to exponentiate	$a^b=c$	base	exponent	power
to take root	$\sqrt[b]{a}=c$	radicand	index	radical

We write the **opposite** of a real number r as $-r$, defined by the sum $r+(-r)=0$. We typeset the **reciprocal** of a nonzero real number r as $\frac{1}{r}$ or r^{-1}, defined by the product $r \cdot r^{-1}=1$.

We define **subtraction** as equivalent to adding the opposite: $a-b=a+(-b)$. We define **division** as equivalent to multiplying with the reciprocal: $a:b=a \cdot b^{-1}$.

When we mix operations we need to apply priority rules for them. There is a fixed priority list 'PEMDAS' in performing mixed operations in $\mathbb{R}$ that can easily be memorized by 'Please Excuse My Dear Aunt Sally'.

- ▷ First process all that is delimited in between Parentheses,
- ▷ then Exponentiate,
- ▷ then Multiply and Divide from left to right,
- ▷ finally Add and Subtract from left to right.

Now we discuss the **distributive law** ruling within $\mathbb{R}$, which we define as threading a 'superior' operation over an 'inferior' operation. Conclusively, distributing requires two *different* operations.

Hence we distribute *exponentiating* over *multiplication* as in $(a \cdot b)^3 = a^3 \cdot b^3$. Likewise rules *multiplying* over *addition* as in $3 \cdot (a+b) = 3 \cdot a + 3 \cdot b$.

However we should never stumble on this 'Stairway of Distributivity' by going too fast:

$$(a+b)^3 \neq a^3 + b^3,$$

$$\sqrt{a+b} \neq \sqrt{a} + \sqrt{b},$$

$$\sqrt{x^2 + y^2} \neq x + y.$$

Fractions

A **fraction** is what we call any rational number written as $\frac{t}{n}$ given $t, n \in \mathbb{Z}$ and $n \neq 0$, wherein t is called the **numerator** and n the **denominator**. We define the reciprocal of a nonzero fraction $\frac{t}{n}$ as $\frac{1}{\frac{t}{n}} = \frac{n}{t}$ or as the power $\left(\frac{t}{n}\right)^{-1}$. We define the opposite fraction as $-\frac{t}{n} = \frac{-t}{n} = \frac{t}{-n}$. We summarize fractional arithmetics:

sum	$\frac{t}{n} + \frac{a}{b} = \frac{t \cdot b + n \cdot a}{n \cdot b}$,
difference	$\frac{t}{n} - \frac{a}{b} = \frac{t \cdot b - n \cdot a}{n \cdot b}$,
product	$\frac{t}{n} \cdot \frac{a}{b} = \frac{t \cdot a}{n \cdot b}$,
division	$\frac{\frac{t}{n}}{\frac{a}{b}} = \frac{t}{n} \cdot \frac{b}{a}$,
exponentiation	$\left(\frac{t}{n}\right)^m = \frac{t^m}{n^m}$,
singular fractions	$\frac{1}{0} = \pm\infty$ infinity,
	$\frac{0}{0} = ?$ undetermined.

Powers

We define a **power** as any real number written as g^m, wherein g is called its **base** and m its **exponent**. The opposite of g^m is simply $-g^m$. The reciprocal of g^m is $\frac{1}{g^m} = g^{-m}$, given $g \neq 0$.

According to the exponent type we distinguish between:

$$g^3 = g \cdot g \cdot g \qquad 3 \in \mathbb{N},$$
$$g^{-3} = \tfrac{1}{g^3} = \tfrac{1}{g \cdot g \cdot g} \qquad -3 \in \mathbb{Z},$$
$$g^{\frac{1}{3}} = \sqrt[3]{g} = w \Leftrightarrow w^3 = g \qquad \tfrac{1}{3} \in \mathbb{Q},$$
$$g^0 = 1 \qquad g \neq 0.$$

Whilst calculating powers we may have to:

multiply	$g^3 \cdot g^2 = g^{3+2} = g^5$,
divide	$\frac{g^3}{g^2} = g^3 \cdot g^{-2} = g^{3-2} = g^1$,
exponentiate	$(g^3)^2 = g^{3 \cdot 2} = g^6$ them.

We insist on avoiding typesetting radicals like $\sqrt[7]{g^3}$ and strongly recommend their contemporary notation using radicand g and exponent $\frac{3}{7}$, consequently exponentiating g to $g^{\frac{3}{7}}$. We recall the fact that all square roots are non-negative numbers, $\sqrt{a} = a^{\frac{1}{2}} \in \mathbb{R}^+$ for $a \in \mathbb{R}^+$.

As well knowing the above exponent types as understanding the above rules to calculate them are inevitable to use powers successfully. We advise memorizing the integer squares running from $1^2 = 1$, $2^2 = 4$, ..., up to $15^2 = 225$, $16^2 = 256$ and the integer cubes running from $1^3 = 1$, $2^3 = 8$, ..., up to $7^3 = 343$, $8^3 = 512$ in order to easily recognize them.

Recall that the only way out of any power is exponentiating with its reciprocal exponent. For this purpose we need to exponentiate both left hand side and right hand side of any given relation (see also paragraph 1.2).

Example: Find x when $\sqrt[7]{x^3} = 5$ by exponentiating this power.

$$x^{\frac{3}{7}} = 5 \Longleftrightarrow \left(x^{\frac{3}{7}}\right)^{\frac{7}{3}} = (5)^{\frac{7}{3}} \Longleftrightarrow x \approx 42.7494.$$

We emphasize the above strategy as the only successful one to free base x from its exponent, yielding its correct expression numerically approximated if we like to.

Example: Find x when $x^2 = 5$ by exponentiating this power.

$$x^2 = 5 \Longleftrightarrow (x^2)^{\frac{1}{2}} = (5)^{\frac{1}{2}} \text{ or } -(5)^{\frac{1}{2}} \Longleftrightarrow x \approx 2.23607 \text{ or} -2.23607.$$

We recall the above double solution whenever we free base x from an *even* exponent, yielding their correct expression as accurate as we like to.

Mathematical expressions

Composed mathematical expressions can often seem intimidating or cause confusion. To gain transparancy in them, we firstly recall indexed variables which we define as subscripted to count them: $x_1, x_2, x_3, x_4, \ldots, x_{99999}, x_{100000}, \ldots$, and $\alpha_0, \alpha_1, \alpha_2, \alpha_3, \alpha_4, \ldots$. It is common practice in industrial research to use thousands of variables, so just picking unindexed characters would be insufficient. Taking our own alphabet as an example, it would only provide us with 26 characters.

We define finite expressions as composed of (mathematical) operations on objects (numbers, variables or structures). We can for instance analyze the expression $(3a+x)^4$ by drawing its **tree form**. This example reveals a Power having exponent 4 and a subexpression in its base. The base itself yields a sum of the variable x Plus another subexpression. This final subexpression shows the product 3 Times a.

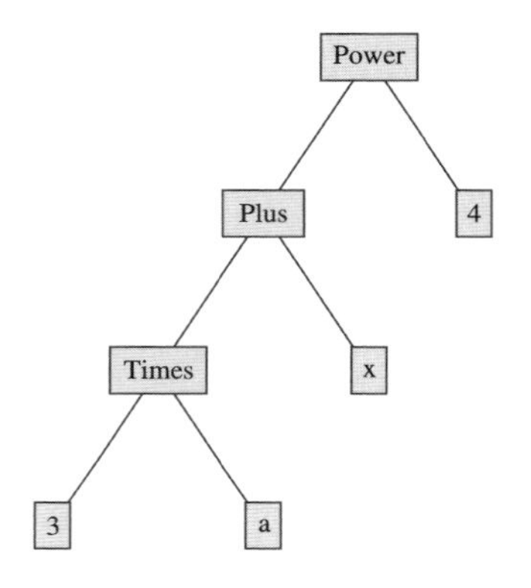

Let us also evaluate this expression $(3a+x)^4$. Say $a = 1$, then we see our expression partly collaps to $(3+x)^4$. If we on top of this assign $x = 2$, our expression then finally turns to the numerical value $(3+2)^4 = 5^4 = 625$.

When we expand this power to its **pure sum expression** $81a^4 + 108a^3x + 54a^2x^2 + 12ax^3 + x^4$, we did nothing but *reshape* its **pure product expression** $(3a+x)^4$.

We warn that trying to solve this expression - which is not a relation - is completely in vain. Recall that inequalities, equations and systems of equations or inequalities are the only objects in the universe we can (try to) solve mathematically.

Relational operators

We also refresh the use of correct terms for inequalities and equations.

We define an **inequality** as any *variable* expression comparing a left hand side to a right hand side by applying the 'is-(strictly)-less-than' or by applying the 'is-(strictly)-greater-than' operator. For example, we can read $(3a+x)^4 \leqslant (b+4)(x+3)$ containing variables *a, x, b*. Consequently we may solve such inequality for any of the unknown quantities a, x or b.

We define an **equation** as any *variable* expression comparing a left hand side to a right hand side by applying the 'is-equal-to' operator. For example $(3a+x)^4 = (b+4)(x+3)$ is an equation containing variables *a, x, b*. Consequently we also may solve equations for any of the unknown quantities a, x or b.

We define an **equality** as a constant relational expression being *true*, e.g. $7 = 7$. We define a **contradiction** as a constant relational expression being *false*, e.g. $-10 > 5$.

Real Polynomials

We elaborate upon the mathematical environment of polynomials over the real numbers in their variable or indeterminate x, a set we denote with $\mathbb{R}_{[x]}$.

▷ Monomials

We define a **monomial** in x as any product ax^n, given $a \in \mathbb{R}$ and $n \in \mathbb{N}$. We can extend this concept to several indeterminates $x, y, z, \ldots$ like the monomials $3(xy)^6$ and $3(x^2y^3z^6)$ are.

We define the **degree** of a monomial ax^n as its natural exponent $n \in \mathbb{N}$ to the **indeterminate part** x. We say constant numbers are monomials of degree 0 and linear terms are monomials of degree 1. We say squares to have degree 2 and cubes to have degree 3, followed by monomials of higher degree.

For instance the real monomial $-\sqrt{12}x^6$ is of degree 6. Extending this concept, the monomial $3(xy)^6$ is of degree 6 in xy and the monomial $3(x^2y^3z^6)^9$ is of degree 9 in $x^2y^3z^6$.

We define **monomials of the same kind** as those having an identical indeterminate part. For instance both $\frac{5}{7}x^6$ and $-\sqrt{12}x^6$ are of the same kind. Extending the concept, likewise $\frac{5}{7}x^3y^5z^2$ and $-\sqrt{12}x^3y^5z^2$ are of the same kind.

All basic operations on monomials emerge simply from applying the calculation rules of fractions and powers.

▷ Polynomials

We define a **polynomial** $V(x)$ as any sum of monomials. We define the **degree** of $V(x)$ as the maximal exponent $m \in \mathbb{N}$ to the indeterminate variable x. For instance the real polynomial

$$V(x) = 17x^2 + \frac{1}{4}x^3 + 6x - 7x^2 - \sqrt{12}x^6 - 13x - 1,$$

is of degree 6.

Whenever monomials of the same kind appear in it, we can simplify the polynomial. For instance our polynomial simplifies to $V(x) = 10x^2 + \frac{1}{4}x^3 - 7x - \sqrt{12}x^6 - 1$.

Moreover, we can sort any given polynomial either in an ascending or descending way according to its powers in x. Sorting our polynomial $V(x)$ in an ascending way yields $V(x) = -1 - 7x + 10x^2 + \frac{1}{4}x^3 - \sqrt{12}x^6$. Sorting $V(x)$ in a descending way yields $V(x) = -\sqrt{12}x^6 + \frac{1}{4}x^3 + 10x^2 - 7x - 1$.

Eventually we are able to evaluate any polynomial, getting a numerical value from it. For instance evaluating $V(x)$ in $x = -1$, yields $V(-1) = -\sqrt{12}(-1)^6 + \frac{1}{4}(-1)^3 + 10(-1)^2 - 7(-1) - 1 = -\sqrt{12} - \frac{1}{4} + 16 = \frac{63}{4} - 2\sqrt{3} \in \mathbb{R}$.

▷ Basic operations

Adding two monomials of the same kind: we add their coefficients and keep their indeterminate part

$$5a^2 - 3a^2 = (5-3)a^2 = 2a^2.$$

Multiplying two monomials of any kind: we multiply both their coefficients and their indeterminate parts

$$-5ab \cdot \frac{7}{4}a^2b^3 = -5 \cdot \frac{7}{4} \cdot a^{1+2}b^{1+3} = \frac{-35}{4}a^3b^4.$$

Dividing two monomials: we divide both their coefficients and their indeterminate parts

$$\frac{-8a^6b^4}{-4a^4} = \frac{-8}{-4}\,a^{6-4}b^{4-0} = 2a^2b^4.$$

Exponentiating a monomial: we exponentiate each and every factor in the monomial

$$\left(-2a^2b^4\right)^3 = (-2)^3(a^2)^3(b^4)^3 = -8a^6b^{12}.$$

Adding or subtracting polynomials: we add or subtract all monomials of the same kind

$$(x^2 - 4x + 8) - (2x^2 - 3x - 1) = x^2 - 4x + 8 - 2x^2 + 3x + 1 = -x^2 - x + 9.$$

Multiplying two polynomials: we multiply each monomial of the first polynomial with each monomial of the second polynomial and simplify all those products to the resulting product polynomial

$$\begin{aligned}(2x^2 + 3y) \cdot (4x^2 - y) &= 2x^2(4x^2 - y) + 3y(4x^2 - y)\\ &= 2x^2 \cdot 4x^2 + 2x^2 \cdot (-y) + 3y \cdot 4x^2\\ &\qquad + 3y \cdot (-y)\\ &= 8x^4 - 2x^2y + 12x^2y - 3y^2\\ &= 8x^4 + 10x^2y - 3y^2.\end{aligned}$$

1.2 Equations in one variable

Anticipating this paragraph we refresh some vocabulary for it. A **solution** is any value assigned to the variable that turns the given equation into an *equality* (being *true*). The **scope** of an equation is any number set in which the equation resides, realizing it will be most likely $\mathbb{R}$. We define the **solution set** as the set containing all legal solutions to an equation. This solution set always is a subset of the scope of the equation.

Linear Equations

A **linear equation** is an algebraic equation of degree one, referring to the maximum natural exponent of the unknown quantity. By simplifying we can always standardize any linear equation to

$$ax + b = 0, \tag{1.1}$$

given $a \in \mathbb{R} \setminus \{0\}$ and $b \in \mathbb{R}$. We cite $3x+7=22, 5x-9d=c$ and $5(x-4)+x=-2(x+2)$ as examples of linear equations, and $3x^2+7=22$ and $5ab-9b=c$ as counterexamples. The adjective 'linear' originates from the Latin word 'linea' meaning (straight) line as referring to the graph of a linear function (see chapter 4).

We solve a linear equation for its unknown part by rewriting the entire equation until its shape exposes the solution explicitly.

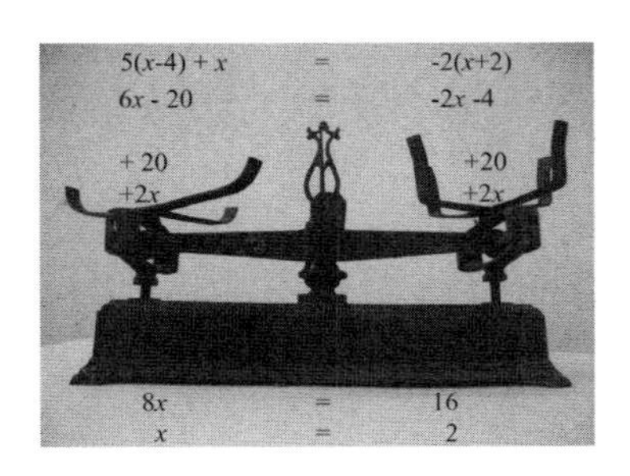

We recall easily the required rules for rewriting a linear equation by the metaphor denoting a linear equation as a 'pair of scales'. This way we should never forget to keep the equation's balance: whatever operation we apply, it has to act on both sides of the equals-sign. If we add (or subtract) to the left hand 'scale' than we are obliged to add (or subtract) the same term to the right hand 'scale'. If we multiply (or divide) the left hand side, than we are likewise obliged to multiply (or divide) the right hand side with the same factor. If not, our equation would loose its balance just like a pair of scales would. We realize that our metaphor covers all usual 'rules' to handle linear equations.

The reason we perform certain rewrite steps depends on which variable we are aiming for. This is called *strategy*. Solving the equation for a different variable implies a different sequence of rewrite steps.

Example: We solve the equation $5(x-4)+x=-2(x+2)$ for x. Firstly, we apply the distributive law: $5x-20+x=-2x-4$. Secondly, we put all terms dependent of x to the left hand side and the constant numbers to the right hand side $5x+x+2x=-4+20$. Thirdly, we simplify both sides $8x=16$. Finally, we find $x=2$ leading to the solution singleton $\{2\}$.

Quadratic Equations

Handling quadratic expressions and solving quadratic equations are useful basics in order to study topics in multimedia, digital art and technology.

▷ Expanding products

We refresh **expanding** a product as (repeatedly) applying the distributive law until the initial expression ends up as a pure *sum* of terms. Note that our given polynomial $V(x)$ itself does not change: we just shift its appearance to a pure sum. We illustrate this concept through $V(x)=(2x-3)(4-x)$.

$$\begin{aligned}(2x-3)(4-x) &= (2x-3)\cdot 4+(2x-3)\cdot(-x)\\ &= (8x-12)+(-2x^2+3x)\\ &= -2x^2+11x?12.\end{aligned}$$

Other examples are

$$5a(2a^2-3b)=5a\cdot 2a^2-5a\cdot 3b=10a^3-15ab$$

and

$$\begin{aligned}4\left(x-\frac{1}{2}\right)\left(x+\frac{13}{2}\right) &= (4x-2)\left(x+\frac{13}{2}\right)\\ &= (4x-2)\cdot x+(4x-2)\cdot\frac{13}{2}\\ &= 4x^2-2x+26x-13=4x^2+24x-13.\end{aligned}$$

▷ Factoring polynomials

We define **factoring** a polynomial as decomposing it into a pure *product* of (as many as possible) factors. Note that our given polynomial $V(x)$ itself does not change: we just shift its appearance to a pure product. Our **trinomial** $V(x)=-2x^2+11x-12$ just shifts its appearance to the pure product $V(x)=(2x-3)(4-x)$ when factored. It merely shows that the product $(2x-3)(4-x)$ is a factorization of the trinomial $-2x^2+11x-12$.

Imagine we had to factor the trinomial $-2x^2+11x-12$ without any hint. This way we realize that factoring generally is a hard job to do. Especially because we do not have any clue about which factors build up the pure product for a polynomial. Many questions arise: *how many* factors to expect, *where* to start from, *what* is the opening step towards factorization?

We observe the need for at least a minimum asset of factoring methods. As an extra motivation to it, we emphasize the importance of factoring as it reveals all essential building blocks of any polynomial. Knowing the roots of a polynomial gives us a deeper insight. We therefore introduce some factoring basics in the next paragraphs.

Common Factor

We show how to separate common factors if they appear.

For instance $6+12x = 6\cdot 1 + 6\cdot(2x) = 6\cdot(1+2x)$ results into a pure product of a number and a linear factor by separating the common factor 6. Another polynomial like $5x+x^2 = 5x+xx = (5+x)\cdot x$ separates into two linear factors by the use of the common factor x. An example expression like $39x+3xy = 3\cdot 13x+3xy = 3\cdot x\cdot(13+y)$ yields a pure product of a number factor, a linear factor in x and a linear factor in y by separating the common factors 3 and x. Occasionally we may have to factor by grouping. For instance

$$\begin{aligned} 1+x+x^2+x^3 &= (1+x)+(x^2+x^3) = (1+x)+(x^2\cdot 1+x^2\cdot x) \\ &= (1+x)+x^2(1+x) = 1\cdot(1+x)+x^2(1+x) \\ &= (1+x^2)\cdot(1+x) \end{aligned}$$

results stepwise into a pure product of a quadratic and a linear factor in x.

Perfect Products

Expanding natural powers of the **binomial** $A+B$ reveals their corresponding pure sum shapes.

$$\begin{aligned} (A+B)^2 &= (A+B)\ (A+B) = A^2+2AB+B^2 \\ (A+B)^3 &= (A+B)^2(A+B) = A^3+3A^2B+3AB^2+B^3 \\ (A+B)^4 &= (A+B)^3(A+B) = A^4+4A^3B+6A^2B^2+4AB^3+B^4 \end{aligned}$$

We define a **perfect product** as any natural exponentiation of a binomial. The important product $(A+B)^2$ we define as the **perfect square** of its binomial $A+B$.

Those perfect products of $A+B$, when ordered to ascending natural exponents, display **Pascal's Triangle** for all $n \in \mathbb{N}$.

$$\begin{array}{c} 1 \\ 1A+1B \\ 1A^2+2AB+1B^2 \\ 1A^3+3A^2B+3AB^2+1B^3 \\ 1A^4+4A^3B+6A^2B^2+4AB^3+1B^4 \\ 1A^5+5A^4B+10A^3B^2+10A^2B^3+5AB^4+1B^5 \\ \vdots \end{array}$$

Notice how a **coefficient** is produced as a sum of its upper two, leading to a symmetric triangle of numbers with the constant '1' on both edges. This 'triangle' is named after its explorer **Blaise Pascal** (1623-1662).

Despite the diminishing need for perfect products in this century of ruling computing power, we do advice to know at least the perfect square by heart.

To put the perfect square *in words*: *'The square of a binomial equals the sum of both squares plus two times the product'*.

$$(A+B)^2 = A^2 + 2AB + B^2. \tag{1.2}$$

We provide a visual aid for better memorizing it. The area of the total square equals $(A+B)\cdot(A+B)$. Alternatively we puzzle this area piece by piece, via adding both white square areas A^2 and B^2 plus the *two* grey rectangular areas AB, jointly equalling the perfect square as the trinomial A^2+B^2+2AB.

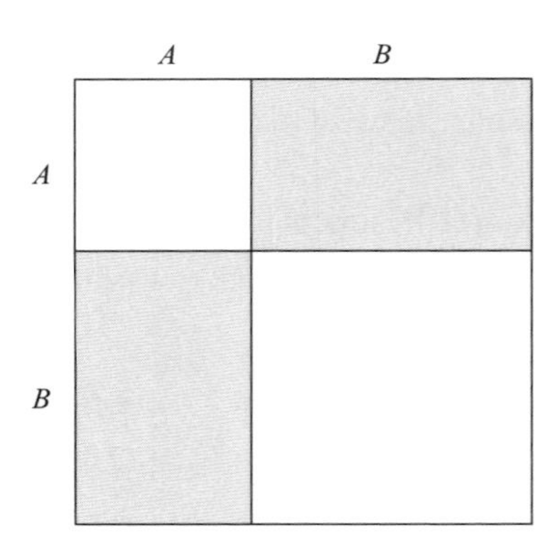

Consequently we now can explore a new factoring method. For instance we intend to factor the trinomial $1-2x+x^2$, whilst we have no guarantee for its pure product shape to even exist.

Strategically we perform two subsequent checks.

1) Verify whether both squares carry the same sign.

2) Then find $2AB$ corresponding correctly to the given A and B.

Only when both checks hold, we are able to shift the given trinomial to its perfect product $(A+B)^2$. We give an example of this strategy to the trionomial $1-2x+x^2$.

We rewrite the trinomial to $+(-1)^2 - 2x + (x)^2$ by assigning $A = -1$ and $B = x$. By substituting A and B into $2AB$, we find $+(-1)^2 + 2(-1)(x) + (x)^2$ equalling our trinomial. Therefore we confirm $A = -1$ and $B = x$, which allows us for the shift $1 - 2x + x^2 = (-1 + x)^2$. We realize that alternatively $(+1 - x)^2$ is a correct factorization as well.

Perfect Quotient

$$(A+B)(A-B) = A^2 - B^2 \tag{1.3}$$

▷ Quadratic formula for quadratic equations

A **quadratic equation** is an algebraic equation of degree two in the unknown quantity x that can be reduced to the default shape

$$ax^2 + bx + c = 0 \tag{1.4}$$

given $a \in \mathbb{R} \setminus \{0\}$ and $b, c \in \mathbb{R}$.

To solve this equation for x we firstly divide both sides of it by a . Dividing by a is valid since $a \neq 0$. In case of $a = 0$ we would have no longer a quadratic but a linear equation.

$$ax^2 + bx + c = 0 \Longleftrightarrow x^2 + \frac{b}{a}x + \frac{c}{a} = 0$$

Secondly we aim for a perfect square by adding and subtracting the special term $\left(\frac{b}{2a}\right)^2$ which is again valid since this is equivalent to adding 0. This way we have created a perfect square $(A+B)^2$, assigning $A = x$ and $B = \frac{b}{2a}$.

$$\begin{aligned}
& x^2 + \mathbf{2}\left(\frac{b}{\mathbf{2}a}\right)x + \left(\frac{\mathbf{b}}{\mathbf{2a}}\right)^{\mathbf{2}} - \left(\frac{\mathbf{b}}{\mathbf{2a}}\right)^{\mathbf{2}} + \frac{c}{a} = 0 \\
\Longleftrightarrow & \left(x^2 + \mathbf{2}\left(\frac{b}{\mathbf{2}a}\right)x + \left(\frac{\mathbf{b}}{\mathbf{2a}}\right)^{\mathbf{2}}\right) - \left(\frac{\mathbf{b}}{\mathbf{2a}}\right)^{\mathbf{2}} + \frac{c}{a} = 0 \\
\Longleftrightarrow & \left(x + \frac{b}{2a}\right)^2 - \left(\frac{b}{2a}\right)^2 + \frac{c}{a} = 0 \\
\Longleftrightarrow & \left(x + \frac{b}{2a}\right)^2 = \left(\frac{b}{2a}\right)^2 - \frac{c}{a}
\end{aligned}$$

The left hand side of this equation is by now a square. Before proceeding we make sure that all denominators of the right hand side are equal.

$$\left(x + \frac{b}{2a}\right)^2 = \frac{b^2}{4a^2} - \frac{c \cdot 4a}{a \cdot 4a} \Longleftrightarrow \left(x + \frac{b}{2a}\right)^2 = \frac{b^2 - 4ac}{4a^2}$$

Arithmetically holds: $L^2 = R \Leftrightarrow L = \sqrt{R}$ or $L = -\sqrt{R}$. Hence we reach two similar solutions to our equation:

$$x + \frac{b}{2a} = \sqrt{\frac{b^2 - 4ac}{4a^2}} \quad \text{or} \quad x + \frac{b}{2a} = -\sqrt{\frac{b^2 - 4ac}{4a^2}}$$

$$\Downarrow$$

$$x + \frac{b}{2a} = \frac{\sqrt{b^2 - 4ac}}{2a} \quad \text{or} \quad x + \frac{b}{2a} = -\frac{\sqrt{b^2 - 4ac}}{2a}$$

$$\Downarrow$$

$$x = -\frac{b}{2a} + \frac{\sqrt{b^2 - 4ac}}{2a} \quad \text{or} \quad x = -\frac{b}{2a} - \frac{\sqrt{b^2 - 4ac}}{2a}$$

The **discriminant** of a quadratic equation $ax^2 + bx + c = 0$, given $a \neq 0$, is the real number to be calculated as

$$D = b^2 - 4ac. \tag{1.5}$$

Furthermore, we solve $ax^2 + bx + c = 0$ for x like this:

if $D < 0$ then there are no solutions in $\mathbb{R}$,

if $D = 0$ we find one real root $x_1 = \frac{-b}{2a}$,

if $D > 0$ we have two similar **roots**

$$x_1 = \frac{-b + \sqrt{D}}{2a} \text{ and } x_2 = \frac{-b - \sqrt{D}}{2a}. \tag{1.6}$$

As a spin-off these roots enable factoring the default left hand side as

$$ax^2 + bx + c = a(x - x_1)(x - x_2) \text{ when } D > 0 \tag{1.7}$$

and as

$$ax^2 + bx + c = a(x - x_1)^2 \text{ when } D = 0. \tag{1.8}$$

Examples: Solving the quadratic equation $-2x^2 + 11x - 12 = 0$ for x, we firstly calculate its discriminant $D = 11^2 - 4 \cdot (-2) \cdot (-12) = 25$ to subsequently determine its roots as $x_1 = \frac{-11+\sqrt{25}}{2\cdot(-2)} = \frac{3}{2}$ and $x_2 = \frac{-11-\sqrt{25}}{2\cdot(-2)} = 4$. As a bonus, this allows us to factor $-2x^2 + 11x - 12$ as $(-2)\left(x - \frac{3}{2}\right)(x - 4)$. The solution set is the pair $\{\frac{3}{2}, 4\} \subset \mathbb{R}$.

We solve $25x^2 - 60x + 36 = 0$ for x as a next example. In this case the discriminant equals zero, yielding a unique root of multiplicity 2 to be found as $x = \frac{-(-60)\pm\sqrt{0}}{2\cdot 25} = \frac{6}{5}$ and thus leading to the solution singleton $\{\frac{6}{5}\} \subset \mathbb{R}$.

Finally solving also $25x^2 + 49x + 36 = 0$ for x, we calculate its discriminant as $D = -1199$. We can impossibly find any real root due to the fact $\sqrt{-1199} \notin \mathbb{R}$, which in this case leads to an empty solution set $\{\} \subset \mathbb{R}$.

Equations of higher degree

Solving the polynomial quadratic equation $ax^2+bx+c=0$ for x by means of the Quadratic Formula

$$\frac{-b\pm\sqrt{D}}{2a}$$

dates back to Babylonian and Greek times. The next big leap forward for solving equations of higher degree had to wait until the 16th century, till the time of the Renaissance.

- ▷ Cubic Equations **Geronimo Cardano** (1501–1576) published a similar Cubic Formula for solving polynomial equations of degree three. Despite Cardano publishing it, the Cubic Formula was actually discovered by another Italian. Historians claim this formula to be discovered by the mathematician **Niccolo Fontana** (1499–1557) (nicknamed Tartaglia or '*stutterer*').
- ▷ Quartic Equations Shortly after the former formula, it was again an Italian mathematician and pupil of Cardano, called **Lodovico Ferrari** (1522 – 1565) who found the Quartic Formula to solve polynomial equations of degree four.
- ▷ Quintic Equations For an apotheosis one needed to wait until the 19th century in France: the very young brilliant mathematician **Evariste Galois** (1811–1832) proves the impossibility of finding a similar Quintic Formula for polynomial equations of larger than degree four. Meanwhile as a workaround (among others, Isaac Newton around 1676) we can solve any polynomial equation numerically, yielding approximations for its solutions. Apart from this modern numerical approach, also special subtypes of polynomial equations of larger degree can still be solved exactly by means of formulas of radical expressions.

1.3 Exercises

Exercise 1 Simplify the following expressions.

1) $a^5 \cdot a^5$
2) $a^5 + a^5$
3) $a^5 - a^5$
4) $(a^5)^3$
5) $a^{10} \cdot a^{-3}$
6) $\frac{a^8}{a^2}$
7) $\frac{a^6}{a^{-2}}$
8) $(2a^{-3})^{-2}$

Exercise 2 Tick every correct answer, given $a, b, c, d, m, n \in \mathbb{Z}_0$.

1)	$-(a+b) =$	$-a-b$	$a-b$	$-a+b$	$a+b$
2)	$-\frac{a+b}{c} =$	$\frac{a+b}{-c}$	$\frac{-a+b}{c}$	$\frac{-a-b}{c}$	$\frac{-a-b}{-c}$
3)	$\frac{c}{d} + \frac{3}{4} =$	$\frac{c+3}{d+4}$	$\frac{c+3}{4d}$	$\frac{4c+3d}{4d}$	$\frac{4c+3d}{d+4}$
4)	$\frac{c}{d} \cdot \frac{3}{4} =$	$\frac{c+3}{d+4}$	$\frac{4d}{3c}$	$\frac{3c}{4d}$	$\frac{12c}{4d}$
5)	$\left(\frac{-1}{7}\right)^{-1} =$	$\frac{1}{7}$	-7	$\frac{-1}{7}$	7
6)	$a^2 \cdot a^n \cdot a^m =$	a^{2nm}	a^{2+n+m}	a^{2+nm}	a^{2n+m}
7)	$(-2a^2b^3)^3 =$	$-6a^6b^9$	$-6a^2b^6$	$-8a^6b^9$	$-8a^8b^{27}$
8)	$\frac{c^6d^3}{c^2d} =$	c^3d^3	c^4d^3	c^4d^2	c^3d^2
9)	$\left(\frac{a^5}{b}\right)^{-2} =$	$\frac{-a^7}{b^2}$	$\frac{b^2}{a^{10}}$	$\frac{b^2}{a^{-10}}$	$\frac{-a^{10}}{b^2}$
10)	$a^{12} : a^3 =$	a^9	a^4	a^{-4}	a^{-9}

Exercise 3 Solve both linear equations for x.

1) $2 \cdot \left(\frac{x}{3} + 1\right) = \frac{x}{4} - \frac{1}{2}$
2) $\left(3 + \frac{2}{7}\right) \cdot \left(\frac{x}{3} - 1\right) = \frac{x}{7} + \frac{5}{3}$

Exercise 4 The sum of three successive odd integers is 141. Find these three numbers.

Exercise 5 Simplify the following expressions by using appropriate rules and use exclusively positive exponents to answer. All variables are nonzero rational numbers.

1) $(a^3)^{-4}$
2) $a^{12} : a^{13}$
3) $(a^{-4})^5$
4) $\left(\frac{a^2}{b^3}\right)^3$
5) $b^3 : b^{-5}$
6) $(-2 \cdot a^3 \cdot b^2)^4$
7) $\left((b^5)^5\right)^2$
8) $((a^{-5})^3 \cdot a^{11}) : (a^9)^{-3}$
9) $\frac{(-4 \cdot a^4 \cdot b)^3}{(-2 \cdot a^3 \cdot b)^2}$
10) $(2 \cdot a^2 \cdot b^3)^4 \cdot (-3 \cdot a \cdot b^4)^2$

Exercise 6 Simplify as much as possible.

1) $-4x^2y^2 + (2xy)^2$
2) $-4x^2y^2(+2xy)^2$
3) $-4x^2y^2 : (2xy)^2$
4) $(x^2y)^3(-x^6y^3)$
5) $(x^2y)^3(-x^6y)^3$
6) $(x^2y)^3 - (x^6y^3)$
7) $x(x-1) + (2x+1)(x-3)$
8) $6x(x+5) - 8x(x^2 - 4x + 3)$
9) $(3x-2)(x-4) - 3x^2 + 7x(2x-1)$

Exercise 7 Factor – *if possible* – by means of the Perfect Square.

1) $81x^4 + 90x^2y + 25y^2$
2) $81x^4 + 60x^2y + 25y^2$
3) $25x^2 + 49x + 36$
4) $25x^2 - 60x + 36$

Exercise 8 Solve these quadratic equations for their unknown quantity (variable) within the scope of $\mathbb{R}$.

1) $6\delta^2 + 5\delta = 0$
2) $4x^2 - 25 = 0$
3) $2t^2 - t - 6 = 0$
4) $4x^2 + 5x + 1 = 0$
5) $(t+1)^2 + (t+2)^2 + (t+3)^2 = 2$
6) $t(t+1) = 6$

Exercise 9 Was the polynomial $V(x,y) = 2(x-2y)(a-3b) + 5(a+3b)(2y-x)$ factored? If so, explain why it is a factorization? If not so, then factor $V(x,y)$.

Exercise 10 Factor the quadratic trinomial $K(x) = 1 + 6x + 9x^2$ using the Quadratic Formula.

Chapter 2 · Linear systems

Not to say an inevitable skill, solving linear systems is a more than useful skill for the expert topics in multimedia. The elimination of variables as chosen in this book for solving linear systems is as straightforward to implement as to scale. Scalability is crucial in situations where variables might run into thousands.

2.1 Definitions

We define a **linear system** as a collection of n linear equations involving the same set of m variables, hence it is often called an n by m system.

Example:

$$\begin{cases} 3x+2y=6 \\ 5x+2y=4 \end{cases}$$

Finding the solution to a linear system does not mean separately solving each equation. It implies the *simultaneous* solving of all equations.

We define **equivalent systems** as systems having the same solution.

Example: The 3 by 3 system

$$\begin{cases} -4y+z+7 &= -2x \\ x+y-3z &= 6 \\ -y+3x+z &= 0 \end{cases}$$

is equivalent to

$$\begin{cases} 2x & + & 2y & - & 6z & = 12 \\ & - & 6y & + & 7z & = -19 \\ 3x & - & y & + & z & = 0 \end{cases}$$

and equivalent to

$$\begin{cases} x & + & y & - & 3z & = 6 \\ & & 2y & - & 5z & = 9 \\ & & & & 8z & = -8. \end{cases}$$

We link equivalent systems by inserting the equivalent-sign '$\Longleftrightarrow$'.

$$\begin{cases} -4y+z+7 &= -2x \\ x+y-3z &= 6 \\ -y+3x+z &= 0 \end{cases} \Longleftrightarrow \begin{cases} x & & & = 1 \\ & y & & = 2 \\ & & z & = -1 \end{cases}$$

The solution set to the latter equivalent system is a singleton containing one triple: $\{(1,2,-1)\} \subset \mathbb{R} \times \mathbb{R} \times \mathbb{R} = \mathbb{R}^3$.

We define the **solution to an n by m linear system**, if it exists, as the list of m values that satisfies all n equations when assigned to the variables.

Example: We solve the linear system

$$\begin{cases} 3x+2y=6 \\ 5x+2y=4 \end{cases}$$

as

$$\begin{cases} x=-1 \\ y=\frac{9}{2} \end{cases}$$

often typeset in a **solution set** $\{(-1,\frac{9}{2})\} \subset \mathbb{R}\times\mathbb{R}=\mathbb{R}^2$.

We define a **square system** as any system having as many equations as unknown quantities and often referred to as an n by n system.

Example: This 3 by 3 system

$$\begin{cases} 2x & + & 2y & - & 6z & =12 \\ & - & 6y & + & 7z & =-19 \\ 3x & - & y & + & z & =0 \end{cases}$$

is a square system.

We define a **triangular system** as a system with all of its equations ordered to their variables, showing a triangle of coefficients 0 under the diagonal.

Example:

$$\begin{cases} x & + & y & - & 3z & =6 \\ & & 2y & - & 5z & =9 \\ & & & & 8z & =-8 \end{cases}$$

Notice how this diagonal runs along the ordered variables x,y,z under which exclusively coefficients 0 appear.

We define a linear system as **underdetermined** when its solution contains one (or more) free **parameter(s)**. An underdetermined linear system has a **degree of freedom** equal to the number of free parameters in its solution.

Example:

$$\begin{cases} -3x+2y & =-4 \\ 9x-6y & =12 \end{cases}$$

We notice that both equations are dependent. Dividing both the left hand side and right hand side of the second equation by -3 simplifies the system to

$$\begin{cases} -3x+2y=-4 \\ -3x+2y=-4. \end{cases}$$

It seems that we actually apply just one condition to two variables x and y. Consequently we need to set free one of both variables, either x or y. Let us for instance choose to set free $x \in \mathbb{R}$. The degree of freedom of this system is 1 and we simply proceed by solving this system for y.

$$\begin{cases} y = \frac{-4+3x}{2} \\ x \in \mathbb{R} \text{ (free parameter)} \end{cases} \Longleftrightarrow \left\{ \left(x, \frac{-4+3x}{2} \right) \text{ with } x \in \mathbb{R} \right\} \subset \mathbb{R} \times \mathbb{R}$$

However this answer admits an infinite number of solutions, they are all according to the solution template as dictated by the free parameter.

$$\left\{ \ldots, \left(-1, \frac{-7}{2} \right), (0,-2), \left(1, \frac{-1}{2} \right), (2,1), \left(3, \frac{5}{2} \right), \ldots \right\} \subset \mathbb{R} \times \mathbb{R}$$

To distinguish the free parameter from the ordinary variable x we typeset it by a Greek character such as λ. We summarize the above solution set as

$$\left\{ \left(\lambda, \frac{-4+3\lambda}{2} \right) \text{ with } \lambda \in \mathbb{R} \right\} \subset \mathbb{R} \times \mathbb{R}.$$

We define a linear system as **inconsistent** when two (or more) of its equations cause a conflict. Consequently, inconsistent systems have an empty solution set.

Example:

$$\begin{cases} 3x - 4y &= 2 \\ -6x + 8y &= 6 \end{cases} \Longleftrightarrow \begin{cases} 3x - 4y = 2 \\ 0x + 0y = 10 \end{cases}$$

Note that one of these equations holds a contradiction, leading to an inconsistent system. Hence no value set can ever make this system to be *true* . As solutions cannot exist this system has an empty solution set $\{\} \subset \mathbb{R} \times \mathbb{R}$.

2.2 Methods for solving linear systems

Solving by substitution

Substitution is a method involving systematically replacing the variables of the system. We solve one of the equations for one variable of our choice. We for instance choose to solve the second equation for y.

$$\begin{cases} -2x + y &= -2 \\ x + y &= 10 \end{cases} \Longleftrightarrow \begin{cases} -2x + y &= -2 \\ y &= 10 - x \end{cases}$$

We then replace or 'substitute' y by its expression $y = 10 - x$ in the first equation $-2x + y = -2$. We are now able to solve the resulting equation for x only.

$$\begin{cases} -2x + (10 - x) & = -2 \\ y & = 10 - x \end{cases} \iff \begin{cases} -3x + 10 & = -2 \\ y & = 10 - x \end{cases}$$

$$\iff \begin{cases} -3x & = -12 \\ y & = 10 - x \end{cases}$$

$$\iff \begin{cases} x & = 4 \\ y & = 10 - x \end{cases}$$

Finally we replace the obtained value $x = 4$ back in the second equation.

$$\begin{cases} x & = 4 \\ y & = 10 - x \end{cases} \iff \begin{cases} x & = 4 \\ y & = 10 - 4 = 6 \end{cases}$$

Solving linear systems by substitution has three disadvantages.

1) Substitution becomes digressive for solving linear systems of larger dimensions n by m.
2) As a consequence of the previous, solving by substitution grows error prone.
3) When mistaking strategically, solving by substitution may end up in a vicious circle.

Solving by elimination

We strongly recommend to solve any *linear* system by elimination. This method involves a stepwise renewing of the initial system, dictated by the algorithm of the famous German mathematician **Carl Friedrich Gauss** (1777-1855), until the complete solution appears. Be aware that this method is valid for *linear* systems only. Solving by elimination is as straightforward to use with pen and paper as to implement in software.

Gaussian algorithm

Step 1: We **trim** the initial system for each equation to its variables. Remember to maintain the same order of appearance for the variables in each equation. Leave a blank space in case of an absent variable.

Step 2: Renew this trimmed version to an equivalent **triangular** system. To achieve this, only three operations can be used:

- ▷ swapping two equations,

- ▷ multiplying an equation with a nonzero number,
- ▷ adding two equations.

Strategically we aim at setting coefficients to zero below each targeted variable.

Step 3: Finally we **replace** from the bottom equation to the top equation each determined value for the subsequent variable. At the final stage, we find the complete solution to the linear system.

Example: We solve a 3 by 3 linear system for x, y and z using the above Gaussian algorithm.

$$\begin{cases} -4y+z+7 &= -2x \\ x+y-3z &= 6 \\ -y+3x+z &= 0 \end{cases}$$

Step 1: We trim the initial system.

$$\begin{cases} 2x & - & 4y & + & z & = -7 \\ x & + & y & - & 3z & = 6 \\ 3x & - & y & + & z & = 0 \end{cases}$$

Step 2: We renew the trimmed system to an equivalent triangular system.

Whenever it suits us we may swap equations in order to get an 'easier' one on top. In this example we switch the first and second equation, what we denote by $R_1 \leftrightarrow R_2$.

$$\begin{cases} x & + & y & - & 3z & = 6 \\ 2x & - & 4y & + & z & = -7 \\ 3x & - & y & + & z & = 0 \end{cases}$$

We will use the top equation $x+y-3z=6$ as a tool for eliminating the first variable x in the underlying equations. Meanwhile we will keep copying our unaffected 'tool' $x+y-3z=6$ to the very end. We eliminate x in the second equation by adding -2 times the top equation to it.

$$\begin{cases} x & + & y & - & 3z & = 6 & \mid -2 \\ 2x & - & 4y & + & z & = -7 & \mid 1 \\ 3x & - & y & + & z & = 0 & \end{cases}$$

We denote the above action by $R_2 \to -2R_1 + R_2$. We recall that we keep the top equation and only renew the second equation by our strategic combination. This leads to

$$\begin{cases} x & + & y & - & 3z & = 6 \\ 0x & - & 6y & + & 7z & = -19 \\ 3x & - & y & + & z & = 0. \end{cases}$$

The equation $0x - 6y + 7z = -19$ is the result of adding the multiples (**-2**)($x + y - 3z = 6$) + **1**($2x - 4y + z = -7$).

Likewise, we eliminate x in the third equation via the renewal $R_3 \to -3R_1 + R_3$.

$$\begin{cases} x & + & y & - & 3z & = 6 & \mid -3 \\ 0x & - & 6y & + & 7z & = -19 & \mid \\ 3x & - & y & + & z & = 0 & \mid 1 \end{cases}$$

$$\iff \begin{cases} x & + & y & - & 3z & = 6 \\ 0x & - & 6y & + & 7z & = -19 \\ 0x & - & 4y & + & 10z & = -18 \end{cases}$$

The equation $0x - 4y + 10z = -18$ is the result of adding the multiples (**-3**)($x + y - 3z = 6$) + **1**($3x - y + z = 0$).

We now acquire a decreased 2 by 2 system in y and z. We now make this 2 by 2 system subject to the Gaussian algorithm.

$$\begin{cases} x & + & y & - & 3z & = 6 \\ & - & 6y & + & 7z & = -19 \\ & - & 4y & + & 10z & = -18 \end{cases}$$

After simplifying and omitting obsolete zeros we reach the linear system above. Whenever it suits us we may multiply any equation with a nonzero value in order to simplify it. We multiply the second equation with the factor -1, logging this action as $R_2 \to -R_2$. We likewise multiply the third equation with the factor $-\frac{1}{2}$, logging this action as $R_3 \to -\frac{1}{2}R_3$.

$$\begin{cases} x & + & y & - & 3z & = 6 \\ & & 6y & - & 7z & = 19 \\ & & 2y & - & 5z & = 9 \end{cases}$$

We proceed by swapping $R_2 \leftrightarrow R_3$.

$$\begin{cases} x & + & y & - & 3z & = 6 \\ & & 2y & - & 5z & = 9 \\ & & 6y & - & 7z & = 19 \end{cases}$$

We keep the first equation. The second equation $2y - 5z = 9$ we adopt as the next tool to eliminate y in the underlying equation. Consequently we now keep $2y - 5z = 9$ till the algorithm ends. We eliminate the variable y by adding -3 times $2y - 5z = 9$ to the third equation.

$$\begin{cases} x & + & y & - & 3z & = 6 & \\ & & 2y & - & 5z & = 9 & \mid -3 \\ & & 6y & - & 7z & = 19 & \mid 1 \end{cases}$$

We add (**-3**)($2y-5z=9$) + **1**($6y-7z=19$), logging it as $R_3 \to -3R_2+R_3$, which yields $8z=-8$.

$$\begin{cases} x + y - 3z = 6 \\ 2y - 5z = 9 \\ 8z = -8 \end{cases}$$

We simplify the third equation to $z=-1$.

$$\begin{cases} x + y - 3z = 6 \\ 2y - 5z = 9 \\ z = -1 \end{cases}$$

Step 3: This triangular system allows for substituting values for the variables, from the bottom equation up to the top equation. Substituting $z=-1$ in the above equations, we find the value for y.

$$\begin{cases} x + y - 3(-1) = 6 \\ 2y - 5(-1) = 9 \\ z = -1 \end{cases} \Longleftrightarrow \begin{cases} x + y + 3 = 6 \\ y = 2 \\ z = -1 \end{cases}$$

Substituting also $y=2$ in the above equations, we find the value for x.

$$\begin{cases} x + 2 + 3 = 6 \\ y = 2 \\ z = -1 \end{cases} \Longleftrightarrow \begin{cases} x = 1 \\ y = 2 \\ z = -1 \end{cases}$$

We solved this system yielding the solution set $\{(1,2,-1)\} \subset \mathbb{R}\times\mathbb{R}\times\mathbb{R} = \mathbb{R}^3$.

Verifying the solution is straightforward and can be done in no time:

$$\begin{cases} 2(1) - 4(2) + (-1) = -7 \\ 1 + 2 - 3(-1) = 6 \\ 3(1) - 2 + (-1) = 0. \end{cases}$$

Corresponding to the disadvantages of solving by substitution, solving by elimination has three advantages.

1) Solving by elimination scales properly, even for very large n by m dimensions.

2) This scalability firmly reduces the error rate when performing on large systems.

3) Based upon a transparant algorithm with a clear ending phase (announced by the triangular shape), it is unlikely to end up in a vicious circle.

2.3 Exercises

Exercise 11 Solve this linear system in $\mathbb{R}^4$:

$$\begin{cases} x+y+2z+3v &= 9 \\ 2x+z+4v &= 11 \\ -y+2z+v &= 4 \\ 2x+7y+8z &= 10. \end{cases}$$

Exercise 12 Solve this linear system in $\mathbb{R}^3$:

$$\begin{cases} 2x_2+x_3+x_1 &= 4 \\ x_1+2x_3+2x_2 &= 3 \\ x_3+x_1+6x_2 &= 1. \end{cases}$$

Exercise 13 A house contains four identical wooden windows. In total it took 32 m of wood to construct them. The length of each window equals three times the width. Find the dimensions of each window.

Exercise 14 The sum of two real numbers is 210 and their difference is 36. Find these numbers.

Exercise 15 A father tells to his son: 'Three years ago I used to be four times your age'. 'Indeed', his son answers, 'but next year you will be three times my age'. Find today's ages of both the father and son.

Exercise 16 After finishing exams Adison and Valence compare their marks. Adison says: 'I was graded 50 points more than Valence'. Valence replies: 'If they had only graded me 310 points more, I would have twice the points of Adison'. Find their graded points, both for Adison and for Valence.

Exercise 17 The main component of a computer is its motherboard: it determines the processor type and frequency, and therefore the computer architecture. A hardware constructor uses two different assembly robots for producing 29000 motherboards. The fastest robot assembles 10000 motherboards in 1 hour, the other robot 6000. Due to a reparation the fastest robot remains out of service for the first 30 minutes. Determine how long it then will take to assemble the total batch of 29000 motherboards and how many of them were made by each robot.

Exercise 18 For manufacturing 3 types of laptop components (I, II and III) we need 3 machines (A, B and C).

We detail the involved production times (in minutes):

▷ component I: 3 minutes using machine A, 11 minutes using machine B and 27 minutes using machine C,

▷ component II: 15 minutes using machine A, 5 minutes using machine B and 6 minutes using machine C,

▷ component III: 12 minutes using machine A, 14 minutes using machine B and 5 minutes using machine C.

How many laptop components of each type should we produce in an 8 hour working day in order to fulltime use each machine?

Exercise 19 Teddy and Phiona each cut a rectangular sheet of paper into two equal parts. Phiona has cut her sheet into two rectangles, each having a circumference of 40 cm. The same applies to Teddy, though with a rectangular circumference of 50 cm. We emphasize that both of them cut an identical rectangular sheet of paper. Find the initial rectangular circumference of the sheet both girls cut.

Exercise 20 Two masses, one of 155 kg and one of 264 kg, are put on a pair of scales. When doubling the smallest mass, we need to shift the center on the arm by 7 m in order to restore balance. Find the arm length. Hint: $\text{mass}_1 \cdot \text{distance}_1 = \text{mass}_2 \cdot \text{distance}_2$.

Exercise 21 It takes an airplane 3 hours to fly the 2000 km from Brussels to Athens, benefiting from a tailwind. During the return flight the airplane now faces the same wind as a headwind, which causes a return duration of 4 hours. Find both the speed of the plane and the speed of this wind.

Exercise 22 On a hot summer day Edmond, Edward and Evelyn are joining an outside drink. Due to the heat they first order four beers and two sodas and it is Evelyn who pays 8 EUR for it. Edmond pays 7.5 EUR for the next order of three beers and three sodas. Although there is no price list outside, Edward manages to figure out the price of one beer. Can you?

Exercise 23 Rita wraps a parcel looping it with 16 cm ribbon. Since it is a brick shaped parcel she has three possibilities in doing so and respectively she remains with 2 cm, 5 cm or just 1 cm left for the knot. Find the parcel's dimensions.

Chapter 3 · Trigonometry

Genuine computer animations often deal with variables representing position, distance, angle, ... The major mathematical concepts that cover these quantities are sine and cosine. Moreover, all the other concepts in this chapter return frequently in the subsequent chapters of this book.

Trigonometry originated from measuring our surrounding space and definitely leads to a multitude of applications such as computer animation, coordinate systems, carrier waves for data transmission, ... We initially define what an angle is in order to turn to triangles. These triangles allow us to define the trigonometric concepts which we then extend to the unit circle. We finalize this chapter by studying the inverse trigonometric functions.

3.1 Angles

We consider the cone that connects our eye-pupil to a golf-ball at armlength. Cutting this 3D-cone vertically we obtain a plane angle formed by its two rays or sides, sharing the common endpoint or vertex O (at our eye-pupil). We now try to measure the size of this spacial quantity 'angle' as the inner sector bordered by both intersecting sides.

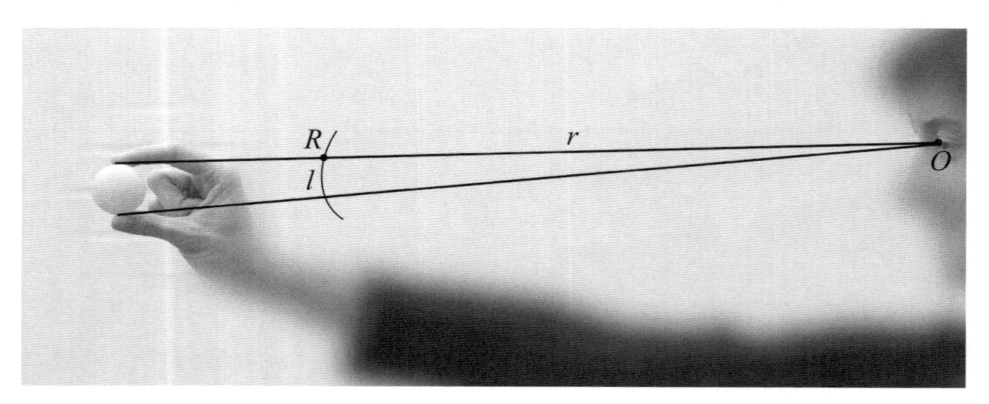

Figure 3.1: Definition of an angle

We catch this sector using a pair of compasses and define an **angle** as the circular arc length l to its radius $|OR| = r$:

$$\alpha = \frac{l}{r}. \tag{3.1}$$

As the above fraction is insensitive to scaling it indeed guarantees a reliable measurement for any angle. When we for instance measure the size of an angle as a circular arc of 6 cm to a radius of 5 cm, then it will remain as a circular arc of 24 cm to its proportionally adjusted radius of 20 cm. We then calculate $\alpha = \frac{6\text{ cm}}{5\text{ cm}} = \frac{24\text{ cm}}{20\text{ cm}} = 1,2$. The unit for this angle

α, based upon the dimensionless fraction $\frac{\text{circular arc length}}{\text{radius}}$, is called **radians** and commonly typeset as rad. Hence the above discussion led to an angle of 1,2 rad.

We might be more familiar with **degrees** as a common unit for angles. This ancient Sumerian measure refers to the approximately 360 days for the Earth to complete its 'circular' orbit around the Sun, stating $\alpha_{\max} = 360°$.

Converting angles from radians to degrees and the other way around can be done easily applying the **Three-Step Rule**. We obtain the maximal plane angle by revolving one ray until $\alpha_{\max} = \frac{l_{\max}}{r} = \frac{\text{circle circumference}}{r} = \frac{2\pi r}{r} = 2\pi$ rad. Hence we can rely on the **full angle** equality 2π rad $= 360°$, given $\pi \approx 3,14$.

Example: Converting between radians and degrees.

from degrees to radians	from radians to degrees
$360° = 2\pi$ rad	2π rad $= 360°$
$1° = \frac{2\pi \text{ rad}}{360}$	$1 \text{ rad} = \frac{360°}{2\pi}$
$30° = \frac{2\pi \text{ rad}}{360} \cdot 30 = \frac{\pi}{6}$ rad	$\frac{\pi}{6} \text{ rad} = \frac{360°}{2\pi} \cdot \frac{\pi}{6} = 30°$

The fractional part of an angle represented in degrees can be expressed in two different base forms. On the one hand we can use the popular DD-form (Duo Decimal) for it, writing the fraction in the decimal number base 10. On the other hand we might use the ancient DMS-form (Degrees Minutes Seconds) for it, writing the fraction in the sexagesimal number base 60 using accents for separating minutes from seconds. Our contemporary time keeping still uses minutes and seconds as parts of one hour. We might for instance express an angle of $180,5°$ as $180°30'00''$ in the latter system. Converting fractional parts of angles from DD to DMS and the other way around goes the same way as converting numbers from decimal base to number base $B = 60$.

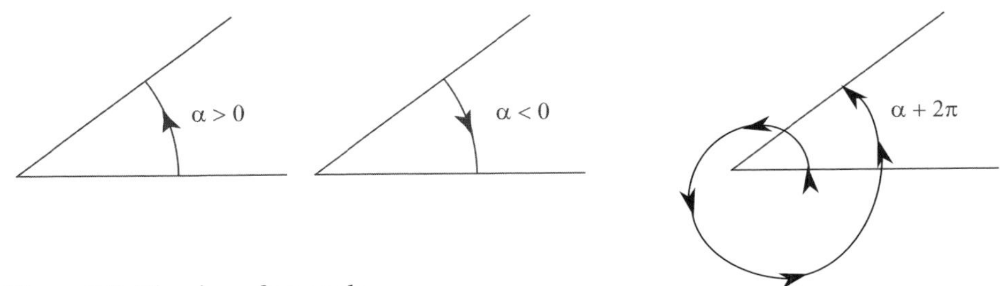

Figure 3.2: The sign of an angle

We can represent the same angle in different ways. We even may add a sign to an angle. We define angles by revolving a ray counter clockwise as positive. Alternatively, we de-

fine angles by revolving a ray clockwise as negative.
Furthermore, all angles α and α', stating $\alpha' = \alpha + k\,2\pi$ with parameter $k \in \mathbb{Z}$ or alternatively put in degrees $\alpha' = \alpha + k\,360°$, are equivalent.

We differentiate the types of angles to some more detail. A **zero** angle equals exactly $0°$. An **acute** angle is larger than $0°$ and smaller than $90°$. A **right** angle equals exactly $90°$ or the quarter of a circle. Both sides of a right angle are said to be orthogonal or perpendicular. An **obtuse** angle is larger than $90°$ and smaller than $180°$. A **straight** angle equals exactly $180°$ or the half of a circle. A **reflex** angle is larger than $180°$ and smaller than $360°$ (full circle).

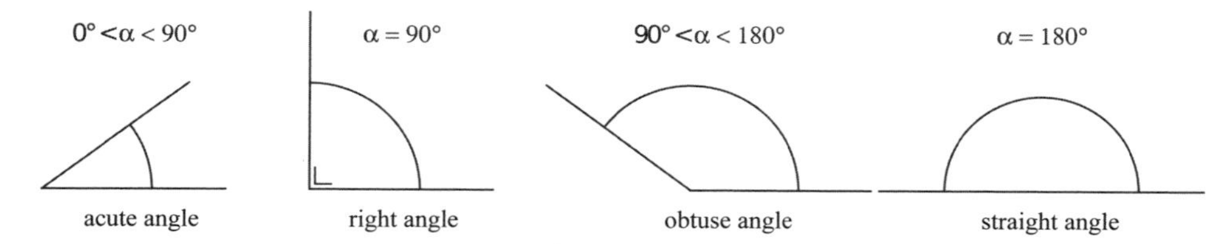

Figure 3.3: Types of angles

3.2 Triangles

Every three points or vertices A, B and C that do not lie on the same straight line, make up a **scalene triangle** ABC when it has no two sides of equal length. The measures of the interior angles of a triangle ABC always add up to the straight angle, $\hat{A} + \hat{B} + \hat{C} = 180° = \pi$ rad.

A **right triangle** is a triangle having one of its interior angles equal to $90° = \frac{\pi}{2}$ rad. Consequently, the sum of both acute angles of any right triangle equals $90° = \frac{\pi}{2}$ rad. Its largest side is the edge opposite the right angle and called the **hypotenuse**.

An **isosceles triangle** is a triangle having two equal sides through the apex and their corresponding base angles having the same measure. An **equilateral triangle** is a triangle having three equal sides, their equal interior angles measuring $60°$.

We recall some geometric concepts as they are defined and used in triangles. An **angle bisector** of a triangle is the straight line through a vertex which cuts the corresponding angle in half. A **median** or **side bisector** of a triangle is the straight line through a vertex and the midpoint of the opposite side. An **altitude** of a triangle is the straight line through a vertex and perpendicular to the opposite side. This opposite side is called the **base** of the altitude and its intersection point is called the **foot**. A **perpendicular bisector** of a triangle is the straight line through the midpoint of a side and being perpendicular to it. In a scalene triangle (see figure 3.4) these four geometric lines differ clearly. In isosceles triangles the four geometric lines through the apex coincide. In equilateral triangles the four geometric lines always coincide.

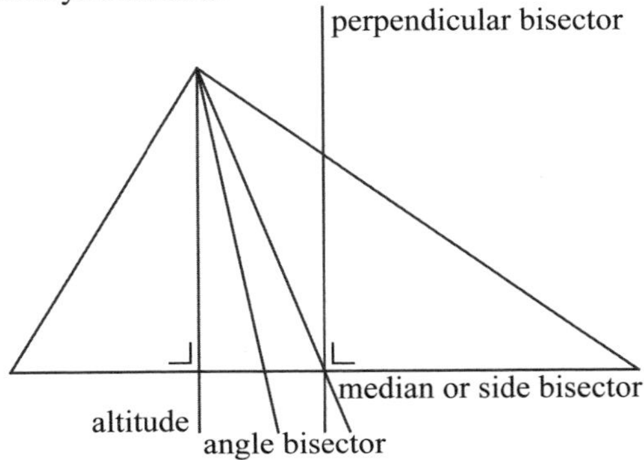

Figure 3.4: Geometric lines in a triangle

Two plane polygons are **similar** in case

- ▷ the ratio of their corresponding edges is constant, and
- ▷ their corresponding angles measure the same size.

We recall more specifically a similarity criterion for triangles: *two triangles are similar whenever two of their corresponding angles measure the same size.*

Before proceeding the discussion on triangles we explain the ruler-and-compass method to construct a perpendicular bisector on a given line segment, as it will be useful in subsequent chapters.

- ▷ We construct the perpendicular bisector m of a segment $[AB]$ by drawing two circles $C(A,r)$ and $C(B,r)$ with the same radius r. These circles intersect, when applying a sufficiently large radius $r > 0$, in two points spanning the perpendicular bisector of the segment $[AB]$. Consequently we acquire the midpoint M of the segment $[AB]$ as an extra bonus.
- ▷ We construct the altitude l from apex D to a base k by drawing a circle $C(D,s)$ which intersects the base k in two points P and Q. Finally we apply the above construction for the perpendicular bisector m of the segment $[PQ]$. Consequently we acquire the foot as the midpoint D' of the segment $[PQ]$.

We recall the **Pythagorean theorem**, named after the Greek **Pythagoras of Samos** (582–507 Before Christ), which states in any right triangle: the square of the length of the

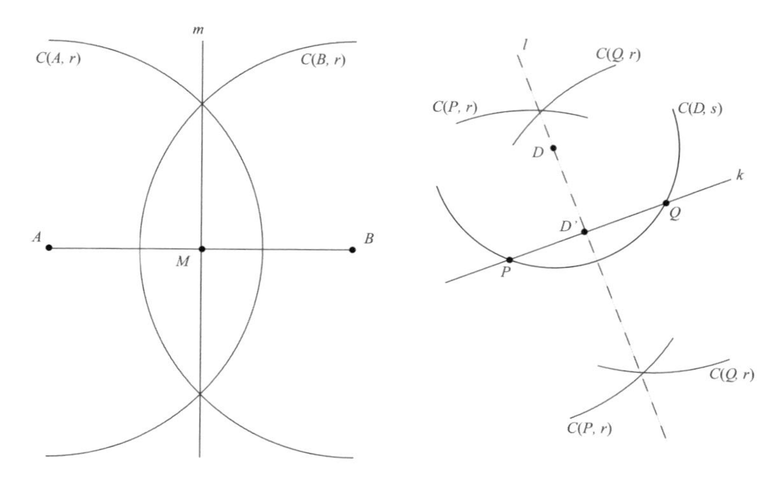

Figure 3.5: Constructions for the perpendicular bisector m and the altitude l

hypotenuse equals the sum of the squares of the lengths of the two other sides.

$$(\text{side}_1)^2 + (\text{side}_2)^2 = (\text{hypotenuse})^2 \tag{3.2}$$

Figure 3.6: Statue of Pythagoras on the isle of Samos

Proof: One way to prove this famous theorem goes via the area of this tiled square. In this square we count four identical right triangles featuring hypotenuse c and other sides a and b.

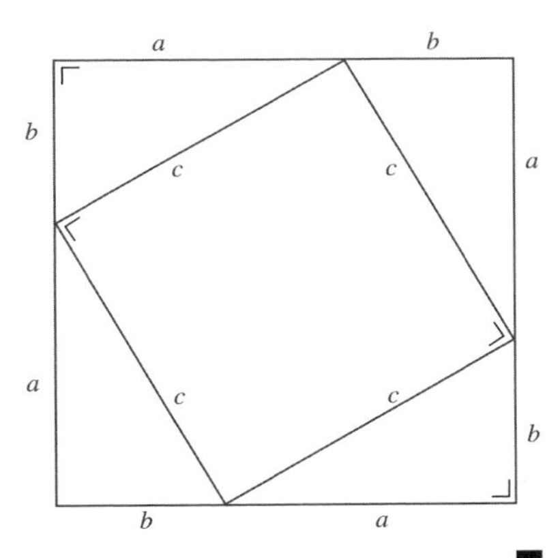

large square area = inner square area + 4 times right triangle area

$$(a+b)^2 = c^2 + 4\frac{ab}{2}$$

$$a^2 + 2ab + b^2 = c^2 + 2ab$$

$$a^2 + b^2 = c^2$$

■

Example: A staircase consists of 17 steps of height 19 cm and depth 15 cm. Find the length of the entire banisters.

The Pythagorean theorem yields $\sqrt{19^2+15^2}$ for the length of the hypotenuse of a single step. Conclusively, for the length of the banisters we multiply $17 \cdot \sqrt{19^2+15^2} = 411,53$ cm.

Game programming often involves calculating the distance between two points on the screen, e.g. between the anchor points of two colliding objects or two interacting personages. The programmed game may for instance respond as soon as its player moves sufficiently near the enemy. We realize now that various screen situations require a fast calculation of the distance between two points. Applying the Pythagorean theorem is the straightforward way to do so.

Assuming we know two points $P(x_1,y_1)$ and $Q(x_2,y_2)$ allows us to draw a right triangle with hypotenuse $[PQ]$. Figure 3.7 reveals (x_2-x_1) and (y_2-y_1) for the lengths of the other sides. Conclusively the Pythagorean theorem states for the distance between P and Q, typeset as $d(P,Q)$, the formula

$$d(P,Q) = \sqrt{(x_2-x_1)^2+(y_2-y_1)^2}. \tag{3.3}$$

Example: Find the distance between the points $A(1,2)$ and $B(5,6)$.
Their distance yields $d(A,B) = \sqrt{(5-1)^2+(6-2)^2} = \sqrt{32}$.

We can extend the plane distance formula to three dimensions. We just need to take the z-coordinate into account to realize it. Consequently, the distance between two spacial points $P(x_1,y_1,z_1)$ and $Q(x_2,y_2,z_2)$ then equals

$$d(P,Q) = |PQ| = \sqrt{(x_2-x_1)^2+(y_2-y_1)^2+(z_2-z_1)^2}. \tag{3.4}$$

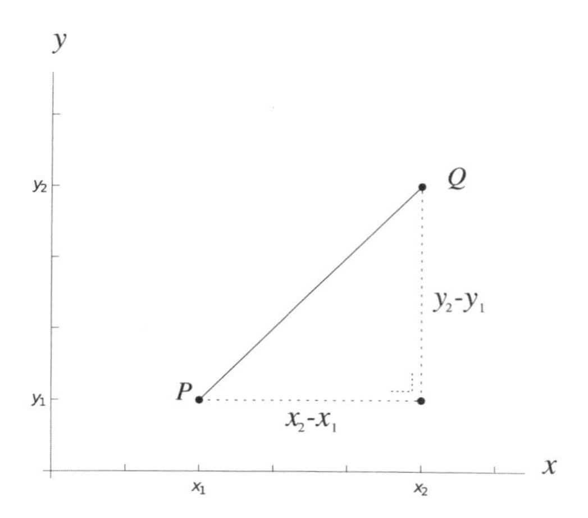

Figure 3.7: Distance between two points

3.3 Right Triangle

We recall measuring the size of angles in radians as defined by the ratio $\alpha = \frac{l}{r}$. Alternatively we define more such ratios as based upon a right triangle. We replace the drawing of a circular arc on radius r (using a pair of compasses) by the drawing of an opposite side perpendicular to r (using a square). We realize the elegance of using a square over the use of a pair of compasses, as the length of a side is way easier to determine than the length of a circular arc.

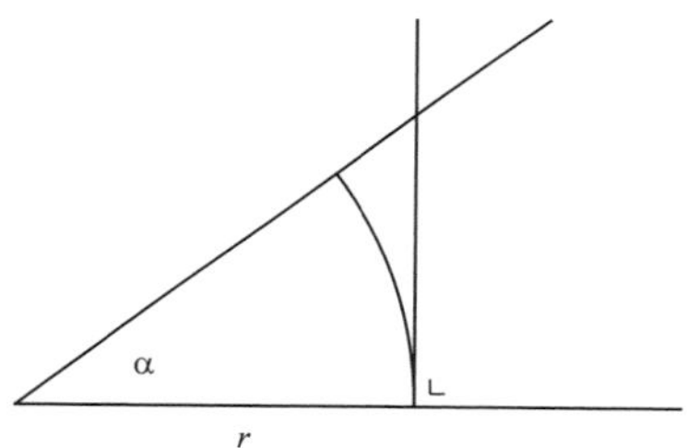

Figure 3.8: Definition trigonometric ratios

This way we acquire a new measure for the interior angle α as the ratio $\frac{\text{opposite side}}{r}$. Such a measure is a 'trigonometric ratio'. We define the above ratio of sides as the **tangent** of the acute angle α. As the above fraction is insensitive to scaling, it again guarantees a reliable measurement for any angle.

The total number of trigonometric ratios is six and they are called **sine**, **cosine**, tangent, **cotangent**, cosecant and secant. They are of fundamental importance to study and practice the more advanced trigonometry. In other words, one should never use them incorrectly!

We hereby list all correct definitions for the ratios sine (sin), cosine (cos) and tangent (tan), as for their reciprocals cosecant (csc), secant (sec) and cotangent (cot).

$$\sin\alpha = \frac{\text{opposite side}}{\text{hypotenuse}} \qquad \csc\alpha = \frac{1}{\sin\alpha}$$

$$\cos\alpha = \frac{\text{adjacent side}}{\text{hypotenuse}} \qquad \sec\alpha = \frac{1}{\cos\alpha}$$

$$\tan\alpha = \frac{\text{opposite side}}{\text{adjacent side}} \qquad \cot\alpha = \frac{1}{\tan\alpha}$$

As a useful mnemonic for the above ratios we make acronyms using their starting characters: 'SOH, CAH, TOA'. We realize that these definitions, as based upon right triangles, are limited to acute angles α only. We recall that each of both acute angles of a right triangle measures a size smaller than $90°$.

Based upon the above trigonometric ratios we immediately discover our first trigonometric formula as

$$\tan\alpha = \frac{\sin\alpha}{\cos\alpha}. \tag{3.5}$$

3.4 Unit Circle

The **unit circle** is a circle within the orthonormal (see paragraph 6.1) frame. The center of the unit circle is the origin O and its radius equals one. We consider the unit circle as being our trigonometric 'dashboard' because for any angle α revolved from the horizontal x-axis (reference side), we find a corresponding point E_α on it (see figure 3.9).

We subdivide this unit circle in four equal parts that we index counter clockwise, starting from the zero angle (on the positive x-axis). Consequently, all acute angles between $0°$ and $90°$ lie in the first quadrant and all obtuse angles are in the second quadrant bordered by $90°$ and $180°$. The reflex angles fall in the third quadrant as bordered by $180°$ and $270°$ or eventually in the fourth quadrant between $270°$ and $360°$.

We simplify the former trigonometric ratios sine, cosine and tangent for a hypotenuse corresponding to the radius of the unit circle, in other words for the hypotenuse set to one. Note that we redefine the trigonometric tangent via a larger right triangle bordering the vertical tangent line to the unit circle at the point $E_0(1,0)$.

$$\cos\alpha = \frac{\text{adjacent side}}{1}$$

$$\sin\alpha = \frac{\text{opposite side}}{1}$$

$$\tan\alpha = \frac{\text{opposite side at the vertical tangent line } x = 1}{1}$$

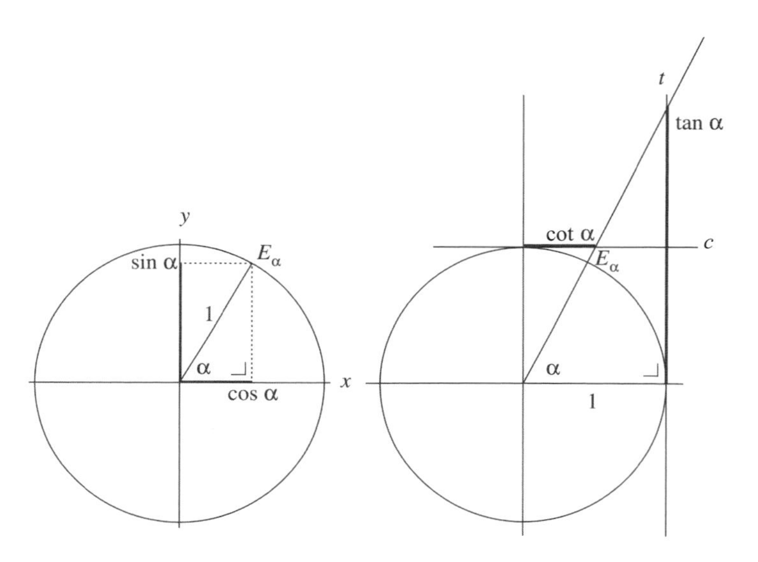

Figure 3.9: Trigonometric ratios in the unit circle

In the unit circle we interpret $\cos\alpha$ as the horizontal shadow on the x-axis caused by the revolving side of angle α. Likewise, we meet $\sin\alpha$ as the vertical shadow on the y-axis caused by the revolving side of angle α. Finally, we encounter $\tan\alpha$ as the opposite side to the angle α on the vertical tangent line $x = 1$. In other words, for the revolving side of an angle α intersecting the unit circle in the corresponding point E_α we define the x-coordinate of E_α as the **cosine** of α and the y-coordinate of E_α as the **sine** of α. Any angle α therefore determines unambiguously its cosine and sine. But for the other way around, every couple of cosine and sine values determines its angle α only to an integer multiple of 2π or 360° (denoted as the periodicity of an angle).

The redefined sine, cosine and tangent in a unit circle overcome their previous limitation to acute angles in such a way that they now are defined for any angle size.

Applying the Pythagorean theorem (3.2) in the unit circle, yields

$$(\sin\alpha)^2 + (\cos\alpha)^2 = 1. \tag{3.6}$$

This important formula linking sine and cosine is known as the trigonometric **Pythagorean Identity**.

The Pythagorean theorem applies to right triangles only. Scalene triangles are ruled by the **Law of Cosines** and the **Law of Sines** to link their interior angles to the length of their sides. For a scalene triangle ABC featuring interior angles α, β and γ at corresponding vertices A, B and C and opposite sides a, b and c, we state both laws (omitting their proofs):

$$\frac{a}{\sin\alpha} = \frac{b}{\sin\beta} = \frac{c}{\sin\gamma} \qquad \text{Law of Sines,} \tag{3.7}$$

$$a^2 = b^2 + c^2 - 2bc\cos\alpha \qquad \text{Law of Cosines.} \tag{3.8}$$

The Law of Cosines for a right angle α implies $\cos\alpha = 0$ and thus simplifies to the Pythagorean theorem.

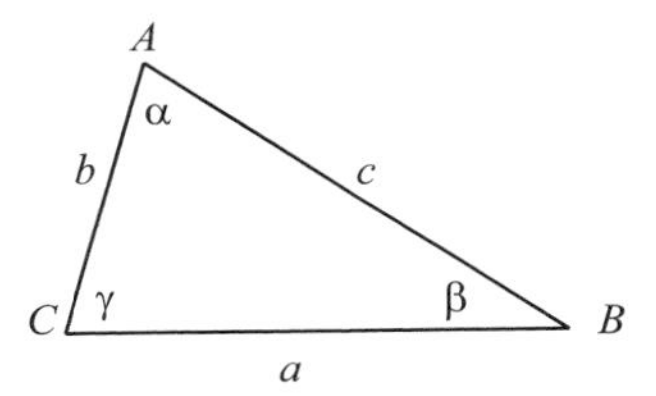

Figure 3.10: Laws of Sines and Cosines in a scalene triangle

3.5 Special Angles

Given their definitions, the trigonometric ratios are limited to these intervals:

$$\sin\alpha \in [-1, 1] \qquad \cos\alpha \in [-1, 1]$$
$$\tan\alpha \in \left]-\infty, +\infty\right[\qquad \cot\alpha \in \left]-\infty, +\infty\right[$$

TRIGONOMETRIC RATIOS FOR AN ANGLE OF $45° = \frac{\pi}{4}$ RAD

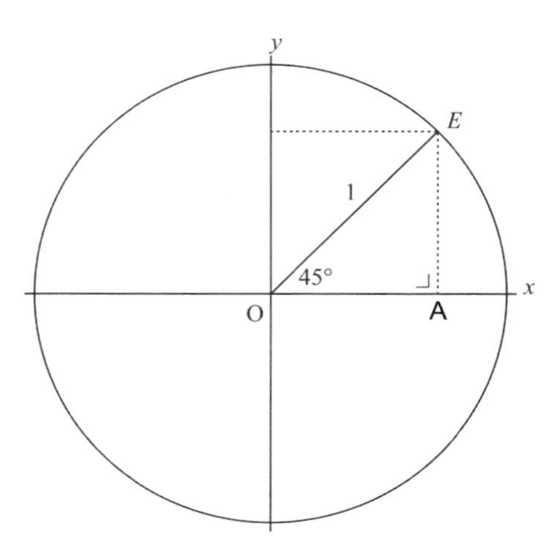

As the interior angles of the triangle OAE add up to 180°, we find the angle $\hat{E}$ equals 45°. As this triangle features equal angles $\hat{O} = \hat{E} = 45°$ we conclude the triangle OAE to be isosceles with apex A and for its corresponding length of sides $|OA| = |AE|$. This geometric reasoning leads finally to $\cos 45° = \sin 45°$. Substituting the above geometric conclusion in the Pythagorean Identity $(\sin 45°)^2 + (\cos 45°)^2 = 1$ yields $(\sin 45°)^2 + (\sin 45°)^2 = 1$. Solving this equality for $\sin 45°$ leads to $\sin 45° = \pm\sqrt{\frac{1}{2}}$. Since the angle of 45° resides in the first quadrant limits $\sin 45° \geqslant 0$, and so we conclude $\sin 45° = \frac{1}{\sqrt{2}} = \frac{\sqrt{2}}{2}$. As an immediate consequence we also conclude $\cos 45° = \frac{\sqrt{2}}{2}$.

TRIGONOMETRIC RATIOS FOR AN ANGLE OF $30° = \frac{\pi}{6}$ RAD

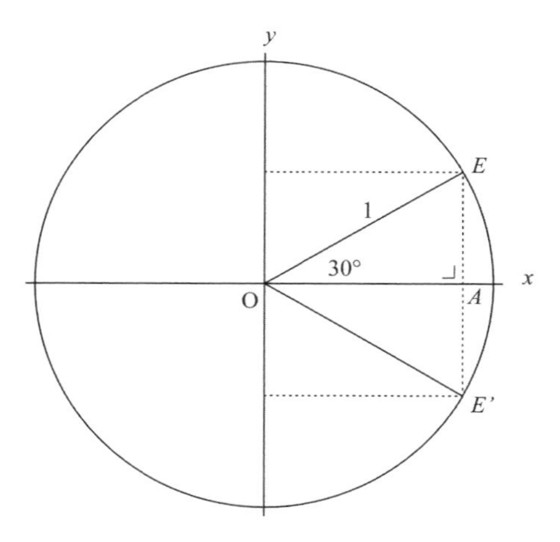

Firstly we reflect the point $E_{30°}$ in the x-axis to its image point E'. Reflections are isometric mappings: they leave distances and angle sizes unchanged. Therefore we conclude the triangle EOE' to have an apex angle $\hat{O} = 60°$ and two equal sides $|OE| = |OE'|$. As the interior angles of the triangle EOE' add up to 180°, we find equal angles $\hat{E} = \hat{E}' = 60°$. Since $\hat{E} = \hat{O} = \hat{E}' = 60°$ we conclude the triangle EOE' to be equilateral with length of sides $|OE| = |OE'| = |EE'| = 1$. Based on the coinciding of all geometric lines in equilateral triangles, we consider the altitude OA on the base $[EE']$ also to be the median that bisects the segment $[EE']$. Therefore we conclude $|EA| = \sin 30° = \frac{1}{2}$. Substituting this geometric conclusion in the Pythagorean Identity straightforwardly yields the corresponding $\cos 30° = \frac{\sqrt{3}}{2}$.

TRIGONOMETRIC RATIOS FOR AN ANGLE OF $60° = \frac{\pi}{3}$ RAD

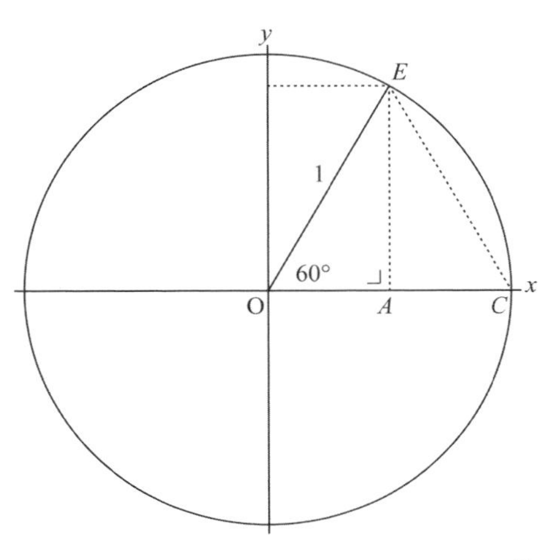

For the special angle of 60° we consider the triangle OEC as the start of a geometric reasoning. Given its vertices E and C lying on the unit circle, we know that $|OE| = |OC| = 1$ leading us to an isosceles triangle OEC with apex O and consequently equal base angles $\hat{E} = \hat{C}$. As the interior angles of the triangle OEC add up to 180°, we find equal angles $\hat{E} = \hat{C} = 60°$. Based on the coinciding of all geometric lines in equilateral triangles, we consider the altitude EA on the base $[OC]$ also to be the median that bisects the segment $[OC]$. Therefore we conclude that $|OA| = \cos 60° = \frac{1}{2}$. Substituting this geometric conclusion in the Pythagorean Identity yields the corresponding $\sin 60° = \frac{\sqrt{3}}{2}$.

OVERVIEW

Given the above sine and cosine values, we can easily calculate the corresponding tangents via their quotient (see formula (3.5)). We hereby draw for some special angles α their corresponding values for sine and cosine within the unit circle and their tangents on the vertical tangent line $x = 1$.

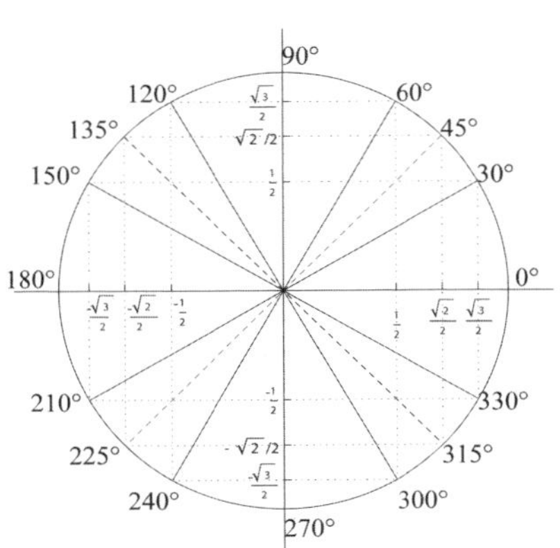

Figure 3.11: Graphical overview via the unit circle

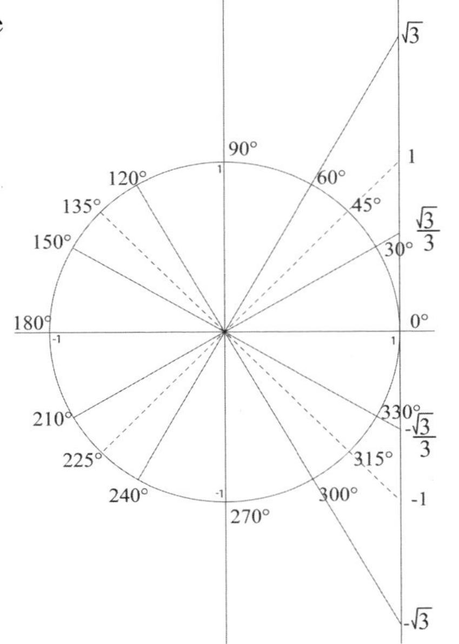

3.6 Pairs of Angles

We briefly explain the properties of two pairs of angles that are useful in this book.

Oppositely signed angles their measures add up to 0°. In other words, if α and β are oppositely signed then $\alpha + \beta = 0^\circ$ or $\beta = -\alpha$. The corresponding figure shows how the cosines of oppositely signed angles remain invariant, while their sines receive opposite signs. This leads to the trigonometric formulas $\cos(-\alpha) = \cos\alpha$ and $\sin(-\alpha) = -\sin\alpha$.

Complementary angles their measures add up to 90°. In other words, if α and β are complementary then $\alpha + \beta = 90^\circ$ or $\beta = 90^\circ - \alpha$. The corresponding figure shows how the sine of α equals the cosine of $90^\circ - \alpha$ and the cosine of α equals the sine of $90^\circ - \alpha$. This leads to the trigonometric formulas $\cos(90^\circ - \alpha) = \sin\alpha$ and $\sin(90^\circ - \alpha) = \cos\alpha$.

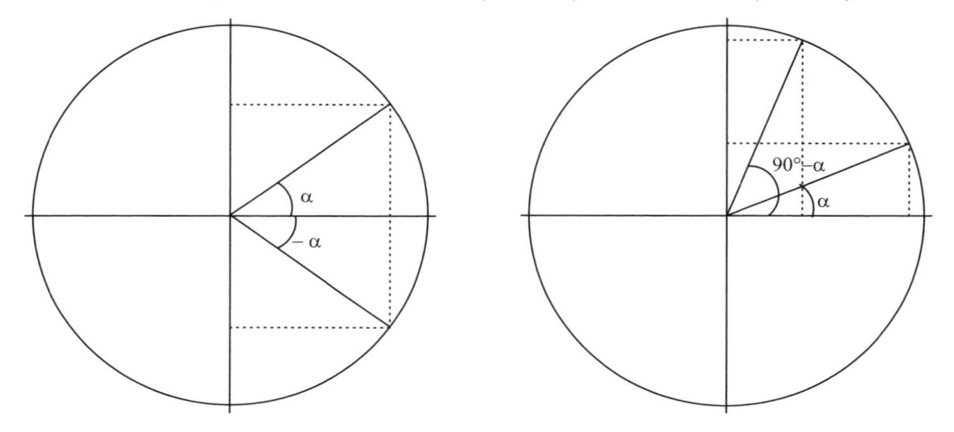

Figure 3.12: Oppositely signed and complementary angles

3.7 Sum Identities

In this paragraph we state and prove all trigonometric ratios of a sum of two angles. We firstly emphasize the non-linearity of all trigonometric ratios: e.g. for the sine we encounter $\sin(\alpha + \beta) \neq \sin\alpha + \sin\beta$. Indeed, e.g. for angles $\alpha = 60^\circ$ and $\beta = 30^\circ$ the value $\sin 90^\circ = 1$ does not equal the sum $\sin 60^\circ + \sin 30^\circ = \frac{\sqrt{3}+1}{2}$. Given the above inequality, we realize the need for the correct formulas which are stated below.

$\sin(\alpha+\beta) = \sin\alpha\cos\beta + \cos\alpha\sin\beta$ $\quad$ $\sin(\alpha-\beta) = \sin\alpha\cos\beta - \cos\alpha\sin\beta$

$\cos(\alpha+\beta) = \cos\alpha\cos\beta - \sin\alpha\sin\beta$ $\quad$ $\cos(\alpha-\beta) = \cos\alpha\cos\beta + \sin\alpha\sin\beta$

$\tan(\alpha+\beta) = \frac{\tan\alpha+\tan\beta}{1-\tan\alpha\tan\beta}$ $\quad$ $\tan(\alpha-\beta) = \frac{\tan\alpha-\tan\beta}{1+\tan\alpha\tan\beta}$

Proof: First of all we prove the formula for the cosine of a subtraction of angles, using the included right triangles and the Law of Cosines. Thereafter we will prove all other Sum Identities based upon this formula for the cosine of a subtraction.

▷ $\cos(\alpha-\beta) = \cos\alpha\cos\beta + \sin\alpha\sin\beta$

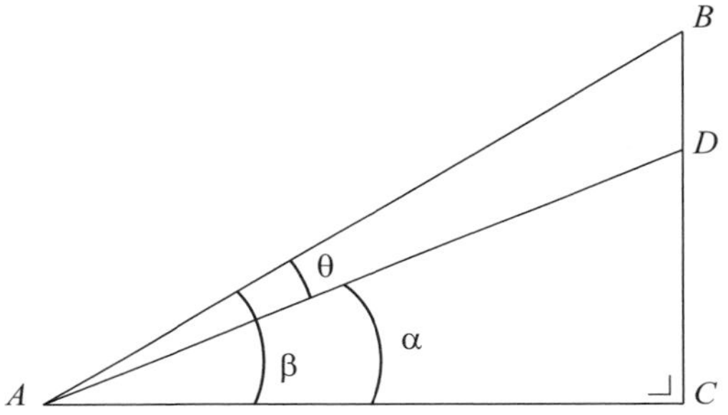

Figure 3.13: Proving the formula for the cosine of a subtraction

The right triangle ACD yields $\cos\alpha = \frac{|AC|}{|AD|}$ and $\sin\alpha = \frac{|DC|}{|AD|}$ and the right triangle ABC yields $\cos\beta = \frac{|AC|}{|AB|}$ and $\sin\beta = \frac{|BC|}{|AB|}$. Substituting the sines and cosines in the right hand side of the formula leads to

$$\begin{aligned}\cos\alpha\cos\beta + \sin\alpha\sin\beta &= \frac{|AC|}{|AD|}\cdot\frac{|AC|}{|AB|} + \frac{|DC|}{|AD|}\cdot\frac{|BC|}{|AB|}\\ &= \frac{|AC|^2 + |DC|\cdot|BC|}{|AD|\cdot|AB|}.\end{aligned}$$

Secondly, we interpret the left hand side $\cos(\alpha-\beta) = \cos(\beta-\alpha)$ as $\cos\theta$ in the scalene triangle BAD. We determine $\cos\theta$ using the Law of Cosines (3.8) for the angle θ and its opposite side $|BD|$ as $|BD|^2 = |AB|^2 + |AD|^2 - 2|AB|\cdot|AD|\cdot\cos\theta$.

To finalize the proof, eliminating $\cos\theta$ from both steps should lead to an equality. Substituting the factor $\cos\theta$ in the second step results in:

$$\begin{aligned}|BD|^2 &= |AB|^2 + |AD|^2 - 2|AB|\cdot|AD|\cdot\left(\frac{|AC|^2 + |DC|\cdot|BC|}{|AD|\cdot|AB|}\right)\\ &= |AB|^2 + |AD|^2 - 2(|AC|^2 + |DC|\cdot|BC|)\end{aligned}$$

We may replace the length $|BD|$ by the difference $|BC|-|DC|$ and apply the Perfect Square:

$$\begin{aligned}
|BD|^2 &= |AB|^2+|AD|^2-2(|AC|^2+|DC|\cdot|BC|)\\
(|BC|-|DC|)^2 &= |AB|^2+|AD|^2-2(|AC|^2+|DC|\cdot|BC|)\\
|BC|^2-2|BC|\cdot|DC|+|DC|^2 &= |AB|^2+|AD|^2-2|AC|^2-2|DC|\cdot|BC|\\
|BC|^2+|DC|^2 &= |AB|^2+|AD|^2-2|AC|^2.
\end{aligned}$$

We finally obtain the equality $(|BC|^2+|AC|^2)+(|DC|^2+|AC|^2)=|AB|^2+|AD|^2$ after grouping terms and applying twice the Pythagorean Theorem.

▷ Based upon $\cos(\alpha-\beta)=\cos\alpha\cos\beta+\sin\alpha\sin\beta$, we have less difficulties in proving the five remaining Sum Identities

$$\begin{aligned}
\cos(\alpha+\beta) &= \cos(\alpha-(-\beta))\\
&= \cos\alpha\cos(-\beta)+\sin\alpha\sin(-\beta)\\
&= \cos\alpha\cos\beta-\sin\alpha\sin\beta \qquad \text{(oppositely signed angles)}
\end{aligned}$$

$$\begin{aligned}
\sin(\alpha-\beta) &= \cos(90^\circ-(\alpha-\beta)) \qquad \text{(complementary angles)}\\
&= \cos((90^\circ-\alpha)+\beta)\\
&= \cos(90^\circ-\alpha)\cos\beta-\sin(90^\circ-\alpha)\sin\beta\\
&= \sin\alpha\cos\beta-\cos\alpha\sin\beta \qquad \text{(complementary angles)}
\end{aligned}$$

$$\begin{aligned}
\sin(\alpha+\beta) &= \sin(\alpha-(-\beta))\\
&= \sin\alpha\cos(-\beta)-\cos\alpha\sin(-\beta)\\
&= \sin\alpha\cos\beta+\cos\alpha\sin\beta \qquad \text{(oppositely signed angles)}
\end{aligned}$$

$$\begin{aligned}
\tan(\alpha\pm\beta) &= \frac{\sin(\alpha\pm\beta)}{\cos(\alpha\pm\beta)}\\
&= \frac{\sin\alpha\cos\beta\pm\cos\alpha\sin\beta}{\cos\alpha\cos\beta\mp\sin\alpha\sin\beta}\\
&= \frac{\frac{\sin\alpha\cos\beta}{\cos\alpha\cos\beta}\pm\frac{\cos\alpha\sin\beta}{\cos\alpha\cos\beta}}{\frac{\cos\alpha\cos\beta}{\cos\alpha\cos\beta}\mp\frac{\sin\alpha\sin\beta}{\cos\alpha\cos\beta}}\\
&= \frac{\tan\alpha\pm\tan\beta}{1\mp\tan\alpha\tan\beta}
\end{aligned}$$

■

When applying the quotient formula of tangent we should never divide by zero! This would occur in case $\tan\alpha=\pm\frac{1}{\tan\beta}$ or equivalently whenever $\alpha\pm\beta=90^\circ+k180^\circ$ given $k\in\mathbb{Z}$. These angular values would cause $\tan(\alpha\pm\beta)=\pm\infty$.

3.8 Inverse Trigonometric Functions

When an angle α returns $\sin\alpha = \frac{1}{2}$ then we may conclude, apart from the obvious $\alpha = \frac{\pi}{6}$, for infinitely more angular sizes α to return $\sin\alpha = \frac{1}{2}$. This is due to the periodicity of the function sine and the sine values (except for -1 and 1) to appear twice within one period. We have $\sin\alpha = \frac{1}{2}$ if $\alpha = \frac{\pi}{6}$ but also if $\alpha = \frac{5\pi}{6}$ and we may add multiples of 2π to each of them (due to the periodicity of the plane angle α). We avoid that an infinite number of angles α are the solution to $\sin\alpha = \frac{1}{2}$ when we restrict the angle α to the radian interval $[\frac{-\pi}{2}, \frac{\pi}{2}]$. A similar issue also applies for the cosine and for the tangent, solved in a similar way. We agree on the restricted interval $[0,\pi]$ for cosine and on the restricted interval $]\frac{-\pi}{2}, \frac{\pi}{2}[$ for tangent.

For any x being the result of a sine, cosine or tangent, we find one unique angle α in their restricted interval giving $\sin\alpha = x$, $\cos\alpha = x$ or $\tan\alpha = x$ respectively. The corresponding functions to trace this unique angle α are called the **arcsine**, the **arccosine** and the **arctangent**. We can typeset the arcsine either as arcsin, asin or $\sin^{-1}$.

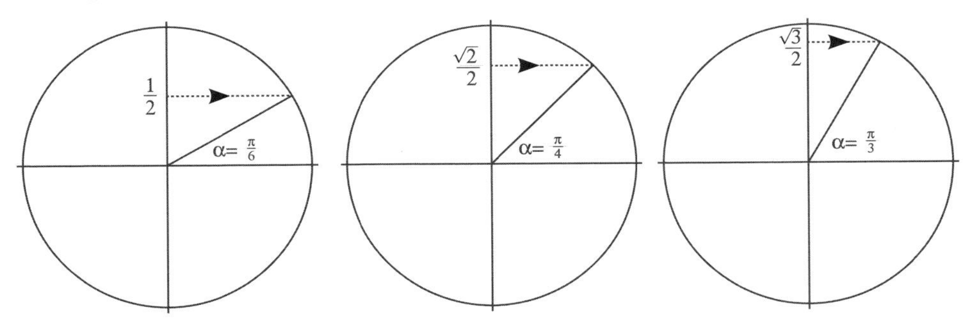

Figure 3.14: The arcsine

Arcsine returns for a given sine value its corresponding circular arc (in radians restricted to quadrants I and IV). For instance $\arcsin\frac{\sqrt{2}}{2} = \frac{\pi}{4}$ because $\sin\frac{\pi}{4} = \frac{\sqrt{2}}{2}$.

$$\alpha = \arcsin x \Leftrightarrow x = \sin\alpha \text{ with } \frac{-\pi}{2} \leqslant \alpha \leqslant \frac{\pi}{2}$$

Similarly, arccosine returns for any given cosine value its corresponding circular arc (in radians restricted to quadrants I and II).

$$\alpha = \arccos x \Leftrightarrow x = \cos \alpha \text{ with } 0 \leqslant \alpha \leqslant \pi$$

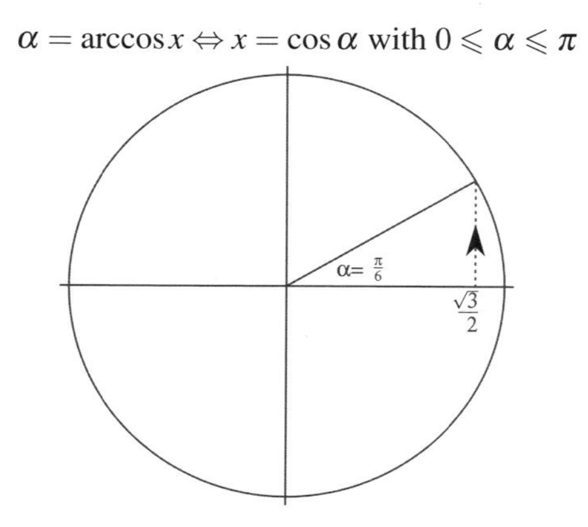

Figure 3.15: The arccosine

Finally, arctangent returns for a given tangent value its corresponding circular arc (in radians restricted to quadrants I and IV).

$$\alpha = \arctan x \Leftrightarrow x = \tan \alpha \text{ with } \frac{-\pi}{2} < \alpha < \frac{\pi}{2}$$

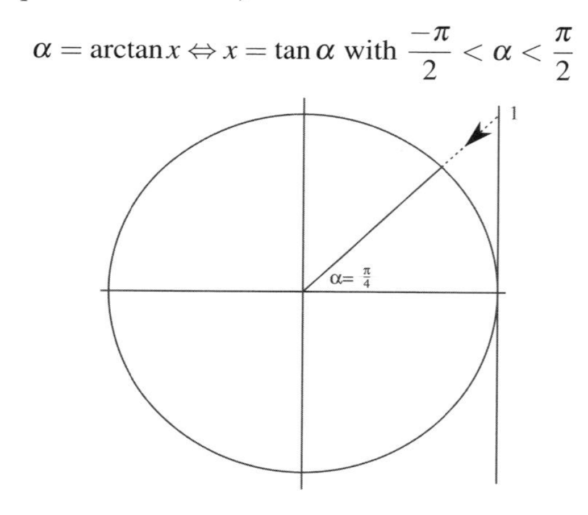

Figure 3.16: The arctangent

Example: What size is the (smallest) angle made by the straight line through the points $O(0,0)$ and $P\left(\frac{1}{2}, \frac{\sqrt{3}}{2}\right)$ and the x-axis?

Drawing the lines shows an opposite side of length $\frac{\sqrt{3}}{2}$ and an adjacent side of length $\frac{1}{2}$. This allows us to calculate the size of angle α made by the line OP and the x-axis via the tangent: $\tan \alpha = \frac{\frac{\sqrt{3}}{2}}{\frac{1}{2}} = \sqrt{3}$. We need to restrict the size of angle α to the radian interval $]\frac{-\pi}{2}, \frac{\pi}{2}[$ for the arctangent to return the corresponding size of the targeted angle as $\alpha = \arctan \sqrt{3} = \frac{\pi}{3}$.

3.9 Exercises

Exercise 24 Calculate the missing side lengths and/or angle sizes in the triangle ABC as typeset in Figure 3.10 for
1) $\alpha = 48°, b = 29$ cm and $\gamma = 90°$;
2) $a = 10$ cm, $\beta = 65°$ and $c = 12$ cm.

Exercise 25 When Martin drives a car on the highway his eyes are 1 meter above groundlevel. He approaches a major traffic sign which is hanging above his head. The traffic sign hangs 6 meter above the road surface and its height measures 3 meter. Find his visual angle when Martin approaches to 20 meter (measured along the road surface) from this traffic sign.

Exercise 26 The Belgian national railroad company renews the railroad from Kortrijk to Bruges. Accidently, ignoring the rail's expansion due to the summer heath, they miscalculated its length. It takes 50 km rail from Kortrijk to Bruges, but in summer this rail suffers an overall expansion of 1m. The railroad authorities decide to incline the track instead of to shorten the rail's length with 1m. Find the track's height at halfway the line from Kortrijk to Bruges.

Exercise 27 A pilot receives flight data from a plane in order to intercept that plane. The situation below was shown on the pilot's display before the data signal was scrambled. Find the distance from the pilot to the point of interception.

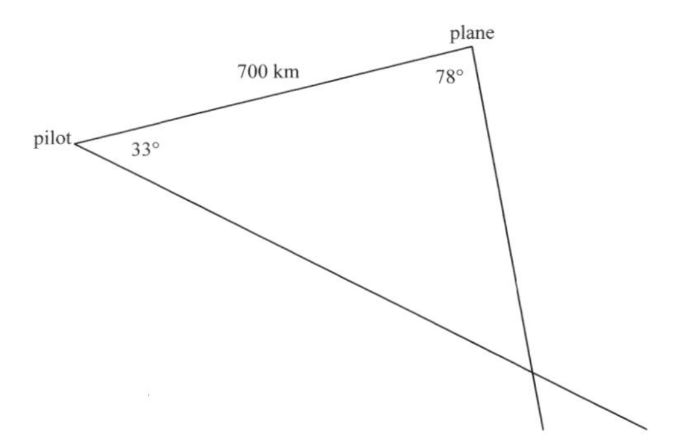

Exercise 28 A scouting plane takes off from a runway at the east coast, heading in the direction of 80°, being measured clockwise from the north direction of 0° onwards. Due to awful weather conditions the flying scout returns to another east coast landing strip at 250 km distance north of its home base. Then the plane takes the direction of 283° for its return flight. What is the total distance this plane will travel?

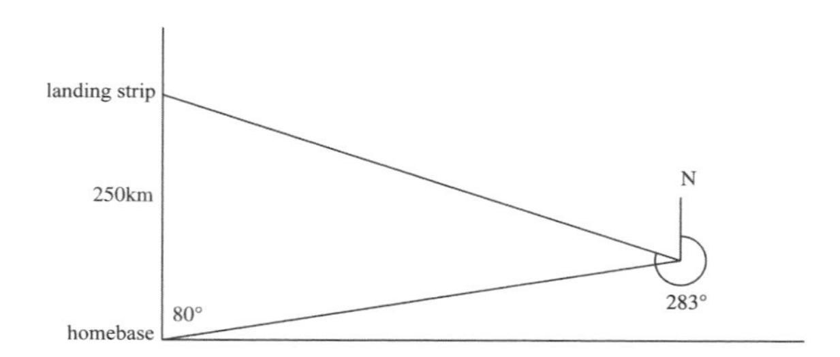

Exercise 29 Two friends work together to determine a plane's height. They keep a distance of 180 meter in the open field and spot the plane simultaneously. Martin measures an angle of 60° and Isaac an angle of 30°. Together they are able to calculate the plane's altitude. Are you?

Exercise 30 Find α in case

1) $\sin\alpha = \frac{\sqrt{3}}{2}$

2) $\tan\alpha = 1$

3) $\cos 2\alpha = \frac{-\sqrt{2}}{2}$

4) $\tan\frac{\alpha}{2} = -\sqrt{3}$

Exercise 31 At each corner of a square field *ABCD* whose side measures 14 meter, a flag-pole is standing. The flag-pole at corner *A* measures 7 meter, the flag-pole at corner *B* has length 10 meter and in corner *C* it is a pole of 15 meter. Somewhere in the square's interior a point *P* is located at equal distance to each flag-pole top.

- ▷ Find the location of this point *P*. Therefore calculate distances x and y, being the horizontal distance from *P* to the side $[AD]$ and the vertical distance from *P* to the side $[AB]$ respectively.

- ▷ Given the distance from *P* to the flag-pole top at corner *D* is the same distance from *P* to the previous tops in corners *A*, *B* and *C*, find the height of the flag-pole in *D*.

Chapter 4 · Functions

How does a computer determine if two objects are colliding, how personages move or what trajectory a canonball takes? It can all be achieved by the use of functions. In this chapter we outline some basic concepts on functions and we discuss the important linear and quadratic functions. We then perform collision detection via intersecting functions. Finally we present the trigonometric functions for they simulate periodical effects.

4.1 Basic concepts on real functions

- ▷ We define a **function** as a mapping f that for each argument x returns at most one image $f(x)$.
- ▷ We define the **domain** of a function as the set of arguments x which have exactly one image $f(x)$.
- ▷ We define the **range** of a function as the set of all images $f(x)$ returned by the function f.
- ▷ We define a **root** of a function f as each argument x_0 that maps to $f(x_0) = 0$.

All four of the above definitions can be summarized in an explicit formula or recipe f : *domain* $\rightarrow$ *range* : $x \mapsto y = f(x)$. The real function $f : \mathbb{R} \rightarrow \mathbb{R} : x \mapsto y = f(x)$ maps arguments x of the domain in $\mathbb{R}$ onto images $f(x)$ of the range in $\mathbb{R}$. Therefore we can easily visualize the function in an $\mathbb{R} - \mathbb{R}$ graph.

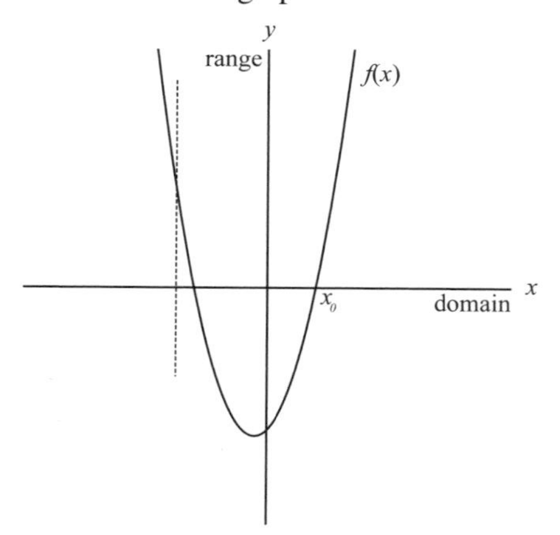

Figure 4.1: The graph of a function $f(x)$

We revisit the former concepts approaching them from the graph of the function f.

- ▷ We verify the defining property graphically via running a vertical line horizontally through the graph of f: at most one intersection point has to occur for f being a function.
- ▷ We recognize the domain of a function f graphically as the area on the x-axis where the function f is defined.
- ▷ We recognize the range of a function f graphically as the set of all returned images $f(x)$ on the y-axis.
- ▷ We graphically interpret a root of a function f as an intersection point x_0 of the graph of the function and the x-axis.

4.2 Polynomial functions

We define a **polynomial function** V via the recipe $y = V(x)$ given $V(x)$ a real polynomial. These functions have domain $\mathbb{R}$, a possibly limited range in $\mathbb{R}$ and up to maximal n roots as solutions to the equation $V(x) = 0$. We call n the degree of the polynomial $V(x)$ and we call n the degree of the polynomial function V as well.

Linear functions

We define a **linear function** via the recipe

$$y = ax + b. \tag{4.1}$$

These functions have domain $\mathbb{R}$, a range in $\mathbb{R}$ which is limited only in case of a horizontal line and a (most likely) unique root as solution to the linear equation $ax + b = 0$.

We define the **slope** $\boldsymbol{a}$ as the coefficient to x in the recipe. The slope meaning the inclination of the linear graph to the horizontal x-axis. A horizontal run of k steps on the x-axis corresponds to a rise of $a \cdot k$ steps on the y-axis when $a > 0$ and to a descent of $a \cdot k$ steps on the y-axis when $a < 0$.

If both axes are equally calibrated then the slope a equals the tangent of the inclination angle α of the linear graph to the x-axis:

$$\text{slope} = \frac{\Delta y}{\Delta x} = \tan \alpha. \tag{4.2}$$

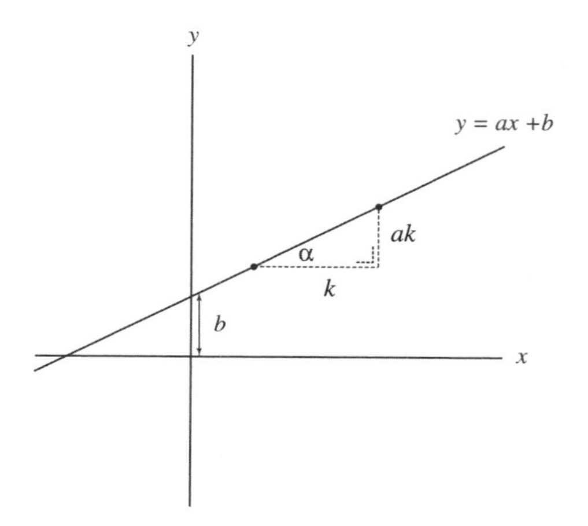

Figure 4.2: The linear function $y = ax + b$

The capital Greek character 'delta' Δ refers to 'd' as the starting character of 'difference'. Hence Δx denotes the difference of two x-values while Δy typesets the difference of two y-values.

In case $a < 0$ we have a descending line, in case $a = 0$ a horizontal line and in case $a > 0$, then the linear graph is ascending. Any two lines $y = ax + b$ and $y = mx + n$ are parallel if their slopes are equal: $a = m$. Any two lines $y = ax + b$ and $y = mx + n$ are perpendicular if their slopes multiply to minus one: $a \cdot m = -1$.

We define the **intercept *b*** as the constant term in the recipe. The intercept meaning the position where the linear graph intersects the y-axis.

Any two points $Q(x_Q, y_Q)$ and $R(x_R, y_R)$ determine the recipe of the linear function they are spanning.

Based upon the formula (4.2) defining the slope, for any point $P(x, y)$ on a non-vertical straight line holds

$$\text{slope} = \frac{\Delta y}{\Delta x} = \frac{y_R - y_Q}{x_R - x_Q} = \frac{y - y_Q}{x - x_Q}.$$

As a consequence of the above formulas we derive an expression for the straight line stretching from point $Q(x_Q, y_Q)$ to point $R(x_R, y_R)$.

$$\begin{aligned} y - y_Q &= \frac{y_R - y_Q}{x_R - x_Q} \cdot (x - x_Q) && (4.3) \\ y - y_Q &= \text{slope} \cdot (x - x_Q). \end{aligned}$$

Example: We determine the linear function $f(x)$ connecting two points $R(0,2)$ and

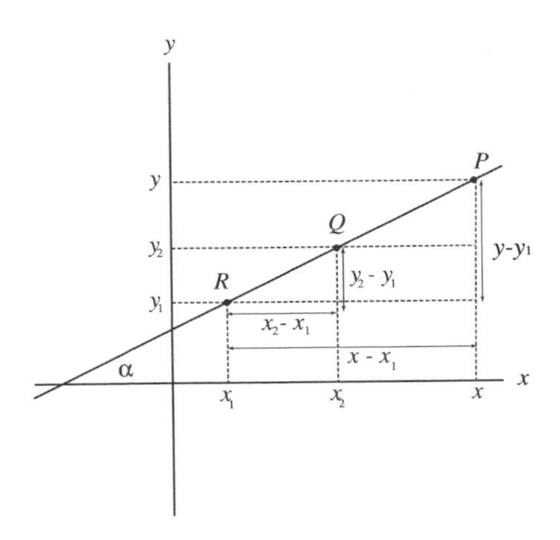

Figure 4.3: The straight line connecting points Q and R

$Q(11,23)$.

Formula (4.3) yields $y-2=\frac{23-2}{11-0}(x-0)$ which we simplify to the recipe $y=\frac{21}{11}x+2$.

As a bonus we also aim for the recipe of the line being perpendicular to the line $y=\frac{21}{11}x+2$ and passing through the point $(2,-7)$. We therefore make use of the property of the slopes of perpendicular lines multiplying to -1. Hence the slope of the perpendicular line we wish to construct equals $-\frac{11}{21}$. Eventually this perpendicular line is $y+7=-\frac{11}{21}(x-2)$ which we simplify to the recipe $y=-\frac{11}{21}x-\frac{125}{21}$.

Quadratic functions

We define the real **quadratic function** as a mapping f which returns for any argument x *maximal* one image $y=ax^2+bx+c$, given a,b,c real coefficients with $a\neq 0$.

Quadratic functions take $\mathbb{R}$ for their domain, limit their range to one side and can have up to two roots x_1 and x_2 which are traceable by the quadratic formulas (see page 26). We call the graph of the quadratic function a parabola.

Example: We hereby present the elementary prototype

$f:\mathbb{R}\longrightarrow\mathbb{R}:x\mapsto x^2$, alternatively $y=x^2$.

We verify this parabola for being a function by running a vertical line horizontally through the (x,y)-frame and counting for the number of return values $f(x)$ corresponding to each argument x.

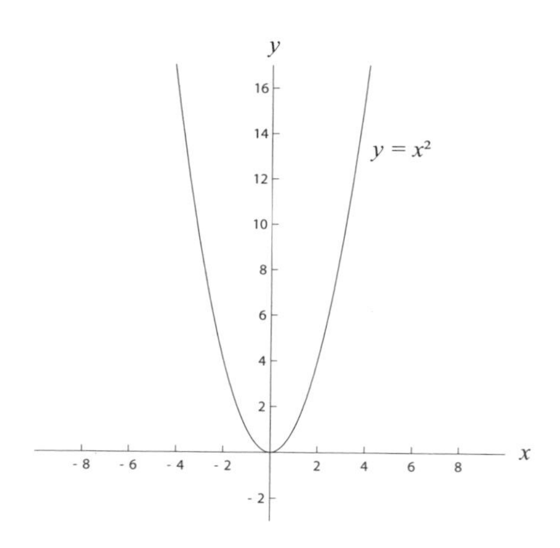

Figure 4.4: The graph of the elementary function $y = x^2$

We see the **parabola opening up** in case $a > 0$ or the **parabola opening down** for $a < 0$. We call the lowest or highest point on it the **vertex of a parabola**. We calculate the x-coordinate of the vertex of the parabola as $x_{vertex} = \frac{-b}{2a}$. The vertical line through the vertex is the axis of symmetry of the parabola. The coefficient a measures the width of the parabola: the larger the absolute value of a the steeper the parabola. The coefficient b is the slope of the tangent line to the parabola in its intersection point on the y-axis. The coefficient c determines the parabola's intersection point on the y-axis, as c gives the y-shift from zero.

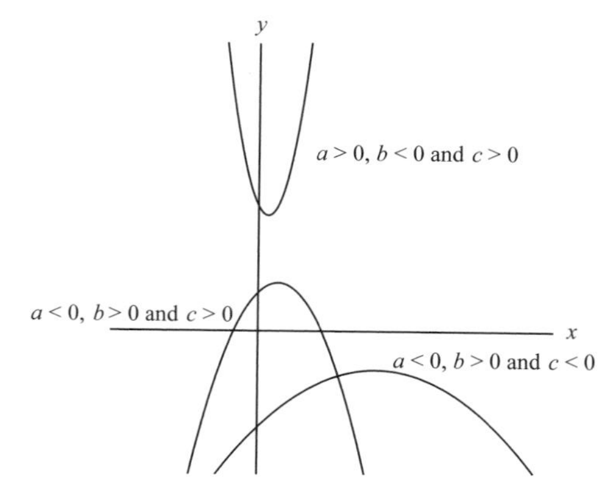

Figure 4.5: The effect of the coefficients a, b and c of the quadratic function $y = ax^2 + bx + c$

We also look at the prototype of the **horizontal parabola** as it is defined by $y^2 = x$. By reducing this equation to zero we can define this horizontal parabola equivalently by $y^2 - x = 0$. Such a horizontal parabola is clearly *no function* as running a vertical line horizontally through the (x,y)-frame reveals *two* return values y for each argument x. Instead we call a horizontal parabola generally a **curve**.

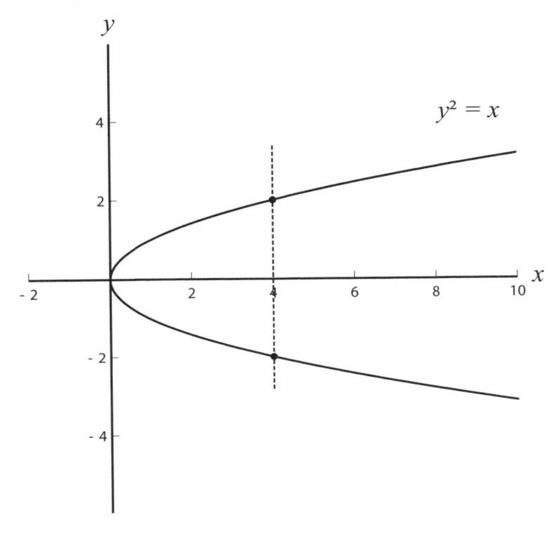

Figure 4.6: The graph of the curve $y^2 = x$

4.3 Intersecting functions

Sometimes animation programming requires the intersection of two lines. For this occasion those lines often model the edge of a building, the horizon or the trajectory of a moving object.

Let us take an easy start aiming for the intersection of just two straight lines. Two different straight lines in the same plane will either intersect in one point or they appear to be parallel. How can we in the former case find their intersection point? We can do this by tracing the x-value that leads to identical return values. Because we try to locate where both straight lines *equal each other*, we have to compare their recipes in one equation.

Example: We intersect the two straight lines $f(x) = 2x - 2$ and $g(x) = -x + 10$. Therefore, we compare their recipes in one equation $2x - 2 = -x + 10$ which we solve for x. Doing so yields $x = 4$ as the unique solution. Substituting this x-value in one of both recipes yields the corresponding y-value, $y = f(4) = g(4) = 6$. Conclusively we find the intersection point $(x,y) = (4,6)$.

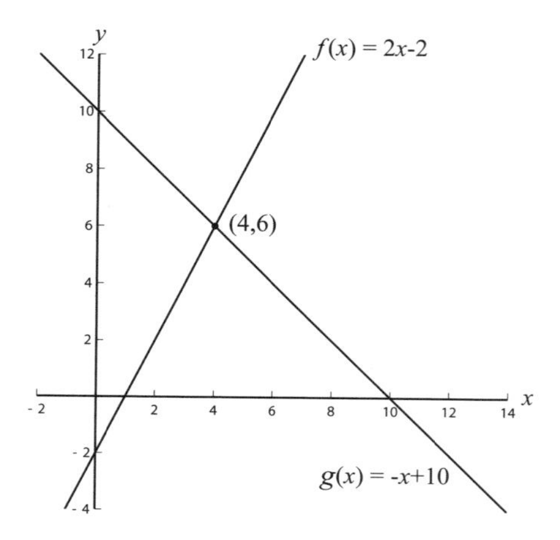

Figure 4.7: Intersecting two straight lines

Alternatively we can find the intersection point of these straight lines by solving the 2 by 2 system (see page 34)

$$\begin{cases} y = 2x - 2 \\ y = -x + 10. \end{cases}$$

Two straight lines may not always have one unique intersection point. They can be parallel without having any intersection point or they can coincide having an infinite number of them.

We apply the same method to determine whether a straight line and a parabola intersect or even to find out whether two parabolas intersect.

Example: Consider two quadratic functions $f(x) = 2x^2 - 10x + 5$ and $g(x) = -x^2 + 2x - 4$. We impose $f(x) = g(x)$ for the x-coordinate of an intersection point, consequently $2x^2 - 10x + 5 = -x^2 + 2x - 4$ after substituting their recipes and equivalently $3x^2 - 12x + 9 = 0$. Solving this quadratic equation using the quadratic formulas (see page 26) yields $x_1 = 1$ and $x_2 = 3$. Calculating $f(1) = g(1) = -3$ and $f(3) = g(3) = -7$ we obtain the coordinates of both intersection points of these parabolas as $(1, -3)$ and $(3, -7)$.

We generalize the above method of 'comparing' recipes. In order to possibly trace the intersection points of two different functions $f(x)$ and $g(x)$, we solve the equation $f(x) = g(x)$ for the horizontal coordinate x. Equivalently we may solve the to zero reduced equation $f(x) - g(x) = 0$ for x. Substituting any found x-value in the recipe of either $f(x)$ or $g(x)$, yields the vertical coordinate y of the corresponding intersection point.

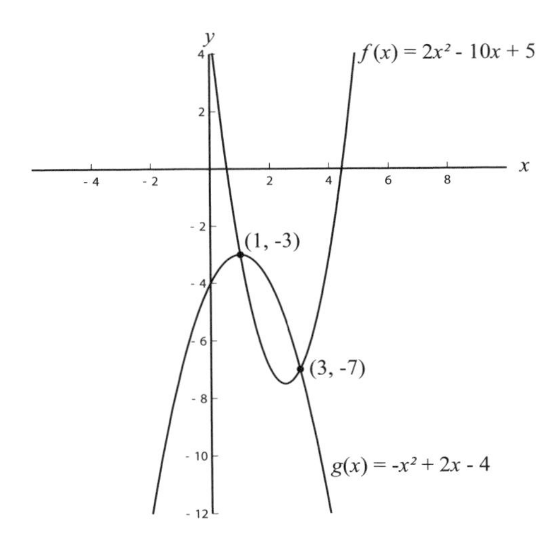

Figure 4.8: Determine intersection points

4.4 Trigonometrical functions

In a previous chapter we met the trigonometric ratios. Amongst them, the trigonometric sine function rules all professional applications which exhibit periodicity (electromagnetic waves, sound, but also certain animations). We will need a more technical insight on how the sine evolves over a time interval. Whenever we input an (angle) argument t, which is variable in time, we call it a sine function $\sin t$.

Elementary sine function

We first of all study the elementary sine function $f : \mathbb{R} \to [-1,1] : x \mapsto \sin x$. Running through the unit circle we return for each angle its sine value.

The domain of this function is the horizontal x-axis or in other words the set $\mathbb{R}$. Due to the between -1 and 1 limited sine values, the range of this function is the vertical window $[-1,1]$. The roots of the elementary sine function are given by $k\pi$, for any $k \in \mathbb{Z}$. Since we can keep running through the unit circle, this function inherits the periodicity of its angle argument, showing an elementary period of 2π.

Generalized sine function

By adding some extra parameters we obtain the generalized sine function. This extended

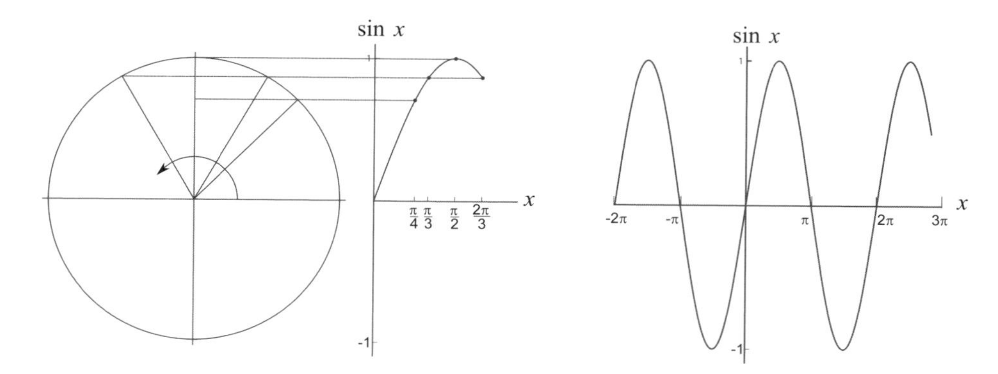

Figure 4.9: The sine function

function looks like $f(x) = a\sin\Big(b(x-c)\Big) + d$. Wave phenomena, oscillations and vibrations are often modeled by such a generalized sine function.

▷ The **amplitude** or **elongation** $\boldsymbol{a}$ is the maximal displacement from the equilibrium. We use the amplitude as the factor to stretch/squeeze the sine function along the y-axis. The distance from the minimum to the maximum return value of the generalized sine function equals $2|a|$. In figure 4.10(a) we show the effect of an amplitude that was brought to 4.

▷ We recall the periodicity of the elementary sine function, its period equal to the full circle: $\sin x = \sin(x+2\pi)$. This equality holds, even after giving x a general coefficient b like this $\sin(bx) = \sin(bx+2\pi)$. We show how this extra coefficient b changes the elementary period by $\sin(bx) = \sin\Big(b(x+\frac{2\pi}{b})\Big)$. For the **generalized period** $\boldsymbol{T}$ we conclude $T = \frac{2\pi}{b}$. After solving this formula for b, we obtain it equivalently in the different shape $b = \frac{2\pi}{T}$ which is better to remember as we can read below.

  The **pulsation** $|\boldsymbol{b}|$ is the inner coefficient that allows us to stretch/squeeze the sine function along the x-axis. Due to $b = \frac{2\pi}{T} = 2\pi f$, given f the frequency in Hertz, we may call the pulsation also **angular frequency**. In figure 4.10(b) we show the effect of a pulsation that was brought to 2 and thus equivalently led to a generalized period of π.

▷ We define the **phase** $\boldsymbol{c}$ as the argument of the 'first incoming crossing' on the baseline of the sine function. In other words, the phase c of the generalized sine is the length by which the elementary graph is shifted to the right ($c > 0$) or to the left ($c < 0$). We may consider the phase c as the horizontal shift. In figure 4.10(c) the elementary graph is shifted by $\frac{\pi}{2}$ to the right. This is the effect of the phase c

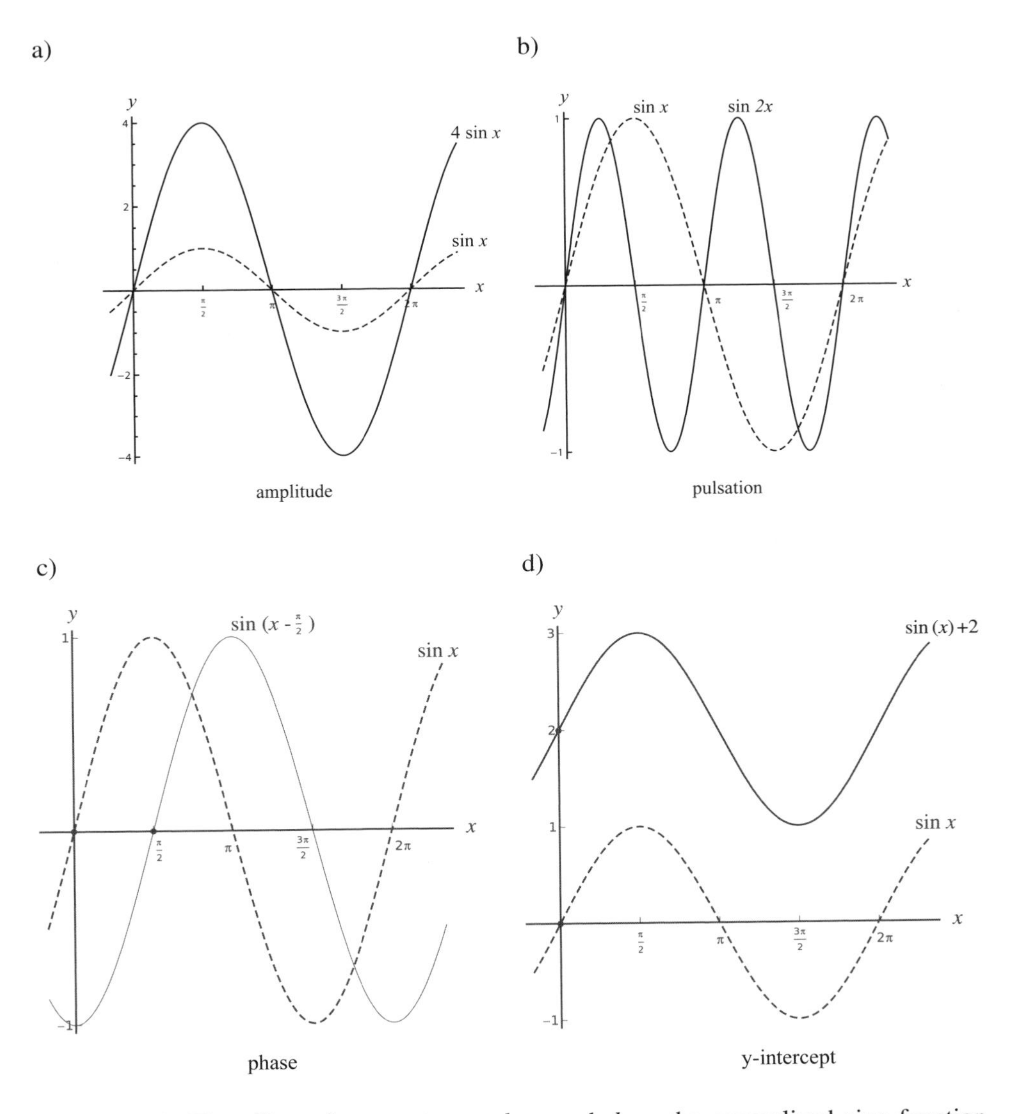

Figure 4.10: The effect of parameters a, b, c and d on the generalized sine function $a\sin\Big(b(x-c)\Big)+d$

brought to $\frac{\pi}{2}$.

According to their phase difference one sine function can be 'ahead of the other'. Sine functions with phase difference zero are said to be 'in-phase'. Sine functions with a phase difference equal to the straight angle ($\Delta c = \pi$) are said to be in 'antiphase' .

▷ The **y-intercept *d*** is the constant term added to the generalized sine function. In case $d \neq 0$ the roots of the generalized sine function are called baseline 'crossings'. In other words, the y-intercept d is the length by which the elementary graph is shifted upwards ($d > 0$) or downwards ($d < 0$). We may consider the y-intercept d as the vertical shift. In figure 4.10(d) the elementary graph is shifted by 2 upwards, showing the effect of a y-intercept brought to 2.

Example: Aiming for the recipe of the generalized sine function as portrayed in figure 4.11, we deduce the four parameters a, b, c and d from it.

The minimal return value is 1 and the maximal is 5. We calculate the amplitude by solving $2a = 5 - 1$ to $a = 2$. Consequently this sine function will oscillate around its horizontal baseline $y = \frac{1+5}{2} = 3$. This means that the elementary graph is shifted by 3 upwards, concluding for a y-intercept of $d = 3$. On the horizontal baseline we spot the 'first incoming crossing' as being shifted by 1 to the right, showing a phase of $c = 1$. Eventually we deduce the period, for instance measured from $x = 1$ to $x = 3$, as $T = 2$. Equivalently we calculate the pulsation as $b = \pi$. Putting it all together we find the function $f(x) = 2\sin\left(\pi(x-1)\right) + 3$.

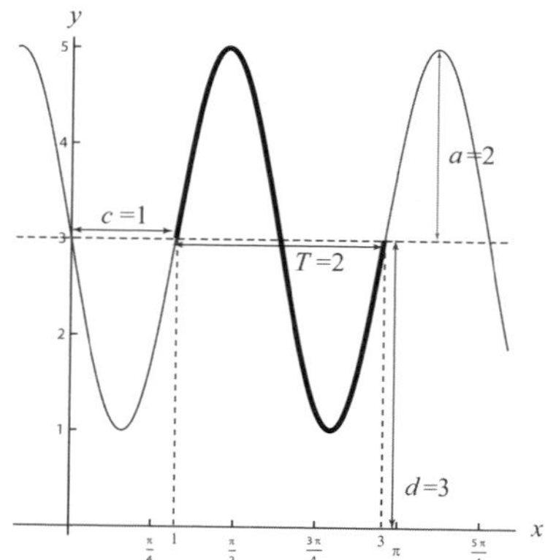

Figure 4.11: Graph of the function $f(x) = 2\sin(\pi(x-1)) + 3$

4.5 Inverse trigonometrical functions

We define the inverse trigonometrical functions (also called cyclometric functions) based on their former definitions (see page 57).

Logically, the real range $[-1,1]$ of the elementary sine function becomes the domain of its inverse function, called arcsine. We recall the periodicity of the elementary sine function (given its natural period of 2π). Due to this periodicity the arcsine would not be a function, as to every argument between -1 and 1, there corresponds an infinite number of returned angles. For this reason we restrict the range of the arcsine to the real interval $\left[\frac{-\pi}{2}, \frac{\pi}{2}\right]$.

$$\begin{array}{llll} \sin: & \mathbb{R} & \longrightarrow & [-1,1] \\ & x \text{ (in rad)} & \mapsto & y = \sin x \end{array}$$

$$\begin{array}{llll} \arcsin: & [-1,1] & \longrightarrow & \left[\dfrac{-\pi}{2}, \dfrac{\pi}{2}\right] \\ & x & \mapsto & y \text{ (in rad) } = \arcsin x \end{array}$$

Note that sine acts as a function on angles (in radians) to return values in the real interval $[-1,1]$, while arcsine acts as a function on numbers (between -1 and 1) to return their corresponding angle in radians (between $\frac{-\pi}{2}$ rad and $\frac{\pi}{2}$ rad).

Example: We calculate the arcsine return for some arguments:

$$\begin{array}{llll} \arcsin: & [-1,1] & \longrightarrow & \left[\dfrac{-\pi}{2}, \dfrac{\pi}{2}\right] \\ & x & \mapsto & y \text{ (in rad) } = \arcsin x \\ & \text{-2} & \mapsto & \arcsin(-2) \text{ is not defined because } -2 \notin [-1,1] \\ & -1 & \mapsto & \arcsin(-1) = \dfrac{-\pi}{2} \\ & -\dfrac{1}{2} & \mapsto & \arcsin\left(-\dfrac{1}{2}\right) = \dfrac{-\pi}{6} \\ & 0 & \mapsto & \arcsin 0 = 0 \\ & \dfrac{1}{2} & \mapsto & \arcsin\left(\dfrac{1}{2}\right) = \dfrac{\pi}{6} \end{array}$$

$$\frac{\sqrt{2}}{2} \mapsto \arcsin\left(\frac{\sqrt{2}}{2}\right) = \frac{\pi}{4}$$

$$\frac{\sqrt{3}}{2} \mapsto \arcsin\left(\frac{\sqrt{3}}{2}\right) = \frac{\pi}{3}$$

Doing so for all arguments x running through the real interval $[-1,1]$ leads to the graph of the arcsine function.

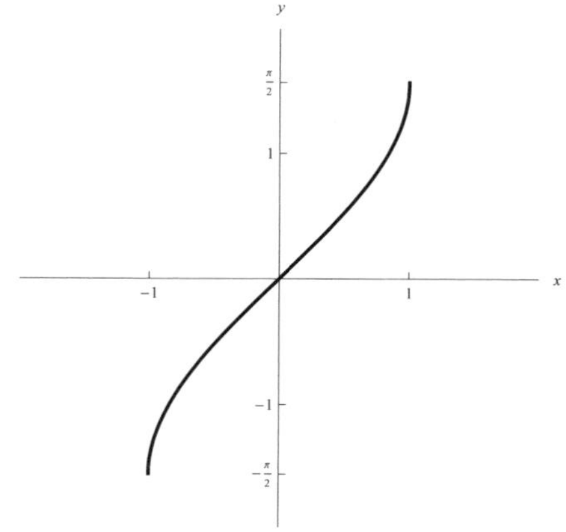

Figure 4.12: The arcsine function

In the same way arccosine is the inverse function of cosine which returns for any given argument (between -1 and 1) its corresponding angle (restricted to the interval $[0,\pi]$) in radians.

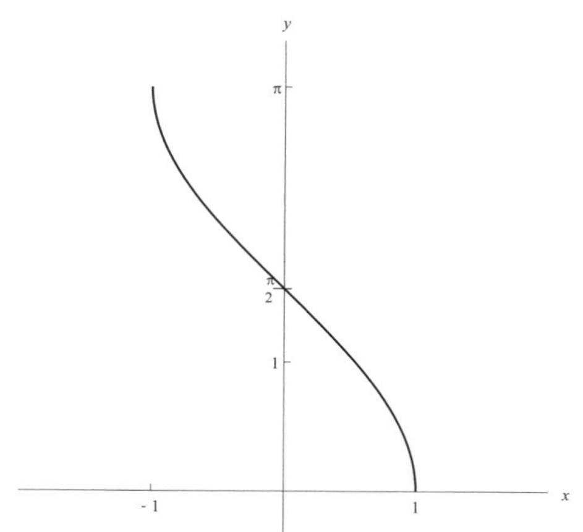

Figure 4.13: The arccosine function

And eventually arctangent is the inverse function of tangent which returns for any given argument its corresponding angle (restricted to the open interval $]\frac{-\pi}{2}, \frac{\pi}{2}[$) in radians.

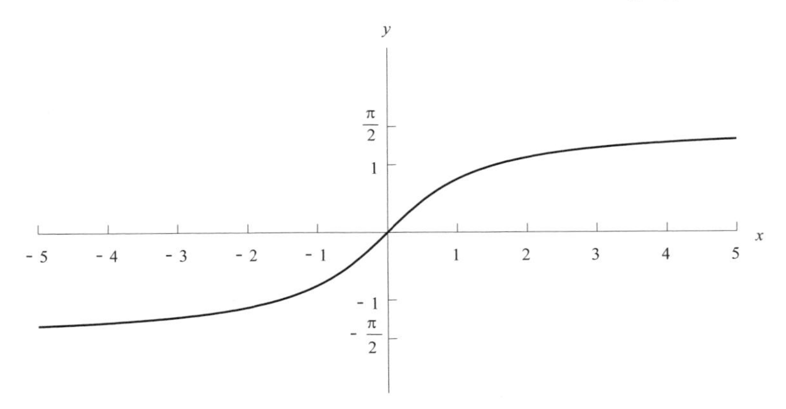

Figure 4.14: The arctangent function

4.6 Exercises

Exercise 32 Give the equation of a straight line

- ▷ without roots,
- ▷ with an infinite number of roots.

Exercise 33 The equation of a parabola is written as $y = ax^2 + bx + c$. Determine the sign of each coefficient a, b and c for each parabola type shown below.

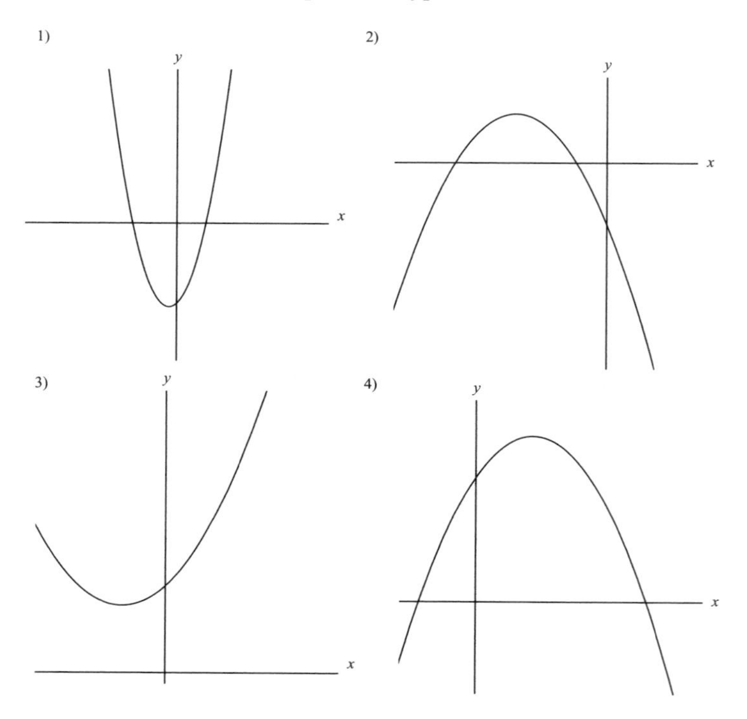

Exercise 34 During a computer game an object is running on a straight line from the point A with coordinates $(0, 20)$ to the point B pinned by $(15, 30)$. Find the equation of this straight line AB. When the object is crossing the point $(30, 40)$, the gamer sends it 90° to the left on another straight line. Find the equation of this new straight line. Draw both lines, the former line AB and the latter new one, in one (x, y)-frame.

Exercise 35 A car is driving on a straight line $3x+5y=8$. On another straight line $x+3y=4$ a brick wall is built. If the car keeps driving on its straight line, will it ever hit this wall? In case it does, in what point exactly? Draw both lines in one (x,y)-frame.

Exercise 36 Find the intersection points of the functions $f(t)=12t-24$ and $g(t)=3t^2+3t-36$.

Exercise 37 A military airplane descends according to the straight line $h_1(t)=-2t+100$, with h referring to the altitude in hectometer. The running argument t is the time being expressed in minutes. A defence rocket orbiting on the parabola $h_2(t)=-t^2+3t+106$ will destroy the plane. Determine the moment and the altitude for the rocket to strike the plane.

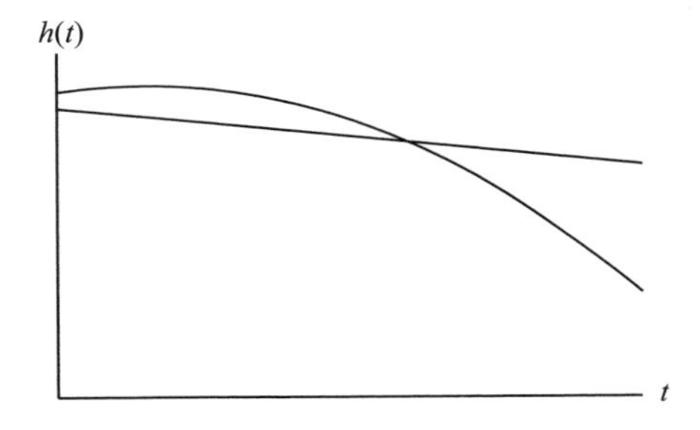

Exercise 38 Give the recipes for the functions as shown below.

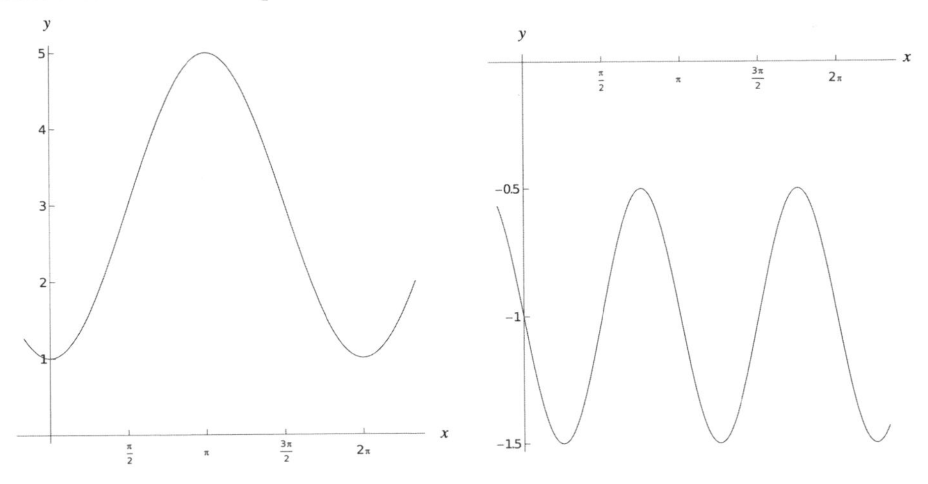

Chapter 5 · The Golden Section

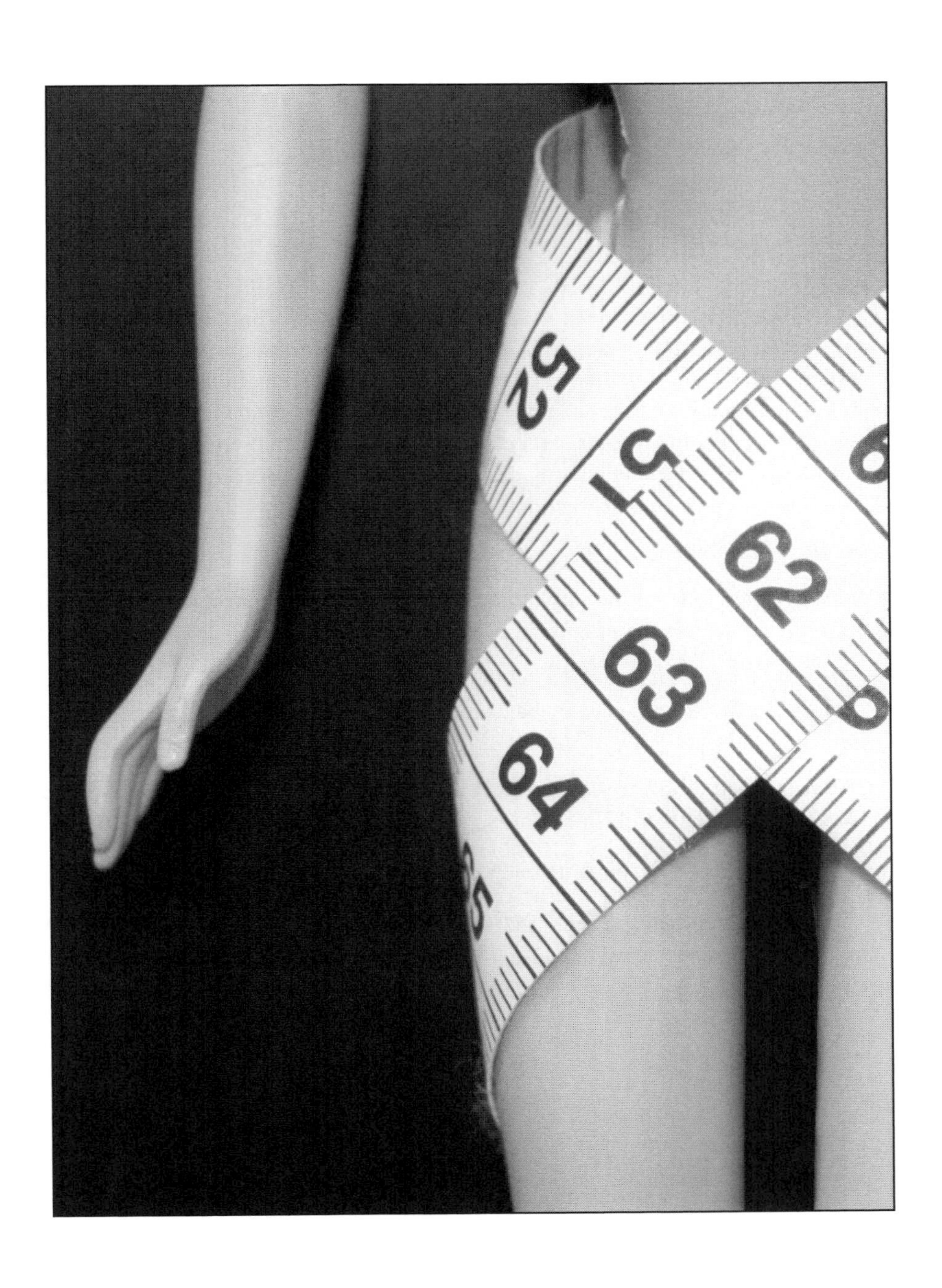

Eras ago, the **golden number** $\frac{1+\sqrt{5}}{2}$ was constructed. Presumably, since Antiquity, dating already 2500 years back into time. Ever since, it is kept in use as a famous irrational number from Antiquity. We easily find illustrations of it in nature as well as applications of it around the world. Throughout this chapter we promote the professional use of the golden section in graphical and digital arts. A pair of compasses, a ruler and a calculator are the required tools for this aesthetic chapter.

5.1 The Golden Number

Around 300 Before Christ the Greek geometer **Euclid** (approx 300 - 250 Before Christ) outlined a ruler-and-compass construction to cut a line segment into **mean proportional parts**, in volume II (11th proposition of the Arithmetical Geometry) and volume VI (30th proposition of the Theory of Ratio) of his thirteen-piece geometric standard 'The Elements'. Euclid stated we can cut a line segment into two parts in such a way that the ratio of the bigger part over the smaller part equals the ratio of the total segment over its bigger part. Ever since this true statement is called the **golden section** for which we outline a modern ruler-and-compass construction.

Construction: the golden section.

Firstly, the construction of $\frac{1+\sqrt{5}}{2}$ implies the drawing of $\sqrt{5}$ in an exact way. We logically base the construction of square roots such as $\sqrt{5}$ on the Pythagorean Theorem (3.2).

1) We draw a right triangle ABC having its right angle in A and its lengths of sides $|AC| = 2|AB|$. Hence, the Pythagorean Theorem dictates the corresponding hypotenuse to be $|BC| = \sqrt{5}|AB|$.

2) We then extend the line segment $[AB]$ with the constructed segment $[BC]$ in such a way that the length of this new line segment becomes $|AD| = (1+\sqrt{5})|AB|$.

3) We finally cut $[AD]$ through its midpoint M which yields $|AM| = \frac{1+\sqrt{5}}{2}|AB|$.

So far, the ratio of the final segment $[AM]$ over the segment $[AB]$ results in $\frac{|AM|}{|AB|} = \frac{1+\sqrt{5}}{2}$. Secondly, we still need to prove that also the ratio of the bigger part $[AB]$ over the smaller part $[BM]$ simplifies to $\frac{1+\sqrt{5}}{2}$.

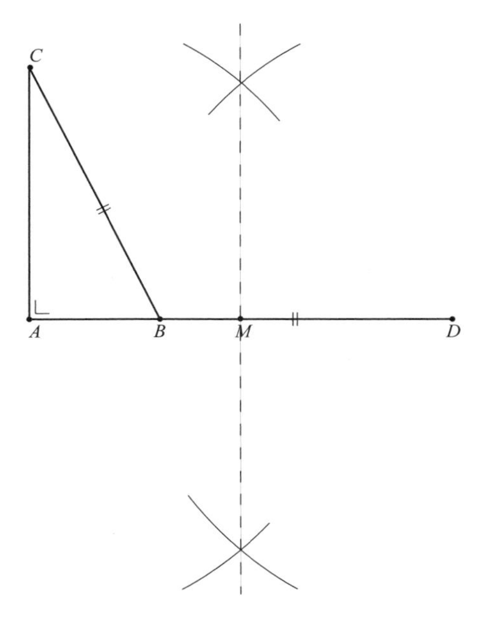

Figure 5.1: A construction of the golden section

Proof:

Setting $|AB| = a$ leads to

$$|BM| = |AM| - |AB| = \frac{1+\sqrt{5}}{2}a - a = \left(\frac{1+\sqrt{5}}{2} - 1\right)a = \frac{\sqrt{5}-1}{2}a.$$

We conclude

$$\frac{|AB|}{|BM|} = \frac{a}{\left(\frac{\sqrt{5}-1}{2}\right)a} = \frac{2}{\sqrt{5}-1},$$

since $a \neq 0$. We get rid of the square root in the denominator by multiplying the fraction with $1 = \frac{\sqrt{5}+1}{\sqrt{5}+1}$ and applying the Perfect Quotient (see formula (1.3)) in such a way that

$$\frac{|AB|}{|BM|} = \left(\frac{2}{\sqrt{5}-1}\right)\left(\frac{\sqrt{5}+1}{\sqrt{5}+1}\right) = \frac{2(\sqrt{5}+1)}{5-1} = \frac{1+\sqrt{5}}{2}.$$

■

The segment $|AB|$ is called the mean proportional part between $|AM|$ and $|BM|$, also known as geometric mean of $|AM|$ and $|BM|$:

$$\frac{|AM|}{|AB|} = \frac{|AB|}{|BM|} \iff |AM| \cdot |BM| = |AB|^2.$$

We call this constant ratio the golden number which we typeset as

$$\Phi = \frac{1+\sqrt{5}}{2} \approx 1.61803. \tag{5.1}$$

This constant proportion Φ, named after the first character of the Greek sculptor **Phidias** (490 - 430 Before Christ), equals 1.618 only approximatively because of its infinite amount of unpredictable fractional digits, just like all **irrational numbers** in $\mathbb{R}$ do.

5.2 The Golden Section

The Golden Triangle

We define the **golden triangle** as any isosceles triangle with an apex angle of 36°.

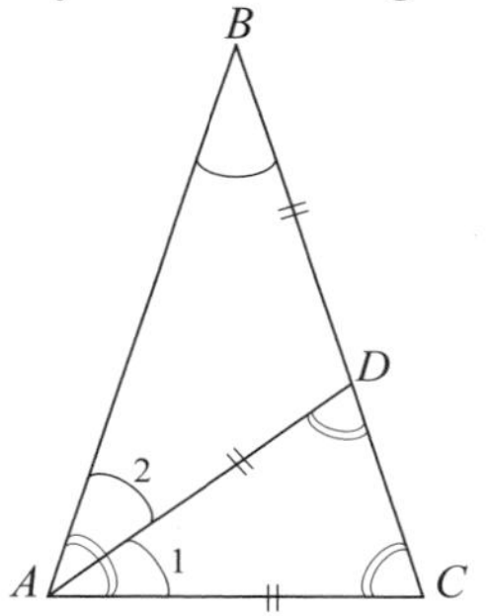

Figure 5.2: The Golden Triangle

The isosceles triangle ABC with apex B featuring angle $\hat{B} = 36°$ must have equal base angles $\hat{A} = 72° = \hat{C}$, because of its interior angles adding up to the straight angle. Measuring its base $[AC]$ as the radius of the circle $C(A, |AC|)$ up to the intersecting point D on segment $[BC]$, creates a new isosceles triangle CAD. This latter triangle features base angles $\hat{C} = 72° = \hat{D}$ and must have an apex angle $\hat{A}_1 = 36°$, because of its interior angles adding up to the straight angle.

The similarity of the triangles ABC and CAD (see page 45) leads to the geometric ratios $\frac{|AB|}{|AC|} = \frac{|AC|}{|CD|} = \Phi$, hence their title of 'golden triangle'.

The Golden Rectangle

Amongst the wide variety of rectangles, we tend to appreciate only one 'ideal rectangle'. It will not be a too stretched rectangle, nor will it be a square (which is also a valid example of a rectangle), but a typical shape somewhere in between. We define this unique 'ideal rectangle' in words: 'it is the rectangle which after cutting away a square, is similar to the remaining rectangle'.

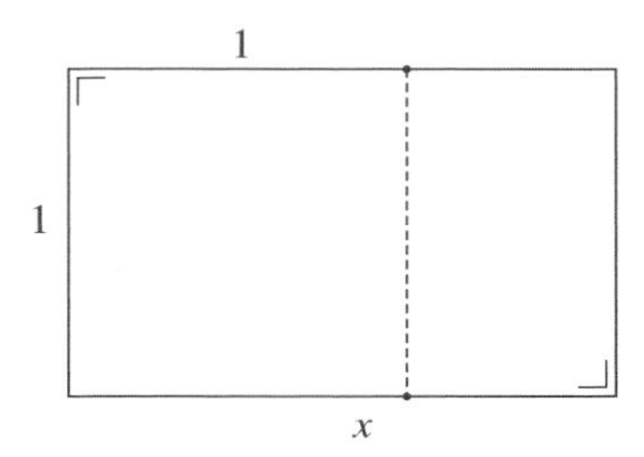

Figure 5.3: The Golden Rectangle

We formalize the above definition based on the similarity in its description:

$$\frac{\text{length}}{\text{width}} \text{ of the golden rectangle} = \frac{\text{length}}{\text{width}} \text{ of its 'remaining rectangle'}.$$

Setting the golden rectangulars' width to 1 and its length to x, we can express this similarity as an equation $\frac{x}{1} = \frac{1}{x-1}$. Solving this equation for x will lead us to the 'ideal ratio' we are searching for. To do so, we rewrite our expression to a standard quadratic equation

$$x^2 - x - 1 = 0, \tag{5.2}$$

subsequently applying the quadratic formula (see formula (1.6)) for it. We calculate its discriminant $D = 5$ and both roots as $x_1 = \frac{1+\sqrt{5}}{2}$ and $x_2 = \frac{1-\sqrt{5}}{2}$.

Given $\sqrt{5} \approx 2.23607$ and the fact that lengths have to be non-negative, we can only allow the x_1-value $\Phi = \frac{1+\sqrt{5}}{2} \approx 1.61803$ as the physical solution to the equation.

Alternatively we should not ignore the negative mathematical solution $x_2 = \frac{1-\sqrt{5}}{2}$ which we therefore distinguish with an accent mark as

$$\Phi' = \frac{1-\sqrt{5}}{2} \approx -0.61803. \tag{5.3}$$

Conclusively, we again rediscovered the ratio corresponding to the golden section as the only meaningful solution and therefore rename the 'ideal rectangles' to **golden rectangles**.

The Golden Spiral

We start from a golden rectangle *ABCD* for which $|AD| = \Phi \cdot |AB|$.

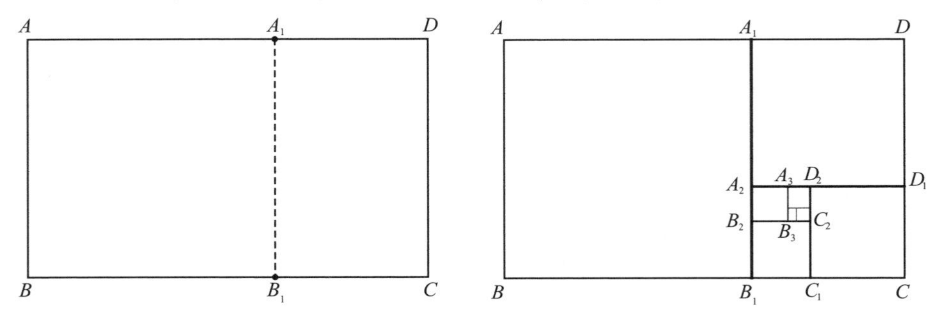

Figure 5.4: Construction of the Golden Spiral in a Golden Rectangle

On segment $[AD]$ we choose the point A_1 and on segment $[BC]$ we choose the point B_1 in such a way that $|AA_1| = |AB| = |BB_1|$. This way we cut away a square from the golden rectangle. We therefore realize $|A_1D| = |AD| - |AB| = \Phi \cdot |AB| - |AB| = (\Phi - 1) \cdot |AB|$. Multiplying both sides of this equality with the factor Φ yields $\Phi \cdot |A_1D| = (\Phi^2 - \Phi) \cdot |AB|$. Due to the former quadratic equation $\Phi^2 - \Phi = 1$ for the golden number Φ, we simplify the latter equality to $\Phi \cdot |A_1D| = |AB|$. Given the equal widths $|AB| = |CD|$ of the original rectangle, we find $|CD| = \Phi \cdot |A_1D|$. The latter proportion proves also the remaining rectangle A_1B_1CD to be a golden rectangle.

We proceed cutting away a square from this new golden rectangle A_1B_1CD, which will again lead to a smaller remaining golden rectangle. Iterating this process yields a sequence of progressively shrinking golden rectangles: $ABCD$, DA_1B_1C, $CD_1A_2B_1, B_1C_1D_2A_2$, $\ldots$.

We observe that the sequence of points $B, A_1, D_1, C_1, B_2, A_3, \ldots$ makes up a spiral. We can show this spiral in a non-exact way by drawing a quarter of a circle inside the progressively shrinking square cutaways, shifting the circle center accordingly. Mathematically this socalled 'spiral' is just a chain of quarters of a circle.

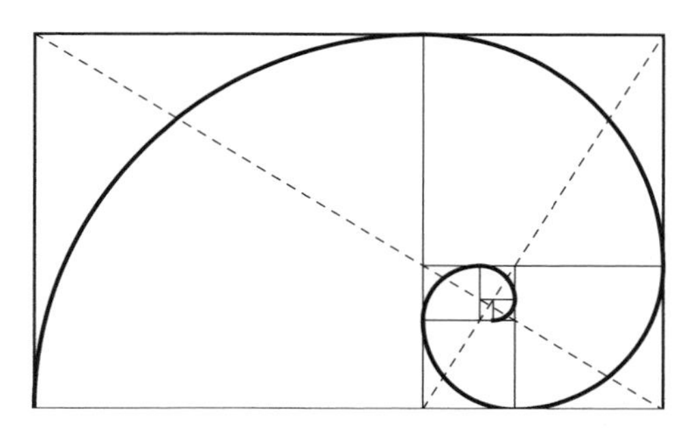

Figure 5.5: The Golden Spiral in a Golden Rectangle

In a similar way we discovered a **golden spiral** based on the sequence of progressively shrinking golden rectangles, we can sketch a golden spiral based on the golden triangle.

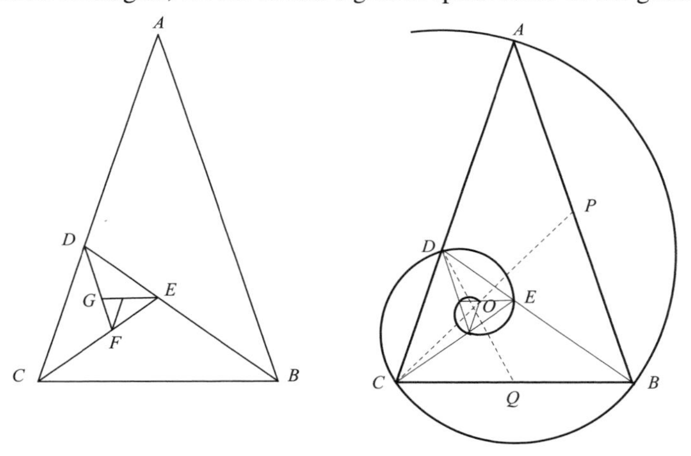

Figure 5.6: The Golden Spiral in the Golden Triangle

The isosceles triangles $ABC, BCD, CDE, DEF, EFG, \ldots$ are similar (see page 45) due to the constructed bisector of angles $\hat{B}$, $\hat{C}$, $\hat{D}$, $\hat{E}$, $\hat{F}$, $\hat{G}, \ldots$. The starting point O for the golden spiral is the intersection point of the medians of the initial golden triangle and its remaining golden triangle. To some more extent, the starting point is the intersection point of the medians CP and DQ of the golden triangles ABC and BCD respectively.

The Golden Pentagon

The diagonals of a pentagon intersect according to the golden section.

Proof:

In this pentagon the triangle SBG is similar to the triangle ABS because they have two equal interior angles (see page 45). The angle in the point B is already shared and based on the symmetry of the pentagon we obtain $\hat{A} = \hat{S}$. Besides, $|AG| = |TR| = |SB|$ which instantly leads to

$$\frac{|AB|}{|BS|} = \frac{|SB|}{|BG|} \Rightarrow \frac{|AB|}{|AG|} = \frac{|AG|}{|BG|}.$$

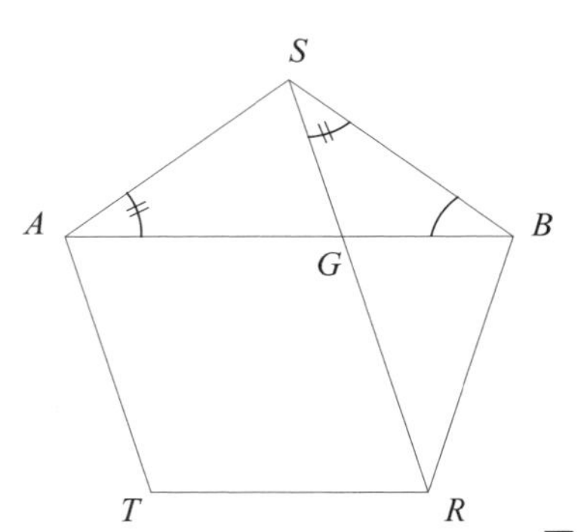

Conclusively, the point G cuts the intersecting diagonal $[AB]$ according to the golden section. ■

Due to the appearance of the golden section Φ on its diagonals, we give the pentagon its alias of **golden pentagon**.

The Golden Ellipse

We consider two concentric circles around center O with radii a and b in such a way that $a > b$. The area of the ring in between both circular disks $C(O,a)$ and $C(O,b)$ equals the difference of both circle areas $\pi a^2 - \pi b^2 = \pi \cdot (a^2 - b^2)$.

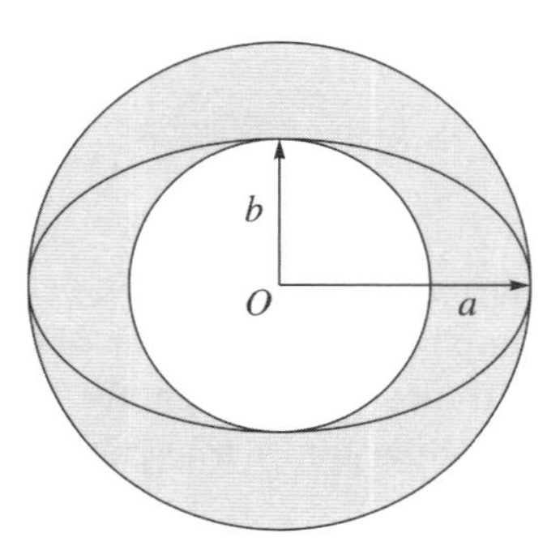

Figure 5.7: The Golden Ellipse

The inner ellipse centered on O featuring **semi-major axis** a and **semi-minor axis** b has an area equal to $\pi \cdot a \cdot b$.

Only in case the area of this ellipse equals the area of the ring, we conclude

$$\pi \cdot (a^2 - b^2) = \pi \cdot ab \Leftrightarrow a^2 - b^2 = ab \Leftrightarrow a^2 - a \cdot b - b^2 = 0.$$

Dividing both sides of the latter equation by b^2, since the existence of the ring assures $b \neq 0$, leads to

$$\frac{a^2}{b^2} - \frac{a \cdot b}{b^2} - \frac{b^2}{b^2} = 0 \Leftrightarrow \left(\frac{a}{b}\right)^2 - \frac{a}{b} - 1 = 0$$

or in other words, to the algebraic expression of the famous numbers Φ and Φ' (5.2).

Only in case of those equal areas, the ratio of the semi-axes $\frac{a}{b} = \Phi$ entitles it as **golden ellipse**.

5.3 Golden arithmetics

Golden Identities

We prove the following golden identities easily by substituting $\Phi = \frac{1+\sqrt{5}}{2}$ and applying some elementary arithmetics.

First Identity

$$\Phi^2 = \Phi + 1. \tag{5.4}$$

Proof:

We substitute $\Phi = \frac{1+\sqrt{5}}{2}$ in both sides and then simplify them until we reach an equality.

For the left hand side we find

$\Phi^2 = \left(\frac{1+\sqrt{5}}{2}\right)^2 = \left(\frac{1+\sqrt{5}}{2}\right)\left(\frac{1+\sqrt{5}}{2}\right) = \frac{1+2\sqrt{5}+5}{2 \cdot 2} = \frac{6+2\sqrt{5}}{4} = \frac{2 \cdot (3+\sqrt{5})}{4} = \frac{3+\sqrt{5}}{2}$,

while for the right hand side we reach $\Phi + 1 = \frac{1+\sqrt{5}}{2} + 1 = \frac{1+\sqrt{5}}{2} + \frac{2}{2} = \frac{3+\sqrt{5}}{2}$.

■

Second Identity

$$-\frac{1}{\Phi} = \Phi'. \tag{5.5}$$

Proof:

$$-\frac{1}{\Phi} = \frac{-1}{\frac{1+\sqrt{5}}{2}}$$

$$= \frac{-2}{1+\sqrt{5}} = \left(\frac{-2}{1+\sqrt{5}}\right) \cdot \left(\frac{1-\sqrt{5}}{1-\sqrt{5}}\right) \quad \text{via the Perfect Quotient (1.3)}$$

$$= \frac{-2(1-\sqrt{5})}{-4} = \frac{1-\sqrt{5}}{2}$$

$$= \Phi'.$$

■

The Fibonacci Numbers

There is a famous number sequence which is prominently related to the golden section, called the **Fibonacci sequence** as mentioned in the novel '*The Da Vinci Code*'. We generate them as the sum of both previous numbers in their sequence by taking 1 and 1 as the two starting numbers: 1, 1, 2, 3, 5, 8, 13, 21, 34, 55, 89, ... Let us study those number sequences to a deeper extent.

In general we define a number sequence u_n for which every element equals the sum of both previous elements:

$$u_{n+1} = u_n + u_{n-1} \tag{5.6}$$

as **Lucas sequences**, named after the French mathematician **Edouard Lucas** (1842-1891). As soon as we initialize both starting numbers u_1 and u_2 for a Lucas sequence, we define it completely.

Example 1: Setting $u_1 = 2, u_2 = 1$, we generate the Lucas sequence

$$2, 1, 3, 4, 7, 11, 18, 29, \ldots$$

and we call its elements **Lucas numbers**.

Example 2: Setting $u_1 = 5, u_2 = 2$, we create the Lucas sequence

$$5, 2, 7, 9, 16, 25, 41, 66, 107, 173, 280, 453, 733, \ldots, 13153, 21282, \ldots.$$

Let us, for incrementing index values n, divide elements u_n of this Lucas sequence by their previous element u_{n-1} and display those fractions $\frac{u_n}{u_{n-1}}$ decimally.

$\frac{u_n}{u_{n-1}}$	ratio	decimally
$\frac{u_5}{u_4}$	$\frac{16}{9}$	1.77778...
$\frac{u_{12}}{u_{11}}$	$\frac{453}{280}$	1.61786...
$\frac{u_{13}}{u_{12}}$	$\frac{733}{453}$	1.61810...
$\frac{u_{20}}{u_{19}}$	$\frac{21282}{13153}$	1.61803...
$\vdots$		

We notice how those ratios $\frac{u_n}{u_{n-1}}$ converge to $\Phi = 1.61803\ldots$ at incrementing indices.

Example 3: Setting $u_1 = 1, u_2 = 1$, we build the Lucas sequence

$$1, 1, 2, 3, 5, 8, 13, 21, 34, 55, 89, 144, 233, \ldots, 17711, 28657, \ldots.$$

Lucas called the elements of this sequence **Fibonacci numbers**, named after the Italian **Leonardo of Pisa** (1175 – 1250, alias Fibonacci or Filius Bonaccio meaning 'son of Bonaccio'). This name was based on Fibonacci's popular 'rabbit population problem' published in his book 'Liber Abaci', as the above numbers were the solution to it.

The British author **Henry Dudeney** (1857–1930) brought a more realistic version of it, known as the 'cow population problem' described in his book '536 Puzzles and Curious Problems'. It also generates the Fibonacci sequence: *'If a cow in her second year produces her first calf (no bull) and continues like this annually, how many cows would this make in twelve years assuming no animal died?'*

$\frac{f_n}{f_{n-1}}$	ratio	decimally
$\frac{f_5}{f_4}$	$\frac{5}{3}$	1.66667...
$\frac{f_{12}}{f_{11}}$	$\frac{144}{89}$	1.61798...
$\frac{f_{13}}{f_{12}}$	$\frac{233}{144}$	1.61806...
$\frac{f_{23}}{f_{22}}$	$\frac{28657}{17711}$	1.61803...
$\vdots$		

Again, for incrementing index values n, we divide elements f_n of this Fibonacci sequence by their previous element f_{n-1} and display those fractions $\frac{f_n}{f_{n-1}}$ decimally. Again we notice how the ratios $\frac{f_n}{f_{n-1}}$ converge to $\Phi = 1.61803\ldots$ at incrementing indices.

We assume the above property to be true for all Lucas sequences. Consequently we try to prove that *for all* Lucas sequences their corresponding sequence of fractions $\frac{u_n}{u_{n-1}}$ will converge to Φ.

Proof:

1) Proving for a sequence of fractions to converge to one constant number is rather hard. This phenomenon is called **convergence**, for which we kindly refer to the literature on real calculus, skipping this first part of the complete proof.

2) Referring to step 1 of this proof, we assume the sequence of fractions $\frac{u_n}{u_{n-1}}$ to tend to one constant limit $L \in \mathbb{R}$. We continue to use the definition of the Lucas sequence (5.6). Dividing both sides of it by u_n yields

$$\frac{u_{n+1}}{u_n} = \frac{u_n}{u_n} + \frac{u_{n-1}}{u_n},$$

which brings the definition of the Lucas sequence to

$$\frac{u_{n+1}}{u_n} = 1 + \frac{u_{n-1}}{u_n}.$$

We assume in step 1 - omitting its proof - that in its limit we may replace the fraction $\frac{u_{n+1}}{u_n}$ by the constant L. Consequently we replace the inverse fraction $\frac{u_{n-1}}{u_n}$ by the inverse limit value $\frac{1}{L}$. After these replacements we encounter the equation

$$L = 1 + \frac{1}{L}$$

or after multiplying both sides with L and rearranging terms, the standardized equation

$$L^2 - L - 1 = 0.$$

This means that in its limit, we can rewrite the definition of Lucas sequences to the quadratic equation (5.2) valid for the golden number Φ.

Conclusively, we either find $L = \Phi$ or $L = \Phi'$. Based on the non-negative fractions $\frac{u_{n+1}}{u_n} > 0$, we inevitably remain with the limit value $L = \Phi > 0$. ■

5.4 The Golden Section worldwide

Probably the **golden section** is already for about two millennia famous as a mathematically modelled shape, discovered in nature. The golden section is an 'ideal proportion' featured during growth and the biological development of people, animals and plants.

Hence, the golden section was and is often implemented by people as an 'ideal measure' for their buildings, paintings, sculptures, musical instruments and compositions.

Though, we need to stay sceptic and cannot refer each proportion around us to Φ. A few prominent counterexamples are the A4 paper size, the Parthenon in Athens or the Great Pyramid of Giza. Despite our sound scepticism we hereby draft a representative list of various golden section implementations.

About eighteen centuries beyond Euclid we credit **Luca Bartolomeo de Pacioli** (1445 - 1517) to rediscover the 'divine' proportion Φ (1509, *De Divina Proportione*). Pacioli used to be a Franciscan munk who published revolutionary books on mathematics. The human body proportions were then successively related to the golden number by a famous pupil of Pacioli, **Leonardo da Vinci** (1490, *Vitruvian Man* as portrayed on the Italian 1€ coins), continued by both Germans **Henry Cornelius Agrippa** (1510, *Pentagram Man*) and **Adolf Zeising** (1854, *Neue Lehre von den Proportionen des menschlichen Körpers*).

Figure 5.8: Statue of the 'Vitruvian Man' in Vinci (Italy)

How did you think Barbie™ is tempting already generations of girls? The golden section applies in unimaginable many ways to the perfect human body. Realizing the average person is far from perfect, let us adopt the Barbie™ fashion doll as the ideal body. The length from her belly button to her feet is called the mean proportional part between her total length and her remaining length. We measure the same golden ratio on her arms, legs, hands and feet.

The golden section is also ruling architecture for ages, from the temples of Antiquity and medieval cathedrals to the anthropometric standard called the 'modulor' as designed by the French-Swiss architect **Le Corbusier** (1887–1965). And we should not forget to add the Islamic (Great Mosque of Kairouan), the Buddhist (Borobudur, Java) and the Spanish architecture being as eager to implement the golden section on many levels.

Many musical instruments are constructed according to Φ; especially the traditional violin. Even musical compositions and performances can contain the golden number; it seems at least to be the case for Beethoven, Haydn, Schubert, Bartók and Satie.

We should not forget painters: Seurat (*Bathers at Asnieres*), Juan Gris (cubism), Rembrandt (*The Night Watch*) and again Leonardo da Vinci, each of them applying the golden section in their art.

Eventually it goes without saying that graphical design makes use of the golden section. For instance the logo of the Belgian railroad company, as designed by the Belgian **Henry Van de Velde** (1863 - 1957), is completely based on the golden section. The same applies to digital art, animation and web design. Talking about the latter, how about a nice two-column design for a webpage? Assuming a screen resolution of 1024×768, we can start from a webpage of width 1000 pixels. Dividing its width by Φ yields 618 pixels for our main column, leaving 382 for the second column. Even this straightforward approach guarantees automatically an aesthetic result.

5.5 Exercises

Exercise 39 Given the construction

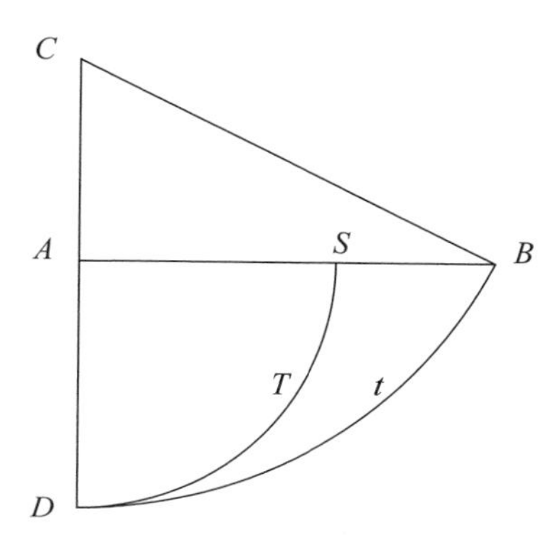

featuring $|AC| = \frac{1}{2} \cdot |AB|$ and the circular arc $\hat{t}$ by the circle $C(C, |BC|)$ intersecting the straight line CA at the side of A in point D. Next to this we have a second circular arc $\hat{T}$ by the circle $C(A, |AD|)$.

Prove that the point S intersects the segment $[AB]$ according to the golden section. In other words, setting length $|AB| = 1$ should imply length $|AS| = \frac{1}{\Phi} \approx 0.618$.

Hint: write the length of $[AD]$ in terms of the length of $[AB]$ and then calculate their ratio $\frac{|AS|}{|AB|}$.

Exercise 40 Given the construction

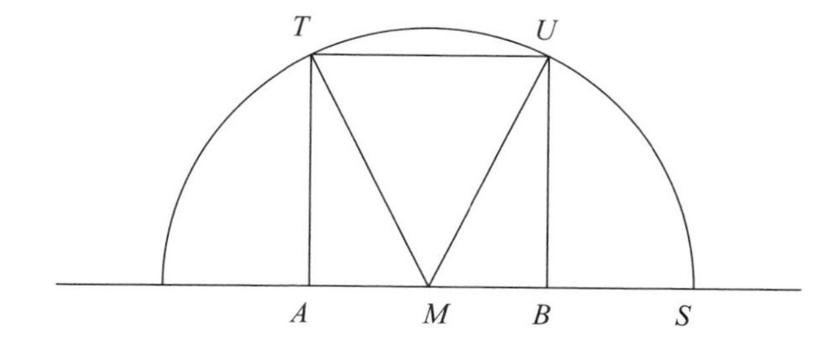

featuring the circular arc by the circle $C(M, |MT|)$, the square $ABUT$, a midpoint causing $|AM| = |MB|$ and S the intersection point of the circular arc $C(M, |MT|)$ and the straight line AB.

Prove that the point B intersects the segment $[AS]$ according to the golden section. In other words, setting length $|AB| = 1$ should imply length $|AS| = \Phi \approx 1.618$.

Hint: write all lengths $|MU|$, $|MS|$, $|AS|$, $|AB|$ in terms of the length of $[AM]$ and then calculate the ratio $\frac{|AS|}{|AB|}$.

Exercise 41 Prove $\Phi^2 + \frac{1}{\Phi^2} = 3$ by substituting $\Phi = \frac{1+\sqrt{5}}{2}$.

Exercise 42 Prove $\frac{1}{\Phi} = \Phi - 1$ by substituting $\Phi = \frac{1+\sqrt{5}}{2}$.

Exercise 43 Prove $\Phi + \frac{1}{\Phi} = \sqrt{5}$ by substituting $\Phi = \frac{1+\sqrt{5}}{2}$.

Exercise 44 Prove by substituting, that $x_1 = -1, x_2 = -\Phi$ and $x_3 = -\Phi'$ are the solutions to the cubic equation $x^3 + 2 \cdot x^2 - 1 = 0$.

Exercise 45 The sequence $1, \Phi, \Phi^2, \Phi^3, \Phi^4, \Phi^5, \ldots$ is a Lucas sequence. Transform this Lucas sequence (up to 6 elements) to a number sequence linear in Φ (such as $1, \Phi, \Phi + 1, \ldots$) by means of the quadratic equation defined by the golden number (5.2).

Exercise 46 In Fibonacci's footsteps anno 1202, we go for a make-over of the solving of his famous rabbit population problem. How many couples of rabbits do we breed in one year, given the following conditions?

- ▷ Every rabbit couple is pubescent at the age of two months.
- ▷ Every rabbit couple that is pubescent produces monthly one new rabbit couple.
- ▷ No rabbit dies.

We hereby ignore their incest problem and their offsprings being constantly one male plus one female. Start in month 0 using one fresh rabbit couple.

Exercise 47 The **silver number** is the irrational number

$$\delta = 1 + \sqrt{2} \approx 2.41421$$

according to its original definition (128 Anno Domini, *Ostiahuis*).

1) Find a ruler-and-compass construction based upon the Pythagorean Theorem to construct the **silver section**.
2) We define the **silver rectangle** in words: 'it is the rectangle which after cutting away two squares, is similar to the remaining rectangle'. Find the meaningful solution δ to the quadratic equation corresponding to the above definition in words.
3) We define the indexed elements $p_n \in \mathbb{N}$ as a number sequence of **John Pell** (1611-1685) by $p_{n+1} = 2p_n + p_{n-1}$. Their ratios $\frac{p_n}{p_{n-1}}$ converge to the limit $\delta = 2.41421\ldots$ for incrementing indices n. Prove this.

Chapter 6 · Coordinate systems

Did you ever wonder how to start programming computer animation? How does the computer keep track of moving objects? This chapter elaborates upon two different $2D$ coordinate systems, in order to label each point (each pixel) efficiently. Subsequently it will also allow us to describe various lines and sets of points. Apart from the popular cartesian coordinate system, professional graphical design and computer animation involve the use of polar coordinates. For instance rotation requires far less computing power (number of operations) and loss of precision (for which we refer to the specialized literature) compared to the cartesian approach. Finally, using polar curves pushes digital design and computer art to a higher level.

6.1 Cartesian coordinates

When we locate objects in a frame we spontaneously make use of the cartesian coordinates. **Cartesian** means rectangular and was named after its inventor, the French mathematician **René Descartes** (1596-1650). So, cartesian coordinates are the popular rectangular (x, y)-coordinates and often orthonormal, which means their x-axis and y-axis are orthogonal and equally scaled. Logically, straight lines can be most efficiently expressed in this rectangular system.

As soon as the origin O is defined as the intersection point of the x- and y-axis, we are able to label any point P in $2D$ by two coordinate numbers: the horizontal x-coordinate called **abscissa** and the vertical y-coordinate called **ordinate**.

6.2 Parametric curves

We define a **parametric curve** as the locus of a moving point $P(x,y)$ with $x = x(t)$ and $y = y(t)$ both being functions of a parameter t, where the parameter t runs through a real interval.

We define the **parametric form** of such a curve as the coordinate functions

$$\begin{cases} x = x(t) \\ y = y(t) \end{cases} . \tag{6.1}$$

Example 1: Whenever we eliminate the parameter t of this ‘curve’

$$\begin{cases} x = t \\ y = t \end{cases}$$

from abscissa x and ordinate y, we obtain its cartesian form $y = x$.

Example 2: When in the parametric form

$$\begin{cases} x = t \\ y = \sqrt{t^2} \end{cases}$$

we substitute abscissa x in ordinate y via the parameter t, we obtain the **absolute value-function** $y = \sqrt{x^2}$. We typeset this absolute value-function either as $y = \text{abs}(x)$ or as $y = |x|$.

$$\text{abs: } \mathbb{R} \longrightarrow \mathbb{R}^+ : x \mapsto \sqrt{x^2} = \begin{cases} x & \text{if } x \geqslant 0 \\ -x & \text{if } x < 0 \end{cases} \tag{6.2}$$

We see on the graph its domain $\mathbb{R}$, range $\mathbb{R}^+$ and the solitary root $x_0 = 0$. This graph shows very clearly that all return values are non-negative.

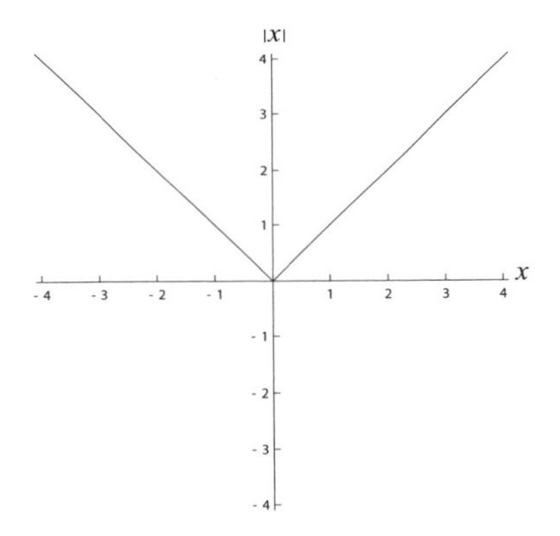

Figure 6.1: Absolute value-function $|x|$

Example 3: The squaring and side to side addition of

$$\begin{cases} x = \cos t \\ y = \sin t \end{cases}$$

eliminates t via the trigonometric Pythagorean Identity (3.6) to $(\cos t)^2 + (\sin t)^2 = 1^2$. This way we obtain the cartesian form $x^2 + y^2 = 1$ of the **unit circle** $C(O,1)$ centered on the origin O and having radius $r = 1$.

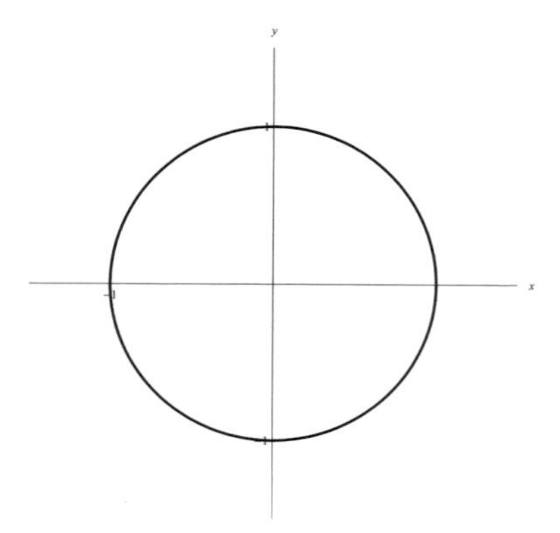

Figure 6.2: Unit circle in the cartesian coordinate frame

Example 4: We eliminate the parameter t from the parametric form

$$\begin{cases} x = a\cos t \\ y = b\sin t \end{cases}$$

via the trigonometric Pythagorean Identity as

$$\begin{cases} \dfrac{x}{a} = \cos t \\ \dfrac{y}{b} = \sin t \end{cases} \Longrightarrow \begin{cases} \left(\dfrac{x}{a}\right)^2 = (\cos t)^2 \\ \left(\dfrac{y}{b}\right)^2 = (\sin t)^2 \end{cases} \Longrightarrow \left(\frac{x}{a}\right)^2 + \left(\frac{y}{b}\right)^2 = 1.$$

This is the cartesian form of an **ellipse** centered on the origin O and having semi-major axis $a > 0$ and semi-minor axis $b > 0$.

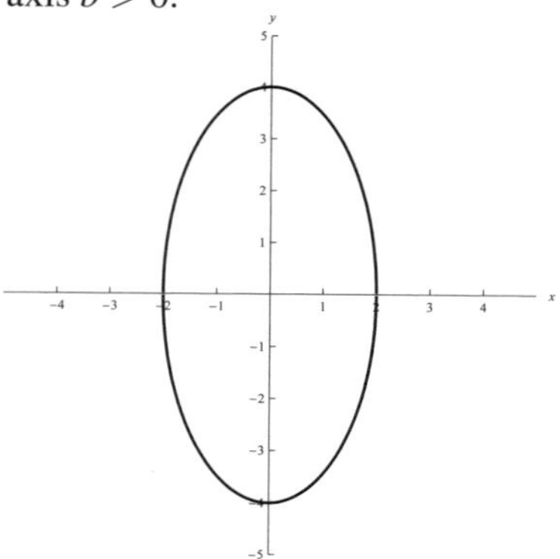

Figure 6.3: Ellipse centered on O, having semi-axes $a = 2$ and $b = 4$

6.3 Polar coordinates

On the other hand, we can express curved and circular lines more efficiently in a non-rectangular system. We alternatively can locate a point P in the plane by referring it to the pole O and the horizontal axis. The **pole** O is the central reference in the polar plane. The **polar axis** is the horizontal axis through the pole O.

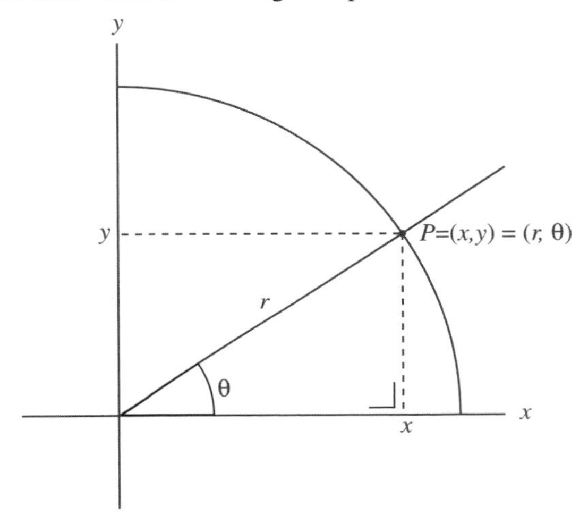

Figure 6.4: Cartesian and polar coordinates of a point P

Whenever we apply a coordinate system to the plane, we are able to locate a point P. This is the case for instance via its unique cartesian label (x,y), but we can locate it also by using *two* alternative coordinate numbers as well. For polar coordinates we make use of the distance $r = d(O,P)$ from P to the pole O, and the angle $\theta \in [0, 2\pi[$ made by the line OP and the positive x-axis. We call them the **radial coordinate** r and the **polar angle** θ respectively. We label a point P in the polar plane uniquely by (r, θ). We note that the radial coordinate r is non-negative and that the angular coordinate θ is described in radians measured counterclockwise.

For converting from cartesian to polar coordinates and vice versa, we refer to the previous chapter 3 on trigonometry. Converting is just a matter of applying the basic formulas of the right triangle (see paragraphs 3.2 and 3.3). Be careful in determining the polar angle θ based on the x- and y-number: only a drawing of the location will show the correct value for θ!

We convert the polar coordinates of the point P easily to cartesian coordinates by just applying the definitions of sine and cosine in the right triangle:

$$x = r\cos\theta \qquad \text{and} \qquad y = r\sin\theta. \tag{6.3}$$

We convert the cartesian coordinates of the point P straightforwardly to polar coordinates by implementing the Pythagorean Theorem and the definition of tangent in the right triangle:

$$r = \sqrt{x^2+y^2} \qquad \text{and} \qquad \tan\theta = \frac{y}{x} \text{ in case } x \neq 0. \tag{6.4}$$

For the polar angle θ we conclude that $\theta = \arctan\left(\frac{y}{x}\right) + k\pi$ given $k \in \mathbb{Z}$. In case $x = 0$ we conclude $r = |y|$ for the radial and $\theta = \frac{\pi}{2}$ or $\frac{3\pi}{2}$ for the angular coordinate. The correct value of k is determined by what quadrant the point P is lying in. By taking the constraint on the polar angle θ into account, we conclude in general

$$\theta = \begin{cases} \arctan\frac{y}{x} & \text{if } x > 0 \text{ and } y > 0 \quad \text{(quadrant I)} \\ \arctan\frac{y}{x} + \pi & \text{if } x < 0 \quad \text{(quadrant II and III)} \\ \arctan\frac{y}{x} + 2\pi & \text{if } x > 0 \text{ and } y < 0 \quad \text{(quadrant IV).} \end{cases}$$

Examples: We convert the cartesian labeled point $A(1,-1)_{\text{cc}}$ to its polar coordinates via

$$r = \sqrt{1^2 + (-1)^2} = \sqrt{2}$$
$$\theta = \arctan\left(\frac{-1}{1}\right) = -\frac{\pi}{4} \text{ or } \frac{3\pi}{4}.$$

The unit circle reveals a corresponding angle of $-\frac{\pi}{4}$ radians. But the point A is located in the fourth quadrant, which leads to a polar angle of $-\frac{\pi}{4} + 2\pi = \frac{7\pi}{4}$. We find the polar coordinates of the point A to be $\left(\sqrt{2}, \frac{7\pi}{4}\right)_{\text{pc}}$.

We convert the polar labeled point $B\left(1, \frac{\pi}{2}\right)_{\text{pc}}$ to its cartesian coordinates via

$$x = 1\cos\frac{\pi}{2} = 0$$
$$y = 1\sin\frac{\pi}{2} = 1,$$

typeset as $B(0,1)_{\text{cc}}$.

There are curves we can most efficiently describe by expressing their relation in r and θ.

Amongst those curved and circular lines we can for instance express a 'normal' spiral very efficiently in its polar form $r = \theta$. We realize this relation is no longer a function.

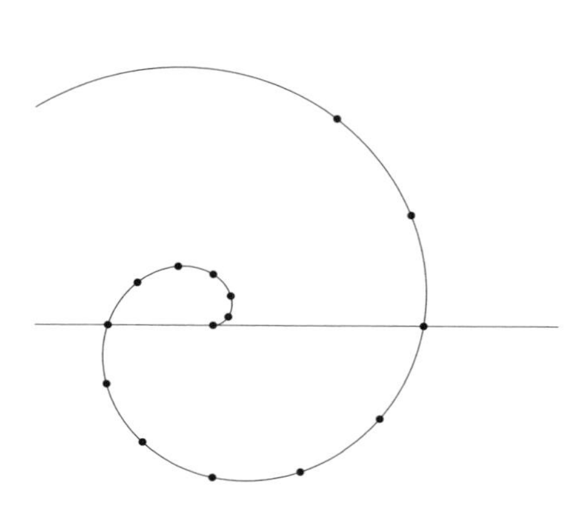

We can rapidly plot its graph by sampling some polar angles θ to calculate their corresponding radial coordinate r.

θ (in rad)	0	$\frac{\pi}{6}$	$\frac{\pi}{3}$	$\frac{\pi}{2}$	$\frac{2\pi}{3}$	$\frac{5\pi}{6}$	$\dots$
r	0	$\frac{\pi}{6} \approx 0.52$	$\frac{\pi}{3} \approx 1.05$	$\frac{\pi}{2} \approx 1.57$	$\frac{2\pi}{3} \approx 2.09$	$\frac{5\pi}{6} \approx 2.62$	$\dots$

Example: Convert the cartesian form of the circle $x^2 + y^2 + 2x = 0$ to its polar form.

We recall for $x^2 + y^2 = r^2$ and substitute the remaining x by $r\cos\theta$. In this way we obtain

$$\begin{aligned} x^2 + y^2 + 2x = 0 &\Leftrightarrow r^2 + 2r\cos\theta = 0 \\ &\Leftrightarrow r(r + 2\cos\theta) = 0 \\ &\Leftrightarrow r = 0 \text{ or } r + 2\cos\theta = 0 \\ &\Leftrightarrow r = 0 \text{ or } r = -2\cos\theta. \end{aligned}$$

Example: Convert the polar form $r = \frac{2}{1-\sin\theta}$ to its cartesian form.

Cancelling out the denominator leads to an equivalent form $r(1-\sin\theta) = 2$. Expanding its left hand side we find $r - r\sin\theta = 2$. Finally we replace r by $\sqrt{x^2+y^2}$ and $r\sin\theta$ by y, which simplifies to

$$\begin{aligned} r - r\sin\theta = 2 &\Leftrightarrow \sqrt{x^2+y^2} - y = 2 \\ &\Leftrightarrow \sqrt{x^2+y^2} = y + 2 \\ &\Leftrightarrow x^2 + y^2 = (y+2)^2 \\ &\Leftrightarrow x^2 = 4y + 4 \\ &\Leftrightarrow y = \frac{1}{4}x^2 - 1. \end{aligned}$$

We finally recognize the cartesian form as the equation of a parabola.

6.4 Polar curves

We define a **polar curve** as the locus of a moving point $P(r,\theta)$ with its radial coordinate $r = f(\theta)$ in terms of the polar angle θ, when θ runs through a real interval.

We define the **polar equation** of a polar curve as the above mentioned relation $r = f(\theta)$ in radial coordinate r and polar angle θ.

Example 1: We define the **crosier curve** of opening $a > 0$ as the polar equation $r = \dfrac{a}{\sqrt{\theta}}$ given its polar angle θ runs through an interval of $\mathbb{R}^+ \backslash \{0\}$.

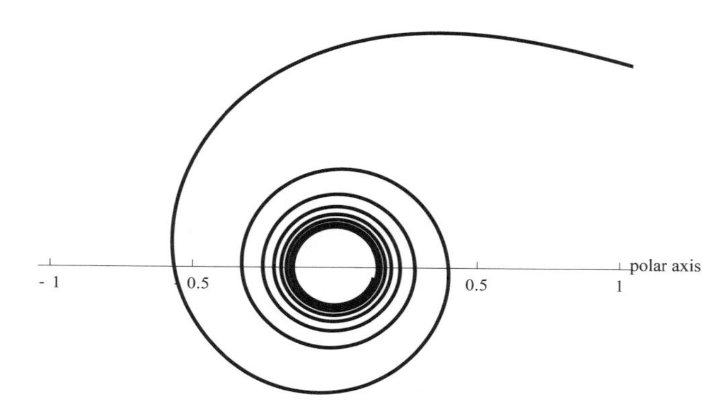

Figure 6.5: A crosier curve of opening $a = 1$

Example 2: We define the odd-petaled **rose** of size $a \neq 0$ as the polar equation $r = a\cos(m\theta)$ for each odd m, given its polar angle θ runs through $[0, 2\pi[$.

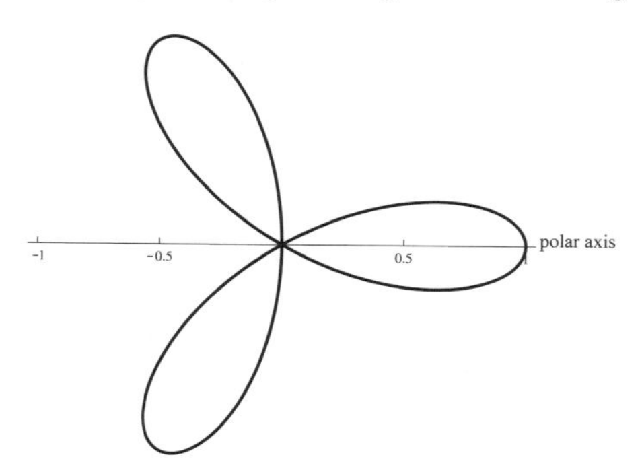

Figure 6.6: A three-petaled rose described by $r = 1\cos(3\theta)$

Example 3: We define the **lemniscate** with semi-axis $a > 0$ as the polar equation $r = a(\cos(2\theta))^{\frac{1}{2}}$ given its polar angle θ runs through $[0, 2\pi[$. Notice how in this case the coefficient of the angle θ stands for the two lobes of the lemniscate, just like it stood for the three petals in case of a rose. Consequently we may interpret the lemniscate as the *even-petaled* rose $r = a(\cos(n\theta))^{\frac{1}{2}}$ featuring the smallest petal number $n = 2$.

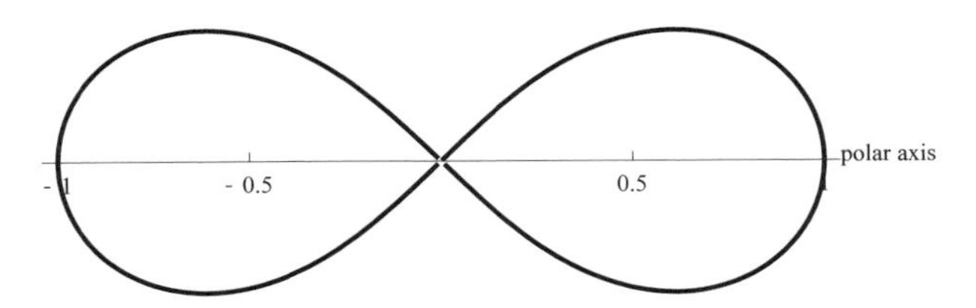

Figure 6.7: Lemniscate with semi-axis $a = 1$

A POLAR SUPERFORMULA

Initially seeking for a mathematical model to fit square bamboo, the Belgian engineer and botanist **Johan Gielis** (1962) developed in 1997 a powerful polar equation capable of plotting almost every polar curve. By implementing six **shaping parameters** Gielis' **superformula** draws circles, polygons, roses, stars, spirals and all kinds of mixtures of them. Due to its unseen combination of computing efficiency (light source code and less computing overhead) and graphical capacity, the superformula was for instance plugged in into Adobe Illustrator®.

We define Gielis' superformula as the polar equation

$$r = \left(\left| \frac{\cos\left(\frac{m}{4}\theta\right)}{a} \right|^{n_2} + \left| \frac{\sin\left(\frac{m}{4}\theta\right)}{b} \right|^{n_3} \right)^{-\frac{1}{n_1}}$$

given its polar angle θ runs through the interval $[0, 2\pi[$. Notice the absolute value over both terms in this formula (see page 97). The superformula implements six shaping parameters:

- the semi-major axis $a > 0$,
- the semi-minor axis $b > 0$,
- the coefficient m of the angle θ implies the number of petals,
- the global shaping exponent n_1,
- the major or horizontal shaping exponent n_2 and

▷ the minor or vertical shaping exponent n_3.

Varying each of the above shaping parameters a, b, m, n_1, n_2 and n_3 reveals an impressive gallery of exotic polar curves.

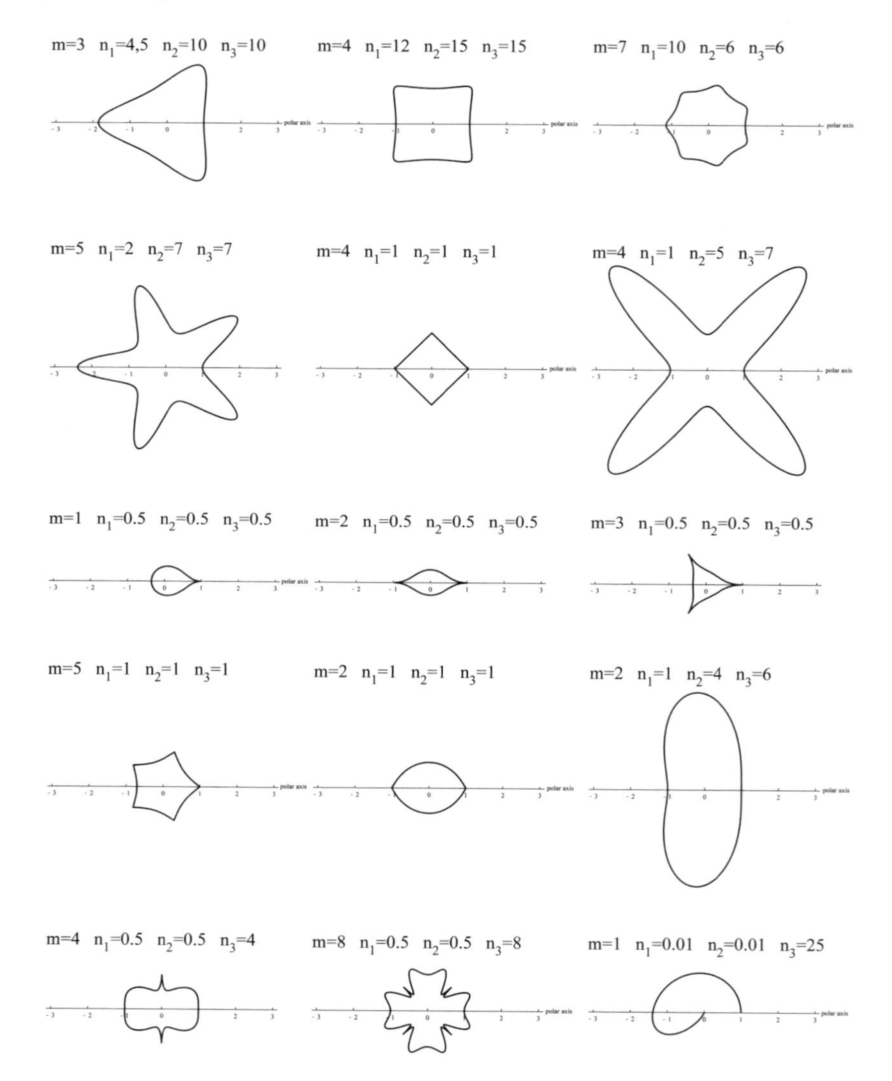

Figure 6.8: Superformula of $a = b = 1$ and varying $m, \ n_1, n_2, n_3$

6.5 Exercises

Exercise 48 Convert from polar to cartesian coordinates. Draw the points in a cartesian frame.

1) $\left(1, \frac{\pi}{2}\right)_{\text{pc}}$
2) $\left(3, \frac{7\pi}{4}\right)_{\text{pc}}$
3) $\left(4, \frac{3\pi}{2}\right)_{\text{pc}}$
4) $\left(2, \frac{5\pi}{4}\right)_{\text{pc}}$

Exercise 49 Convert from cartesian to polar coordinates. Draw the points in the polar plane.

1) $\left(2, 2\right)_{\text{cc}}$
2) $\left(0, -1\right)_{\text{cc}}$
3) $\left(\frac{1}{2}, \frac{\sqrt{3}}{2}\right)_{\text{cc}}$
4) $\left(\sqrt{2}, -\sqrt{2}\right)_{\text{cc}}$

Exercise 50 Rotate the point $\left(\frac{-10}{4}, \frac{5\sqrt{5}}{2}\right)_{\text{cc}}$ counterclockwise $90°$ around the origin O. Hint: first label the point in polar coordinates.

Exercise 51 Convert the following equations to their polar form.

1) $y = 3x$
2) $y = -2x + 3$
3) $x^2 + y^2 + 2x = 0$

Exercise 52 Convert the following equations to their cartesian form.

1) $r = \frac{1}{1 - \cos\theta}$
2) $r = 5\theta$

Exercise 53 Draw with pen and paper the **spiral of Theodore of Cyrene** (5th century Before Christ), taking an isosceles right triangle with sides of length 1 for a start. Put again a side of length 1 perpendicular on its hypotenuse, and keep repeating this for some steps. We hereby picture the initial triangle ABC and the first step.

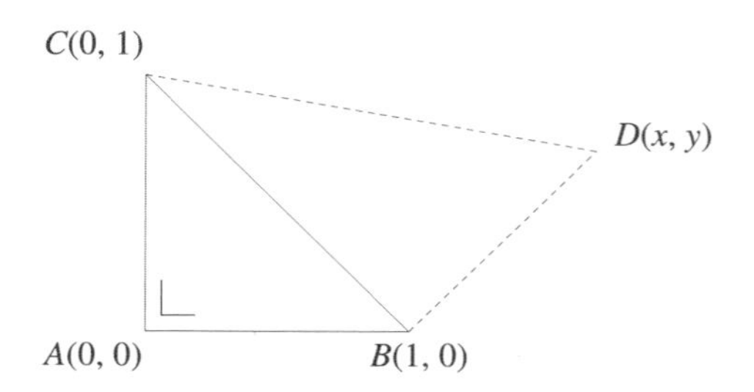

Determine the coordinates of the point D.

Determine the irrational number sequence corresponding to the growing hypotenuses.

Exercise 54 The parametric form of the crosier curve with opening $a > 0$ is

$$\begin{cases} x = \dfrac{a\cos t}{\sqrt{t}} \\ y = \dfrac{a\sin t}{\sqrt{t}} \end{cases} . \tag{6.5}$$

Convert the parametric form to the polar equation of the crosier curve by the subsequent use of the Pythagorean Identity and Theorem (see page 50). Interpret the remaining parameter t as polar angle θ.

Exercise 55 We define the **scarabaeus curve** with coefficients $a \geqslant 0$ and $b \geqslant 0$ as the polar equation $r = b\cos(2\theta) - a\cos(\theta)$ given its polar angle θ runs through the interval $[0, 2\pi[$.

Determine with pen and paper the scarabaeus curve coefficients a and b describing

1) a circle with radius 1.

2) a rose of size 3.

Draw on computer the scarabaeus curves with coefficients

1) $a = 2$ and $b = 3$ and vice versa,

2) $a = 3$ and $b = 2$.

Exercise 56 Determine the shaping parameter(s) a, b, m, n_1, n_2 and n_3 for the superformula to describe

1) the unit circle (see figure 6.2) and try this with pen and paper.

2) the vertical ellipse centered on the origin O on semi-axes of length 2 and 4 (see figure 6.3) using a computer for this.

Chapter 7 · Vectors

We all know numbers. But numbers as such fail to describe motion, because they are unable to express direction. We model the motion and forces in mechanics by the use of vectors. Vectors became increasingly popular through their inevitable use in physics. During the last few decades vectors became even more important as they are applied in robotics, computer animation, computer games, pattern matching, digital security, ...

In this chapter we introduce vectors before studying the basic operations on them: vector addition, scalar multiplication, dot product and cross product. We finally explain some required techniques for 3D animation such as producing a normal vector on polygon surfaces.

7.1 The concept of a vector

Distance and displacement are different measures. If we for instance drive from Brussels to Amsterdam to Brussels, then we travel quite a distance but made eventually no displacement. Distance is just a non-negative number and differs a lot from displacement, which is a vector.

Vectors as arrows

We define **scalar** measures as being completely determined by a numerical value, and often expressed in a certain unit. The definition of a scalar implies it can for instance not determine a direction. Illustrations of scalar measures are *distance* (500 m), *speed* (90 km per hour), temperature (23° C), the length of a person (182.5 cm) or the solutions to an algebraic equation ($2x - 6 = 0 \Leftrightarrow x = 3$), ...

We define a **vector** as a measure determined by a magnitude (or absolute value) and a direction. The definition of a vector implies it can be represented by an arrow when it is in 2D or 3D space. Illustrations of vectors are *displacement*, *velocity*, acceleration, force, pressure, ...

Usually we represent a vector by an arrow with length proportional to the vector's magnitude, with a certain direction and with its initial point meaningful located.

In this figure the origin O of the plane and two points in it A and B, are shown. The location of the points A and B is fully determined by the arrow from O to A and by the arrow from O to B respectively. Such arrows are called vectors. We typeset the vector ending in A using its lowercase character as $\vec{a}$ and the one ending in B likewise as $\vec{b}$. Then there is one exceptional vector which we cannot draw since it starts and ends in the origin O. This special vector is called the **null vector** or **zero vector**: we typeset it accordingly as $\vec{o}$.

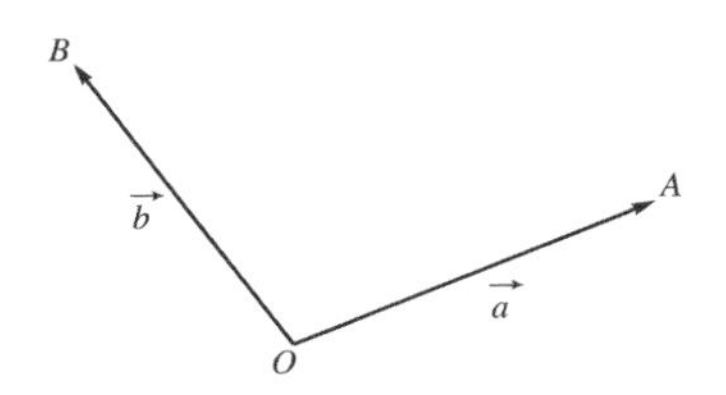

Vectors can even be applied apart from their geometrical use in 2D and 3D space, for holding ordered lists of numbers called arrays. As for this (programming) purpose the number of components often exceeds three, we are no longer able to draw these vectors as arrows. A multicomponent vector $\vec{x}$ can for instance hold four variables x_1 tot x_4 as $\vec{x} = (x_1, x_2, x_3, x_4)$ makes an array (see Quaternions, paragraph 12.5). Multimedial programming performs operations on vectors, which again requires to define vectors as ordered lists of numbers.

Vectors as arrays

We introduced vectors as arrows in the plane or in 3D space. This approach offers a good geometrical insight but is less helpful when it comes to performing operations on vectors. Let us for this reason apply the cartesian coordinate system on the plane and on 3D space, in order to label points and consequently vectors as well.

In this figure the origin O and two points A and B in the plane are shown again. Applying the cartesian coordinate system on this plane, its two orthonormal axes intersecting in O are shown. Consequently point A gets coordinates $(5,2)$ and point B coordinates $(-3,4)$. Subsequently, these coordinates are as labels inherited by their corresponding vectors $\vec{a}$ and $\vec{b}$, and typeset in a column of **vector components**:

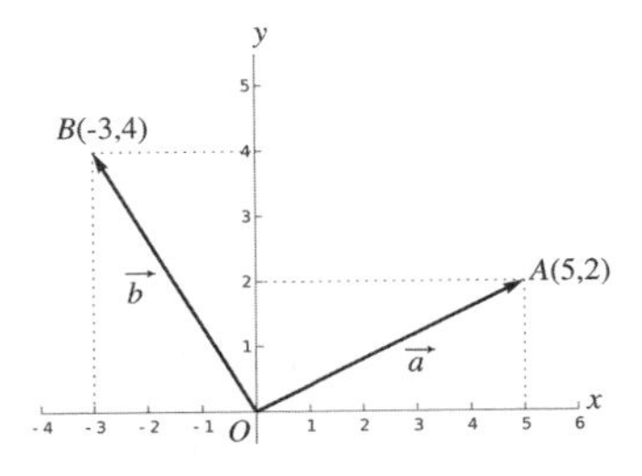

$$\vec{a} = \begin{pmatrix} 5 \\ 2 \end{pmatrix}, \quad \vec{b} = \begin{pmatrix} -3 \\ 4 \end{pmatrix}, \quad \vec{o} = \begin{pmatrix} 0 \\ 0 \end{pmatrix}.$$

We can extend cartesian coordinates to 3D space in which we label a point P by three coordinates x, y and z as located on three perpendicular axes. Here we need to decide whether the z-axis 'disappears in' our sheet or 'appears outwards' from it. In other words, whether ascending z-values escape from or approach us. Both options are valid and accepted. The former is known as a left-handed frame, the latter as a right-handed frame.

We define a **right-handed frame** by aligning all four right-hand fingers in the direction of the positive x-axis and then turn them to the positive y-axis. Doing so, our thumb will point in the direction of the positive z-axis. We call this practical definition the **right-hand rule**. We similarly define a **left-handed frame** by equivalently using our left hand. Alternatively, we can define a right-handed frame by applying the **corkscrew-rule**. We rotate the corkscrew from the x-axis to the y-axis. If the resulting effect is in the direction of the positive z-axis, we are in a right-handed frame.

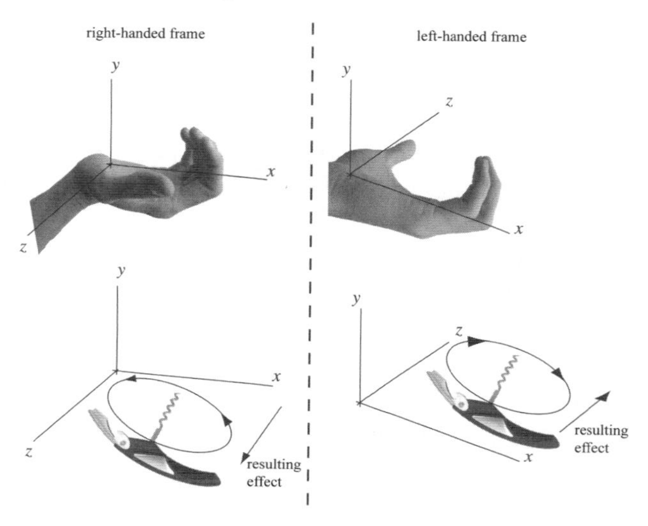

Figure 7.1: Right-handed and left-handed frame

For instance Adobe® Flash® holds a right-handed frame for which the x-axis runs to the right, the y-axis points downwards and the z-axis disappears in their plane. In this book we work in the right-handed frames only.

Applying the cartesian coordinate system to 3D space will accordingly label its vectors from arrows to arrays. We typeset the x-coordinate of a point A in 3D space as a_1, its y-coordinate as a_2 and its z-coordinate as a_3. This point $A(a_1,a_2,a_3)$ as well as its corresponding vector $\vec{a}$ are now uniquely labeled, with $\vec{a} = \begin{pmatrix} a_1 \\ a_2 \\ a_3 \end{pmatrix}$ in components.

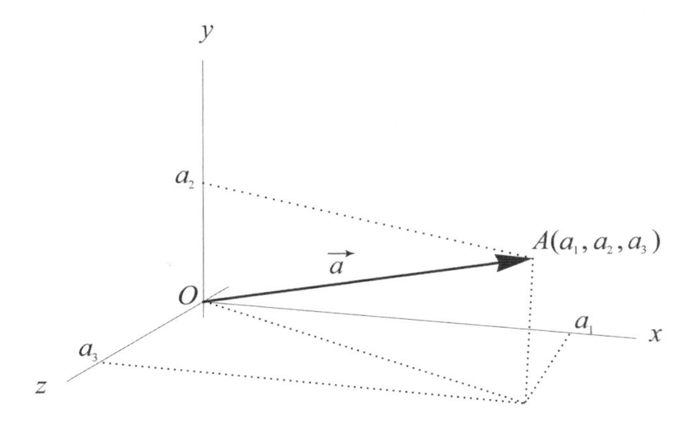

Figure 7.2: Vector in 3D space

A **vector** $\vec{a}$ is as an arrow uniquely determined by its **length** (or **norm** or **magnitude**) and its **direction** (holding an orientation and a sense).

To be more precise, a vector $\vec{a}$ features

- ▷ a non-negative **length** $||\vec{a}|| \in \mathbb{R}^+$;
- ▷ an **orientation** parallel to its arrow;
- ▷ a **sense** given by its arrow-head.

We call the given combination of the vector's orientation and sense, the vector's direction.

We typeset a **vector** $\vec{a} \in \mathbb{R}^n$ as

$$\vec{a} = \begin{pmatrix} a_1 \\ a_2 \\ \vdots \\ a_n \end{pmatrix}. \tag{7.1}$$

We call the numbers $a_1, a_2, \ldots, a_n$ the **components** of the vector $\vec{a}$. We typeset the **length** of this vector as $||\vec{a}|| \in \mathbb{R}^+$.

We calculate the length of a plane vector $\vec{a} = \begin{pmatrix} a_1 \\ a_2 \end{pmatrix}$ via the Pythagorean Theorem (3.2), $||\vec{a}|| = \sqrt{a_1^2 + a_2^2}$.

We easily extend this calculation for multidimensional vectors. We calculate the length of a 3D vector by applying the Pythagorean Theorem twice, leading to: $||\vec{a}|| = \sqrt{a_1^2 + a_2^2 + a_3^2}$.

Logically, we calculate the length or **norm** of an n-dimensional vector $\vec{a} \in \mathbb{R}^n$ via

$$||\vec{a}|| = \sqrt{a_1^2 + a_2^2 + a_3^2 + \cdots + a_n^2}. \tag{7.2}$$

We define **unit vectors** as vectors of length 1. For instance, vector $\vec{a} = \begin{pmatrix} \frac{3}{5} \\ \frac{4}{5} \end{pmatrix}$ is a unit vector because $||\vec{a}|| = \sqrt{\frac{9}{25} + \frac{16}{25}} = 1$.

Free Vectors

We define **location vectors** as vectors having the origin O as initial point or **tail**. We typeset location vectors with a lowercase character referring to their terminal point or **head** A, like $\vec{a} = \overrightarrow{OA}$. We define **free vectors** as vectors whose initial point is not the origin O, but a free floating tail. We typeset a free vector from tail A to head B with an overhead arrow as $\overrightarrow{AB}$.

Two vectors $\overrightarrow{AB}$ and $\overrightarrow{CD}$ are equal ($\overrightarrow{AB} = \overrightarrow{CD}$) only if their length and direction are equal. Two vectors $\overrightarrow{AB}$ and $\overrightarrow{CD}$ are **parallel** ($\overrightarrow{AB} \uparrow\uparrow \overrightarrow{CD}$) if their direction is equal. Practically we create equal vectors by parallel displacing a given vector. Two vectors $\overrightarrow{AB}$ and $\overrightarrow{CD}$ are **antiparallel** ($\overrightarrow{AB} \uparrow\downarrow \overrightarrow{CD}$) if their orientation is equal and their sense is opposite.

Base Vectors

The vectors $\begin{pmatrix} 1 \\ 0 \\ 0 \end{pmatrix}$, $\begin{pmatrix} 0 \\ 1 \\ 0 \end{pmatrix}$ and $\begin{pmatrix} 0 \\ 0 \\ 1 \end{pmatrix}$ are the **base vectors** of 3D space. We typeset these base vectors as $\vec{e}_1, \vec{e}_2$ and $\vec{e}_3$ as they align the x, y and z-axis respectively. Due to these alignments they are perpendicular. Similarly the vectors $\begin{pmatrix} 1 \\ 0 \end{pmatrix}$ and $\begin{pmatrix} 0 \\ 1 \end{pmatrix}$ are the base vectors of $\mathbb{R}^2$. We define base vectors more properly where appropriate in this book (see page 119).

7.2 Addition of vectors

We already introduced a vector $\vec{a}$ as an arrow from the origin O to a terminal point A. Somehow, such a location vector $\vec{a}$ and its head A are holding equivalent information, so we could wonder why we ever introduced vectors? One main reason for it are the operations we perform on vectors, but remain invalid on points. For instance the sum $A+B$ of two points A and B is meaningless, also the product of a point A and a number λ is without meaning. On the contrary the sum of two vectors $\vec{a}$ and $\vec{b}$, and even the product of a vector $\vec{a}$ and a number λ are meaningfully defined.

Vectors as arrows

The figure shows how to add two vectors geometrically. Vector $\vec{c}$ heading the point C is the sum of the vectors $\vec{a}$ and $\vec{b}$, briefly $\vec{c}=\vec{a}+\vec{b}$. We construct the sum $\vec{c}$ by parallel displacing vector $\vec{b}$, putting its tail at the head of the vector $\vec{a}$. We call this the *head-to-tail rule* for vector addition and typeset it as $\vec{a}+\vec{b}$. We could as well have been parallel displacing vector $\vec{a}$, putting its tail at the head of vector $\vec{b}$ to reach $\vec{b}+\vec{a}$. Apparently the order in which we perform vector addition is indifferently leading to the same sum $\vec{c}=\vec{a}+\vec{b}=\vec{b}+\vec{a}$, which we define as the **commutative property** of vector addition.

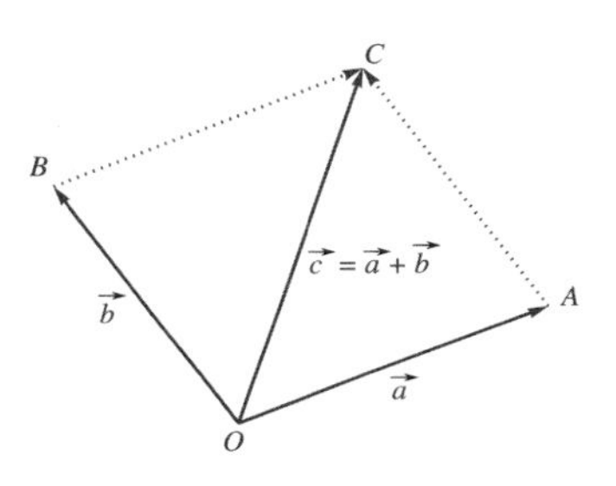

Alternatively, the result vector $\vec{c}$ can be recognized as the diagonal of the parallellogram *AOBC*. This *parallellogram method* applied to two vectors $\vec{a}$ and $\vec{b}$, states that their sum $\vec{a}+\vec{b}$ is the diagonal vector $\vec{c}$ in their spanned parallellogram.

Vectors as arrays

Vectors are most easy to add in cartesian coordinate systems. We add vectors by simply adding their corresponding cartesian components.

$$\vec{a}+\vec{b}=\begin{pmatrix}5\\2\end{pmatrix}+\begin{pmatrix}-3\\4\end{pmatrix}=\begin{pmatrix}5+(-3)\\2+4\end{pmatrix}=\begin{pmatrix}2\\6\end{pmatrix}.$$

Vector addition summarized

- ▷ geometrical: head-to-tail rule or parallellogram method
- ▷ in words: The addition of vectors is defined as the addition of their corresponding cartesian components.

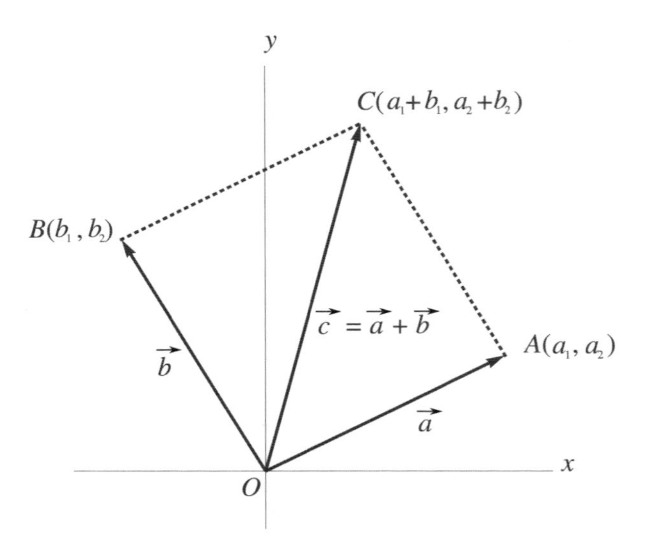

Figure 7.3: Vector addition

▷ in symbols: Given $\vec{a} = \begin{pmatrix} a_1 \\ a_2 \\ \vdots \\ a_n \end{pmatrix}$ and $\vec{b} = \begin{pmatrix} b_1 \\ b_2 \\ \vdots \\ b_n \end{pmatrix}$ are vectors in $\mathbb{R}^n$, then their sum is

$$\vec{a} + \vec{b} = \begin{pmatrix} a_1 + b_1 \\ a_2 + b_2 \\ \vdots \\ a_n + b_n \end{pmatrix} \in \mathbb{R}^n. \tag{7.3}$$

7.3 Scalar multiplication of vectors

We define the **scalar multiplication** of vectors as a number λ times a vector $\vec{a}$, resulting in a scaled vector $\lambda\vec{a}$.

Vectors as arrows

The figure shows some examples of vectors $\vec{a}$ and $\vec{b}$ multiplied with real numbers. We obtain the scaled vector $3\vec{b}$ by adding $\vec{b}+\vec{b}+\vec{b}$. Geometrically we therefore extend $\vec{b}$ via the head-to-tail rule with vector $\vec{b}$ itself, which results in the doubled vector $2\vec{b}$. Extending $2\vec{b}$ again by parallel displacing the tail of the vector $\vec{b}$ to the head of $2\vec{b}$, results in $3\vec{b}$. Hence, the vector $3\vec{b}$ keeps the direction of $\vec{b}$ and has a length of 3 times $\left\|\vec{b}\right\|$.

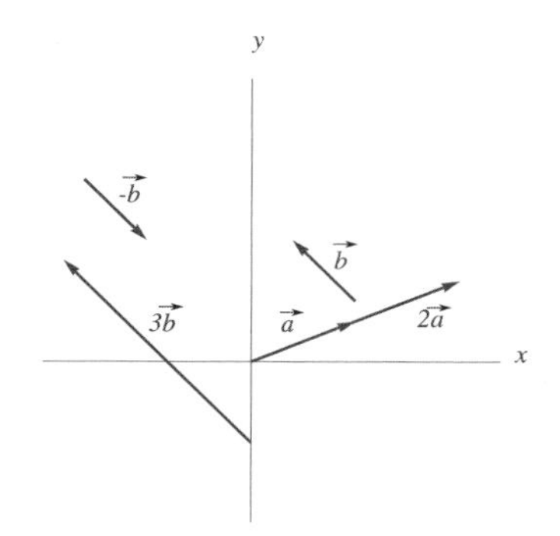

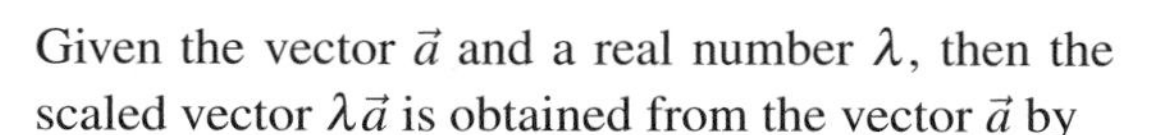

Given the vector $\vec{a}$ and a real number λ, then the scaled vector $\lambda\vec{a}$ is obtained from the vector $\vec{a}$ by

- ▷ multiplying its **length** by the absolute value $|\lambda|$,
- ▷ keeping its **orientation**,
- ▷ adjusting its **sense** according to λ:
  - ▶ in case $\lambda > 0$ then $\lambda\vec{a}$ keeps the sense of $\vec{a}$;
  - ▶ in case $\lambda < 0$ then $\lambda\vec{a}$ gets the opposite sense of $\vec{a}$;
  - ▶ in case $\lambda = 0$ then $\lambda\vec{a} = 0\vec{a} = \vec{o}$. (The zero vector $\vec{o}$ has no direction at all.)

Consequently, the **opposite vector** $-\vec{a} = (-1)\vec{a}$ has the same length and orientation of the vector $\vec{a}$, but an opposite sense. We say the vectors $-\vec{a}$ and $\vec{a}$ are antiparallel. We meanwhile also define the subtraction of vectors as $\vec{a}-\vec{b} = \vec{a}+(-\vec{b})$.

Vectors as arrays

The scalar multiplication of a vector is the scalar multiplication of each of its cartesian components, by for instance the same number $\lambda = 4$.

$$4\vec{a} = \vec{a}+\vec{a}+\vec{a}+\vec{a} = \begin{pmatrix} 3 \\ 1 \end{pmatrix} + \begin{pmatrix} 3 \\ 1 \end{pmatrix} + \begin{pmatrix} 3 \\ 1 \end{pmatrix} + \begin{pmatrix} 3 \\ 1 \end{pmatrix}$$

$$= \begin{pmatrix} 3+3+3+3 \\ 1+1+1+1 \end{pmatrix} = \begin{pmatrix} 4\cdot 3 \\ 4\cdot 1 \end{pmatrix} = \begin{pmatrix} 12 \\ 4 \end{pmatrix}.$$

Scalar multiplication summarized

- in words: We multiply a vector with a number by multiplying each component with that number.
- in symbols: Given $\vec{a} = \begin{pmatrix} a_1 \\ a_2 \\ \vdots \\ a_n \end{pmatrix} \in \mathbb{R}^n$ and $\lambda \in \mathbb{R}$, we obtain

$$\lambda\vec{a} = \begin{pmatrix} \lambda a_1 \\ \lambda a_2 \\ \vdots \\ \lambda a_n \end{pmatrix}. \tag{7.4}$$

Properties

Given $\vec{a}, \vec{b}, \vec{c} \in \mathbb{R}^n$ and $\lambda, \mu \in \mathbb{R}$ we summarize the addition, scalar multiplication and subtraction of vectors arithmetically.

closure	$\vec{a}+\vec{b} \in \mathbb{R}^n$
associative property	$(\vec{a}+\vec{b})+\vec{c} = \vec{a}+(\vec{b}+\vec{c})$
zero vector	$\vec{a}+\vec{o} = \vec{a} = \vec{o}+\vec{a}$
opposite vector	$\vec{a}+(-\vec{a}) = \vec{o} = (-\vec{a})+\vec{a}$
commutative property	$\vec{a}+\vec{b} = \vec{b}+\vec{a}$

We define an **algebraic structure** as a set that takes a meaningful operation on its elements. A structure with the above properties is called a commutative **group**. Hence, we conclude the vector addition is a commutative group.

mixed distributive law	$\lambda(\vec{a}+\vec{b}) = \lambda\vec{a}+\lambda\vec{b}$
	$(\lambda+\mu)\vec{a} = \lambda\vec{a}+\mu\vec{a}$
mixed associative law	$(\lambda \cdot \mu)\vec{a} = \lambda(\mu\vec{a})$

Vector subtraction

▷ Components of free vectors

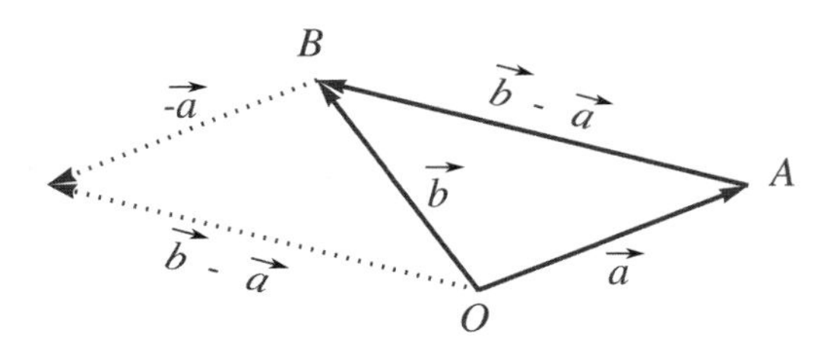

Aiming for the components of a free vector $\overrightarrow{AB}$, we recall

$$\vec{a} + \overrightarrow{AB} = \vec{b} \Longleftrightarrow \overrightarrow{AB} = \vec{b} - \vec{a}. \tag{7.5}$$

Hence we calculate the *components* of the vector $\overrightarrow{AB}$ by subtracting the *coordinates* of its initial point A from those of its terminal point B. We memorize this important rule for instance as 'head minus tail'.

Example: We calculate the components of the vector $\overrightarrow{AB}$, pointing from tail $A(4,1,3)$ to head $B(-1,4,5)$.

$$\overrightarrow{AB} = \overrightarrow{OB} - \overrightarrow{OA} = \begin{pmatrix} -1 \\ 4 \\ 5 \end{pmatrix} - \begin{pmatrix} 4 \\ 1 \\ 3 \end{pmatrix} = \begin{pmatrix} -5 \\ 3 \\ 2 \end{pmatrix}.$$

▷ Distance between two points

We are now able to calculate the distance between two given points A and B straightforward as the length of their free vector $\left\|\overrightarrow{AB}\right\|$ (see formula (7.2)).

▷ Euler's method for trajectories

Especially in game programming, the **trajectory** $\vec{s}(t)$ of a point S due to an acceleration function $\vec{a}(t)$ is often realized via a method named after the Swiss mathematician **Leonard Euler** (1707-1783). Euler estimates the moving point's path, given its acceleration function $\vec{a}(t)$, as an initial value problem. For such a problem we know at time zero t_0:

- the initial location vector $\vec{s}_0$,
- its initial velocity $\vec{v}_0$,

- and the time interval t between two frames.

Taking off with our initial location $\vec{s}_0$ and velocity $\vec{v}_0$, we now try to determine $\vec{s}_1$ and $\vec{v}_1$ straightforward for the next frame. We recall from physics that in case of sufficiently small time intervals t, we may assume the actual velocity and acceleration equal to the average velocity and acceleration:

$$\begin{cases} \vec{v} & \approx & \dfrac{\Delta\vec{s}}{t} \\ \vec{a} & \approx & \dfrac{\Delta\vec{v}}{t}. \end{cases}$$

Accepting this approximation yields a **recursive** formula for $\vec{s}$ and $\vec{v}$ to determine them for a frame $i+1$, based upon their former values in frame i (see page 251).

$$\begin{cases} \vec{v}_i & = & \dfrac{\vec{s}_{i+1}-\vec{s}_i}{t} \\ \vec{a}_i & = & \dfrac{\vec{v}_{i+1}-\vec{v}_i}{t} \end{cases} \Longleftrightarrow \begin{cases} \vec{v}_i t & = & \vec{s}_{i+1}-\vec{s}_i \\ \vec{a}_i t & = & \vec{v}_{i+1}-\vec{v}_i \end{cases} \Longleftrightarrow \begin{cases} \vec{s}_i+\vec{v}_i t & = & \vec{s}_{i+1} \\ \vec{v}_i+\vec{a}_i t & = & \vec{v}_{i+1} \end{cases}$$

In this way the initial values $\vec{s}_0$ and $\vec{v}_0$ lead to $\vec{s}_1$ and $\vec{v}_1$ for the next frame, given the acceleration function $\vec{a}(t)$ can either be statically prescribed within the programme, event-driven by the user or a combination of both and takes off at $\vec{a}_0 = \vec{a}(0)$.

$$\begin{cases} \vec{s}_1 & = & \vec{s}_0+\vec{v}_0 t, \quad \text{see also page 134} \\ \vec{v}_1 & = & \vec{v}_0+\vec{a}_0 t \end{cases}$$

Euler's method to determine parametric trajectories is not the most robust one: after a number of successive frames $0,1,2,3,\ldots$ the difference to the actual locations $\vec{s}(t)$ annex velocities $\vec{v}(t)$ may grow out of bound. But this technique is economic and easy to implement, which is why a lot of physics engines use it.

Decomposition of a plane vector

We are able to calculate the x- and y-component of a vector $\vec{a} \in \mathbb{R}^2$ in 2D space whenever its length $||\vec{a}||$ and its direction α (the angle made by the vector $\vec{a}$ and the positive x-axis) are given. Notice the equivalence of this to the description of a point A in polar coordinates (see chapter 6).

We recall that for the right triangle holds

$$\cos\alpha = \frac{a_1}{||\vec{a}||} \Rightarrow a_1 = ||\vec{a}||\cos\alpha,$$
$$\sin\alpha = \frac{a_2}{||\vec{a}||} \Rightarrow a_2 = ||\vec{a}||\sin\alpha,$$

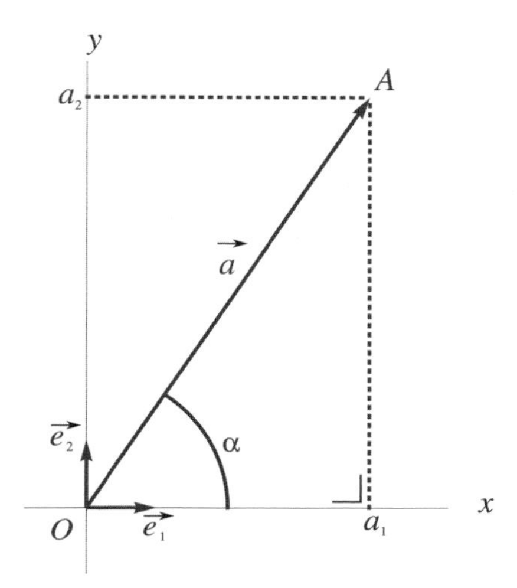

Figure 7.4: Decomposition of a 2D vector

leading to the trigonometrical components of the location vector

$$\vec{a} = \begin{pmatrix} ||\vec{a}|| \cos\alpha \\ ||\vec{a}|| \sin\alpha \end{pmatrix}. \tag{7.6}$$

BASE VECTORS DEFINED

▷ Base vectors are spanning

We obtain any vector as a sum of scalar multiplied base vectors $\vec{e}_i$.

$$\begin{pmatrix} 3 \\ 2 \end{pmatrix} = \begin{pmatrix} 3 \\ 0 \end{pmatrix} + \begin{pmatrix} 0 \\ 2 \end{pmatrix} = 3 \begin{pmatrix} 1 \\ 0 \end{pmatrix} + 2 \begin{pmatrix} 0 \\ 1 \end{pmatrix}$$
$$= 3\,\vec{e}_1 + 2\,\vec{e}_2$$

We call such a sum of scalar multiplied vectors a **linear combination**. Any vector in $\mathbb{R}^n$ can be written as a linear combination of its n base vectors. For this reason, we call base vectors a **spanning set**.

$$\vec{a} = \begin{pmatrix} a_1 \\ a_2 \\ \vdots \\ a_n \end{pmatrix} = a_1 \begin{pmatrix} 1 \\ 0 \\ \vdots \\ 0 \end{pmatrix} + a_2 \begin{pmatrix} 0 \\ 1 \\ \vdots \\ 0 \end{pmatrix} + \cdots + a_n \begin{pmatrix} 0 \\ 0 \\ \vdots \\ 1 \end{pmatrix}$$

$$= a_1\vec{e_1} + a_2\vec{e_2} + \cdots + a_n\vec{e_n} = \sum_{i=1}^{n} a_i\vec{e_i}$$

- ▷ Base vectors are independent

  Base vectors are said to be **linearly independent**: a given base vector can never be a linear combination of the remaining base vectors.

- ▷ Dimension

  We define the **dimension** of a space as its number of base vectors.

7.4 Dot product

The **dot product** or 'inner product' multiplies two vectors and returns a *scalar*. We consistently typeset this product with a dot.

DEFINITION

- ▷ in words: The dot (product) of two vectors multiplies all of their corresponding components, finally adding them to result in a real number.
- ▷ in symbols: Given $\vec{a} = \begin{pmatrix} a_1 \\ a_2 \\ \vdots \\ a_n \end{pmatrix}$ and $\vec{b} = \begin{pmatrix} b_1 \\ b_2 \\ \vdots \\ b_n \end{pmatrix}$ two vectors in $\mathbb{R}^n$, we define their dot product as the real number

$$\begin{aligned} \vec{a} \cdot \vec{b} &= a_1b_1 + a_2b_2 + a_3b_3 + \cdots + a_nb_n \\ &= \sum_{i=1}^{n} a_ib_i. \end{aligned} \tag{7.7}$$

As a constraint, given $\vec{a} \in \mathbb{R}^n$ and $\vec{b} \in \mathbb{R}^m$ with $m \neq n$, then their dot product is not defined.

Example:

$$\vec{a} = \begin{pmatrix} 2 \\ 3 \end{pmatrix} \in \mathbb{R}^2 \qquad \vec{b} = \begin{pmatrix} 6 \\ -1 \end{pmatrix} \in \mathbb{R}^2$$

$$\vec{c} = \begin{pmatrix} 1 \\ 2 \\ -3 \end{pmatrix} \in \mathbb{R}^3 \qquad \vec{d} = \begin{pmatrix} 2 \\ -1 \\ 0 \end{pmatrix} \in \mathbb{R}^3$$

$$\begin{aligned}
&\vec{a} \cdot \vec{b} = 2 \cdot 6 + 3 \cdot (-1) = 9 \in \mathbb{R} \\
&\vec{b} \cdot \vec{a} = 6 \cdot 2 + (-1) \cdot 3 = 9 \in \mathbb{R} \\
&\vec{c} \cdot \vec{d} = 1 \cdot 2 + 2 \cdot (-1) + (-3) \cdot 0 = 0 \in \mathbb{R} \\
&\vec{c} \cdot \vec{a} \text{ is not defined}
\end{aligned}$$

Properties: Given $\vec{a}, \vec{b} \in \mathbb{R}^n$ and $\lambda \in \mathbb{R}$.

independency of a scalar	$\lambda \vec{a} \cdot \vec{b} = \lambda\ (\vec{a} \cdot \vec{b})$
commutative property	$\vec{a} \cdot \vec{b} = \vec{b} \cdot \vec{a}$
length of a vector	$\lVert\vec{a}\rVert = \sqrt{\vec{a} \cdot \vec{a}}$

For the dot products on the base vectors $\vec{e_1} = \begin{pmatrix} 1 \\ 0 \\ 0 \end{pmatrix}$, $\vec{e_2} = \begin{pmatrix} 0 \\ 1 \\ 0 \end{pmatrix}$ and $\vec{e_3} = \begin{pmatrix} 0 \\ 0 \\ 1 \end{pmatrix}$,

we calculate:

$$\begin{aligned}
&\vec{e_1} \cdot \vec{e_1} = \vec{e_2} \cdot \vec{e_2} = \vec{e_3} \cdot \vec{e_3} = 1, \\
&\vec{e_1} \cdot \vec{e_2} = \vec{e_2} \cdot \vec{e_1} = \vec{e_2} \cdot \vec{e_3} = \vec{e_3} \cdot \vec{e_2} = \vec{e_3} \cdot \vec{e_1} = \vec{e_1} \cdot \vec{e_3} = 0.
\end{aligned}$$

Angle between vectors

Animation programming may require angles between two vectors. Applying the dot product on two vectors leads us to their internal angle, expressed in the components of both vectors.

We draw two location vectors $\vec{a}$ and $\vec{b}$ and the base vectors $\vec{e_1}$ and $\vec{e_2}$ in $\mathbb{R}^2$. Vector $\vec{b}$ makes an internal angle θ with vector $\vec{a}$. Vector $\vec{a}$ makes an angle α with base vector $\vec{e_1}$ on the x-axis. We then express both location vectors $\vec{a}$ and $\vec{b}$ according to their trigonometrical

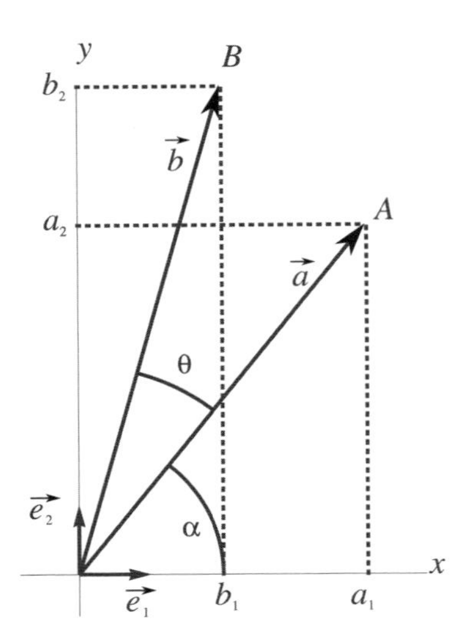

Figure 7.5: The internal angle between two vectors

components (see formula (7.6)) as

$$\vec{a} = \begin{pmatrix} ||\vec{a}|| \cos\alpha \\ ||\vec{a}|| \sin\alpha \end{pmatrix} \text{ and } \vec{b} = \begin{pmatrix} ||\vec{b}|| \cos(\alpha+\theta) \\ ||\vec{b}|| \sin(\alpha+\theta) \end{pmatrix}$$

and calculate their dot product as

$$\begin{aligned} \vec{a}\cdot\vec{b} &= a_1 b_1 + a_2 b_2 \\ &= ||\vec{a}||\ ||\vec{b}||\ \cos\alpha\ \cos(\alpha+\theta) + ||\vec{a}||\ ||\vec{b}||\ \sin\alpha\ \sin(\alpha+\theta) \\ &= ||\vec{a}||\ ||\vec{b}||\ \big(\cos\alpha\ \cos(\alpha+\theta) + \sin\alpha\ \sin(\alpha+\theta)\big) \\ &= ||\vec{a}||\ ||\vec{b}||\ \cos\big(\alpha - (\alpha+\theta)\big) \\ &= ||\vec{a}||\ ||\vec{b}||\ \cos(-\theta) = ||\vec{a}||\ ||\vec{b}||\ \cos\theta. \end{aligned}$$

In $\mathbb{R}^2$ we were able to prove this result rather straightforwardly. We note that its generalization is way harder to prove and we hereby state it without any proof.

The internal angle θ between two vectors $\vec{a}$ and $\vec{b}$ in $\mathbb{R}^n$ is the angle which satisfies

$$\vec{a}\cdot\vec{b} = ||\vec{a}||\ ||\vec{b}|| \cos\theta \tag{7.8}$$

or

$$\cos\theta = \frac{a_1b_1 + a_2b_2 + \cdots + a_nb_n}{\sqrt{a_1^2 + a_2^2 + \cdots + a_n^2}\,\sqrt{b_1^2 + b_2^2 + \cdots + b_n^2}}. \tag{7.9}$$

Examples:

$$\vec{a} = \begin{pmatrix} 3 \\ 4 \\ 2 \end{pmatrix} \quad \text{and} \quad \vec{b} = \begin{pmatrix} 6 \\ -1 \\ 1 \end{pmatrix}$$

$$\begin{aligned} \cos\theta &= \frac{3\cdot 6 + 4\cdot(-1) + 2\cdot 1}{\sqrt{3^2+4^2+2^2}\,\sqrt{6^2+(-1)^2+1^2}} \\ &= \frac{18-4+2}{\sqrt{9+16+4}\,\sqrt{36+1+1}} \\ &= 0.4819 \\ \theta &= \arccos(0.4819) \\ &= 61.1851^\circ \end{aligned}$$

$$\vec{c} = \begin{pmatrix} 3 \\ \sqrt{3} \end{pmatrix} \quad \text{and} \quad \vec{d} = \begin{pmatrix} \sqrt{3} \\ -3 \end{pmatrix}$$

$$\begin{aligned} \cos\theta &= \frac{3\sqrt{3} - 3\sqrt{3}}{\sqrt{3^2+3}\,\sqrt{3+(-3)^2}} = 0 \\ \theta &= \arccos(0) \\ &= 90^\circ \end{aligned}$$

Orthogonality

Two vectors $\vec{a}$ and $\vec{b}$ are orthogonal in $\mathbb{R}^n$ if and only if their dot product is zero.

$$\vec{a} \perp \vec{b} \Longleftrightarrow \vec{a}\cdot\vec{b} = 0 \tag{7.10}$$

We note that the null vector is orthogonal to any vector.

Proof:

$\Rightarrow$	$\Leftarrow$
given: $\vec{a} \perp \vec{b}$ to prove: $\vec{a} \cdot \vec{b} = 0$	given: $\vec{a} \cdot \vec{b} = 0$ to prove: $\vec{a} \perp \vec{b}$
partial proof:	partial proof:
1. $\vec{a} \neq \vec{o} \neq \vec{b}$ $\vec{a} \perp \vec{b} \Rightarrow \theta = \frac{\pi}{2}$ $\vec{a} \cdot \vec{b} = \|\vec{a}\| \; \|\vec{b}\| \cos \frac{\pi}{2}$ $\vec{a} \cdot \vec{b} = \|\vec{a}\| \; \|\vec{b}\| \cdot 0$ $\vec{a} \cdot \vec{b} = 0$ 2. $\vec{a} = \vec{o}$ of $\vec{b} = \vec{o}$ $\vec{a} \cdot \vec{b} = 0$	$\vec{a} \cdot \vec{b} = 0$ $\Downarrow$ $\|\vec{a}\| \; \|\vec{b}\| \cos \theta = 0$ $\Downarrow$ $\|\vec{a}\| = 0$ or $\|\vec{b}\| = 0$ or $\cos \theta = 0$ $\Downarrow$ $\vec{a} = \vec{o}$ or $\vec{b} = \vec{o}$ or $\theta = \frac{\pi}{2}$ $\Downarrow$ $\vec{a} \perp \vec{b}$ ■

Whenever applying this check for orthogonality on a computer, we need to compare the dot product to a tiny threshold value $\varepsilon \approx 10^{-5}$ instead of to zero itself: $\vec{a} \cdot \vec{b} < \varepsilon \Longleftrightarrow \vec{a} \perp \vec{b}$.

In case of a dot product significant different from zero, we are able to interpret its sign (positive or negative). Given θ as the internal angle between vectors $\vec{a}$ and $\vec{b}$, then for

- ▷ $\vec{a} \cdot \vec{b} < 0$ we conclude $90^\circ < \theta < 180^\circ$,
- ▷ $\vec{a} \cdot \vec{b} > 0$ we deduct $0^\circ < \theta < 90^\circ$.

Vector components in 3D

We recall that a vector $\vec{a}$ is completely determined by its length and direction. We obtain its direction by measuring its internal angles to the base vectors $\vec{e_1}, \vec{e_2}$ and $\vec{e_3}$ which are lying on the cartesian axes x, y and z. We call this kind of **direction angles** respectively α, β and γ, and calculate them via the dot product (see formula (7.9)).

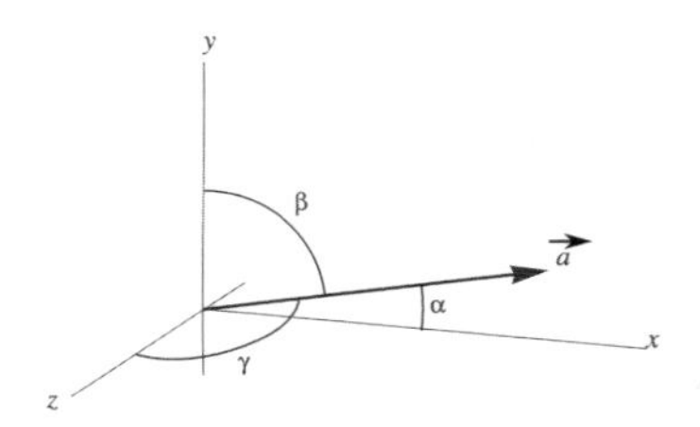

We obtain for instance the angle α between vector $\vec{a} = \begin{pmatrix} a_1 \\ a_2 \\ a_3 \end{pmatrix}$ and the x-axis via

$$\cos\alpha = \frac{\vec{a}\cdot\vec{e_1}}{||\vec{a}||\,||\vec{e_1}||} = \frac{a_1}{||\vec{a}||}.$$

Similarly, we find for both remaining direction angles

$$\cos\beta = \frac{a_2}{||\vec{a}||} \qquad \text{and} \qquad \cos\gamma = \frac{a_3}{||\vec{a}||}.$$

The other way around, knowing its length $||\vec{a}||$ and direction through its angles α,β and γ, determines the components of the vector $\vec{a}$ as

$$a_1 = ||\vec{a}||\cos\alpha, \qquad a_2 = ||\vec{a}||\cos\beta, \qquad a_3 = ||\vec{a}||\cos\gamma.$$

Since α,β and γ are called the direction angles of the vector $\vec{a}$, we call $\cos\alpha, \cos\beta$ and $\cos\gamma$ the **direction cosines** of the vector. We prove

$$\begin{aligned}(\cos\alpha)^2 + (\cos\beta)^2 + (\cos\gamma)^2 &= \frac{a_1^2}{||\vec{a}||^2} + \frac{a_2^2}{||\vec{a}||^2} + \frac{a_3^2}{||\vec{a}||^2} \\ &= \frac{a_1^2 + a_2^2 + a_3^2}{||\vec{a}||^2} \text{ with } ||\vec{a}|| = \sqrt{a_1^2 + a_2^2 + a_3^2} \\ &= 1.\end{aligned}$$

Example: Find the vector $\vec{a}$ given its length is 2 and its direction angles are $\alpha = \frac{\pi}{3}, \beta = \frac{2\pi}{3}$ and $0 \leqslant \gamma \leqslant \frac{\pi}{2}$.

We find γ via the latter formula as $\cos\gamma = \pm\sqrt{1 - (\cos\alpha)^2 - (\cos\beta)^2}$. Given γ is an acute angle, we conclude $\cos\gamma$ to be non-negative. Evaluating α and β yields

$$\cos\gamma = \sqrt{1 - \left(\cos\frac{\pi}{3}\right)^2 - \left(\cos\frac{2\pi}{3}\right)^2} = \frac{1}{\sqrt{2}} = \frac{\sqrt{2}}{2} \Rightarrow \gamma = \frac{\pi}{4}.$$

We calculate the components of the vector $\vec{a}$ as

$$\begin{aligned} a_1 &= ||\vec{a}|| \; \cos\alpha = 2\cos\frac{\pi}{3} = 1, \\ a_2 &= ||\vec{a}|| \; \cos\beta = 2\cos\frac{2\pi}{3} = -1, \\ a_3 &= ||\vec{a}|| \; \cos\gamma = 2\cos\frac{\pi}{4} = \sqrt{2}. \end{aligned}$$

All this results in the vector $\vec{a} = \begin{pmatrix} 1 \\ -1 \\ \sqrt{2} \end{pmatrix}$.

7.5 Cross product

The **cross product** or 'vector product' multiplies two 3D vectors and returns a new 3D vector. We consistently typeset this product with a cross.

DEFINITION

▷ in words: We define the cross product $\vec{a} \times \vec{b}$ of two vectors $\vec{a} \neq \vec{b}$ by determining its three aspects as a vector (see page 111):

1) $\vec{a} \times \vec{b}$ is perpendicular to the plane spanned by $\vec{a}$ and $\vec{b}$ (orientation of $\vec{a} \times \vec{b}$);
2) $\vec{a}, \vec{b}$ and $\vec{a} \times \vec{b}$ make up a right-handed frame (sense of $\vec{a} \times \vec{b}$);
3) $\|\vec{a} \times \vec{b}\| = \|\vec{a}\| \, \|\vec{b}\| \sin\theta$, with θ the internal angle made by $\vec{a}$ and $\vec{b}$ (length of $\vec{a} \times \vec{b}$).

▷ in symbols: Given $\vec{a} = \begin{pmatrix} a_1 \\ a_2 \\ a_3 \end{pmatrix}$ and $\vec{b} = \begin{pmatrix} b_1 \\ b_2 \\ b_3 \end{pmatrix}$ two vectors in $\mathbb{R}^3$, we define their cross product as the vector:

$$\vec{a} \times \vec{b} = \begin{pmatrix} a_2 b_3 - b_2 a_3 \\ -a_1 b_3 + b_1 a_3 \\ a_1 b_2 - b_1 a_2 \end{pmatrix}. \tag{7.11}$$

As a condition, the cross product requires both $\vec{a} \in \mathbb{R}^3$ and $\vec{b} \in \mathbb{R}^3$, then yields $\vec{a} \times \vec{b} \in \mathbb{R}^3$.

1) The first aspect describes the direction of $\vec{a} \times \vec{b}$, which is perpendicular to the vector $\vec{a}$ and to the vector $\vec{b}$.

2) The second aspect can be realized using the right-hand rule or corkscrew-rule (see paragraph 6.1). Applying the right-hand rule, we align our right-hand fingers to vector $\vec{a}$ and turning them to vector $\vec{b}$ makes the thumb point in the direction of the vector $\vec{a} \times \vec{b}$. Applying the corkscrew-rule, turning the corkscrew from the vector $\vec{a}$ to the vector $\vec{b}$ makes it move in the direction of the vector $\vec{a} \times \vec{b}$.

3) The third aspect says that the length of the cross product is given by the area of the parallellogram edged by the vectors $\vec{a}$ and $\vec{b}$.

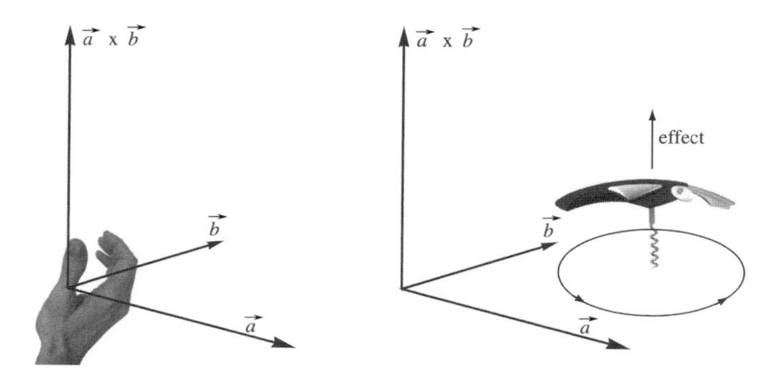

Figure 7.6: The right-hand rule or corkscrew-rule yields the direction of a cross product

Example: Given $\vec{a} = \begin{pmatrix} 5 \\ -6 \\ 0 \end{pmatrix} \in \mathbb{R}^3$ and $\vec{b} = \begin{pmatrix} 1 \\ 2 \\ 3 \end{pmatrix} \in \mathbb{R}^3$.

$$\vec{a} \times \vec{b} = \begin{pmatrix} -6 \cdot 3 - 0 \cdot 2 \\ -5 \cdot 3 + 0 \cdot 1 \\ 5 \cdot 2 - (-6) \cdot 1 \end{pmatrix}$$

$$= \begin{pmatrix} -18 \\ -15 \\ 16 \end{pmatrix} \in \mathbb{R}^3$$

We verify the cross product $\vec{a} \times \vec{b}$ to be perpendicular to the vector $\vec{a}$ by $(\vec{a} \times \vec{b}) \cdot \vec{a} = -18 \cdot 5 + (-15) \cdot (-6) + 16 \cdot 0 = 0$. Similarly we confirm the cross product $\vec{a} \times \vec{b}$ to be orthogonal to the vector $\vec{b}$.

Property: The cross product is **anticommutative**:

$$\vec{a} \times \vec{b} = -(\vec{b} \times \vec{a}).$$

We cross product the three base vectors $\vec{e_1} = \begin{pmatrix} 1 \\ 0 \\ 0 \end{pmatrix}$, $\vec{e_2} = \begin{pmatrix} 0 \\ 1 \\ 0 \end{pmatrix}$ and $\vec{e_3} = \begin{pmatrix} 0 \\ 0 \\ 1 \end{pmatrix}$ to discover how they relate.

$$\vec{e_1} \times \vec{e_1} = \vec{e_2} \times \vec{e_2} = \vec{e_3} \times \vec{e_3} = \vec{o}$$

$$\vec{e_1} \times \vec{e_2} = \vec{e_3}$$
$$\vec{e_2} \times \vec{e_3} = \vec{e_1}$$
$$\vec{e_3} \times \vec{e_1} = \vec{e_2}$$

Given two not null vectors, their cross product is often used to produce a vector perpendicular to the plane they span.

Parallelism

Two vectors $\vec{a}$ and $\vec{b}$ are parallel or antiparallel in $\mathbb{R}^3$ if and only if their cross product is the null vector.

$$\vec{a} \uparrow\uparrow \vec{b} \text{ or } \vec{a} \uparrow\downarrow \vec{b} \Longleftrightarrow \vec{a} \times \vec{b} = \vec{o}$$

We note that the null vector is parallel and antiparallel to any vector.

Proof:

$\Rightarrow$

given: $\vec{a} \uparrow\uparrow \vec{b}$ or $\vec{a} \uparrow\downarrow \vec{b}$
to prove: $\vec{a} \times \vec{b} = \vec{o}$

partial proof:

1. $\vec{a} \neq \vec{o} \neq \vec{b}$

$$\left(\vec{a} \uparrow\uparrow \vec{b} \Rightarrow \theta = 0\right) \text{ or } \left(\vec{a} \uparrow\downarrow \vec{b} \Rightarrow \theta = \pi\right)$$
$$\left\|\vec{a} \times \vec{b}\right\| = \|\vec{a}\| \; \|\vec{b}\| \sin\theta$$
$$\left\|\vec{a} \times \vec{b}\right\| = \|\vec{a}\| \; \|\vec{b}\| \; 0$$
$$\left\|\vec{a} \times \vec{b}\right\| = 0$$

2. $\vec{a} = \vec{o}$ or $\vec{b} = \vec{o}$

$$\left\|\vec{a} \times \vec{b}\right\| = 0$$
$$\Downarrow$$
$$\vec{a} \times \vec{b} = \vec{o}$$

$\Leftarrow$

given: $\vec{a} \times \vec{b} = \vec{o}$
to prove: $\vec{a} \uparrow\uparrow \vec{b}$ or $\vec{a} \uparrow\downarrow \vec{b}$

partial proof:

$$\vec{a} \times \vec{b} = \vec{o}$$
$$\Downarrow$$
$$\left\|\vec{a} \times \vec{b}\right\| = \|\vec{a}\| \; \|\vec{b}\| \sin\theta = 0$$
$$\Downarrow$$
$$\|\vec{a}\| = 0 \text{ or } \|\vec{b}\| = 0 \text{ or } \sin\theta = 0$$
$$\Downarrow$$
$$\vec{a} = \vec{o} \text{ or } \vec{b} = \vec{o} \text{ or } \theta = 0 \text{ or } \theta = \pi$$
$$\Downarrow$$
$$\vec{a} \uparrow\uparrow \vec{b} \text{ or } \vec{a} \uparrow\downarrow \vec{b}$$

Example: $\vec{e_1} \times (\vec{e_2} \times \vec{e_3}) = \vec{o}$

7.6 Normal vectors

All 3D objects are modeled by **polygons**. Polygons are flat surfaces determined by vertices and bordered by straight edges. Triangles are very popular polygons since most render engines require an input sequence of polygons featuring three edges. One single 3D object is often described by thousands of polygons.

Apart from the essential polygon data (the cartesian x-, y- and z-coordinates of its vertices) it is as useful to store the components of the polygon's **normal vector**. The normal vector is perpendicular to the polygon's surface. Normal vectors determine whether the surface should be visible on screen and calculate its shading and reflection, depending on the type of ambient light. The separate sequence of all normal vectors of one single 3D object is called a **normal map**.

We construct for a triangle ABC a normal vector using the 3D cartesian coordinates of its three vertices A, B and C. Each plane in 3D space has a vector which is perpendicular to it. By calculating for instance $\overrightarrow{BA} \times \overrightarrow{BC}$ we immediately gain a vector which is orthogonal to the free vector $\overrightarrow{BA}$ and to its neighbouring vector $\overrightarrow{BC}$, and hence is orthogonal to the plane ABC they edge. In order to be able to calculate *percentages* of reflection later on, it is a common practice to divide a normal vector by its own length $\|\overrightarrow{BA} \times \overrightarrow{BC}\|$ which scales it into a unit vector:

$$\vec{n} = \frac{\overrightarrow{BA} \times \overrightarrow{BC}}{\|\overrightarrow{BA} \times \overrightarrow{BC}\|}. \tag{7.12}$$

We define **normalizing** a vector as scaling it into a unit vector by means of its own length. We call such a normalized normal vector $\vec{n}$ a **unit normal vector**.

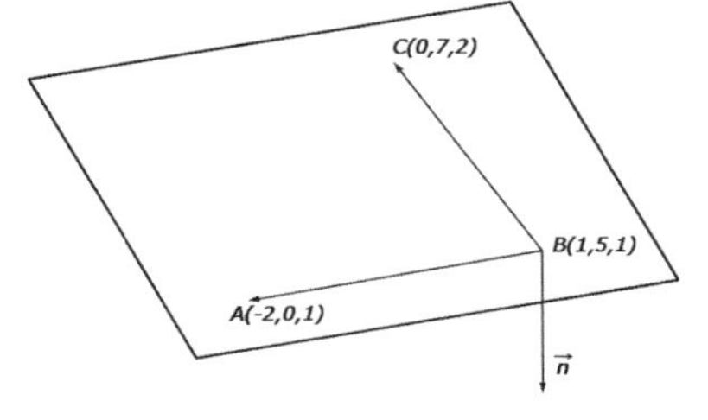

Example: Imagine we need to construct a unit normal vector on the polygon defined by the three vertices: $A(-2,0,1), B(1,5,1)$ and $C(0,7,2)$. The plane containing these three points A, B and C can for our purpose be defined by two vectors each aligning one of the straight edges $[AB], [BC]$ or $[CA]$. We are free to choose for instance the vectors $\overrightarrow{BA}$ and $\overrightarrow{BC}$ for it. We emphasize that also other pairs of non parallel vectors would work.

Firstly, we find the components of our free vectors $\overrightarrow{BA}$ and $\overrightarrow{BC}$.

$$\overrightarrow{BA} = \overrightarrow{OA} - \overrightarrow{OB} = \begin{pmatrix} -2 \\ 0 \\ 1 \end{pmatrix} - \begin{pmatrix} 1 \\ 5 \\ 1 \end{pmatrix} = \begin{pmatrix} -3 \\ -5 \\ 0 \end{pmatrix}$$

$$\overrightarrow{BC} = \overrightarrow{OC} - \overrightarrow{OB} = \begin{pmatrix} 0 \\ 7 \\ 2 \end{pmatrix} - \begin{pmatrix} 1 \\ 5 \\ 1 \end{pmatrix} = \begin{pmatrix} -1 \\ 2 \\ 1 \end{pmatrix}$$

Secondly, we calculate the cross product of $\overrightarrow{BA}$ and $\overrightarrow{BC}$ because this yields a normal vector to the plane defined by A, B and C.

$$\overrightarrow{BA} \times \overrightarrow{BC} = -5\vec{e_1} + 3\vec{e_2} - 11\vec{e_3} = \begin{pmatrix} -5 \\ 3 \\ -11 \end{pmatrix}$$

Finally, to obtain its corresponding unit normal vector, we need to divide this normal vector by its own length $\left\|\overrightarrow{BA} \times \overrightarrow{BC}\right\| = \sqrt{155}$, yielding

$$\vec{n} = \frac{1}{\sqrt{155}} \begin{pmatrix} -5 \\ 3 \\ -11 \end{pmatrix} \approx \begin{pmatrix} -0.40 \\ 0.24 \\ -0.88 \end{pmatrix}.$$

We can easily verify that $\|\vec{n}\| \approx 1$, concluding that $\vec{n}$ indeed is a unit vector.

Verifying that $\vec{n} \cdot \overrightarrow{BA} = 0$ and that $\vec{n} \cdot \overrightarrow{BC} = 0$, proves $\vec{n}$ to be orthogonal to both free vectors $\overrightarrow{BA}$ and $\overrightarrow{BC}$, hence concluding $\vec{n}$ is a normal vector of the plane they define.

7.7 Exercises

Exercise 57 Calculate the length and the direction of the resultant force $\vec{F_1} + \vec{F_2} + \vec{F_3} + \vec{F_4}$ caused by the four forces $\vec{F_1}, \vec{F_2}, \vec{F_3}$ and $\vec{F_4}$, given their magnitudes $||\vec{F_1}|| = 120N$, $||\vec{F_2}|| = 100N$, $||\vec{F_3}|| = 80N$ and $||\vec{F_4}|| = 40N$.

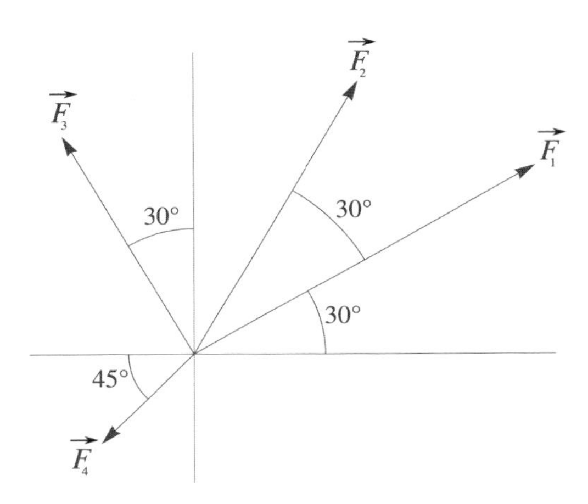

Exercise 58 An airplane is flying at a speed of 800 kilometers per hour and in the direction of 210°. At the same time there is a wind blowing at a speed of 90 kilometers per hour and from the direction of 165° (from the SE to the NW). Find the 'ground speed' and the resulting direction of this airplane. Hint: start finding all vector components for a straightforward solution. Please consider that in navigation directions are measured clockwise, starting from N, which is labeled as 0°.

Exercise 59 Given the location vectors

$$\vec{a} = \begin{pmatrix} 3 \\ 2 \\ -4 \end{pmatrix} \qquad \vec{b} = \begin{pmatrix} -2 \\ 0 \\ 4 \end{pmatrix} \qquad \vec{c} = \begin{pmatrix} -5 \\ 1 \\ 4 \end{pmatrix}$$

Find

1) $-2(\vec{b} + 5\vec{c}) + 5(\vec{a} - 3\vec{b})$
2) $3(\vec{a} \cdot \vec{b})\vec{c} - 5(\vec{b} \cdot \vec{c})\vec{a}$
3) $(\vec{a} - 3\vec{b}) \cdot (4\vec{c})$
4) $(\vec{a} \cdot \vec{b})\vec{c} - \vec{b}$
5) $(-\vec{a} + 2\vec{c}) \times (-\vec{b})$
6) $(2\vec{a}) \times (-\vec{b} + 5\vec{c})$

Exercise 60 A camera is positioned in cartesian coordinate $C(1,4)$ and has a 'focus vector' $\vec{f} = \begin{pmatrix} 5 \\ 3 \end{pmatrix}$ describing its line of sight. Meanwhile we spot an animal roaming in $A(7,2)$. Given our camera swings over an angle of $90°$ to both sides, gaining an eyeshot of $180°$, can it film this animal?

Exercise 61 Some fancy sports cars feature a tiny spoiler at the back of their roof. After simplifying its bodywork, we can model such a sports car frame in 3D cartesian coordinates, as shown in the figure. Assume the spoiler $\overrightarrow{FG}$ has a length 0.3 and is perpendicular to the rear window *ABCD*. Find the coordinates of the point *G*, making use of vectors.

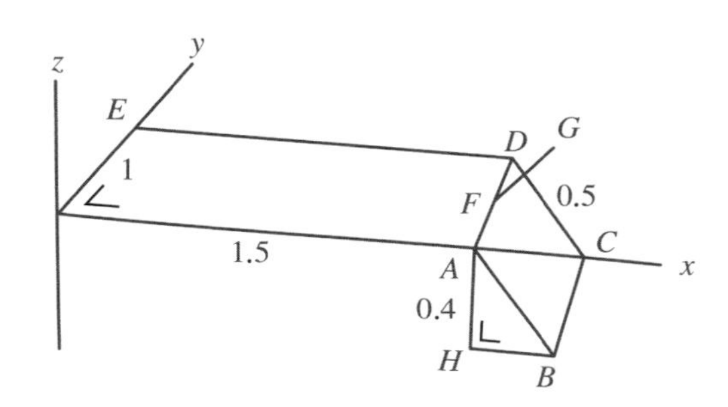

Hints:

- ▷ Calculate the distance $|HB|$ to determine the coordinates of several useful points.
- ▷ Given the spoiler $\overrightarrow{FG}$ is perpendicular to the rear window, it must be a normal vector of the polygon *ABCD*. Hence, the direction of $\overrightarrow{FG}$ can be obtained via the cross product of two vectors which align along the non parallel edges of *ABCD*. Calculate an appropriate cross product.
- ▷ Given the length of the spoiler, determine the coordinates of the point *G*.

Exercise 62 Find the area of the parallellogram spanned by the location vectors $\vec{a} = (4, -10, 5)$ and $\vec{b} = (-3, -1, -3)$.

Exercise 63 Find the internal angle in the point *B*, made by the straight lines connecting the points $A(1,1,1)$, $B(2,3,4)$ and $C(-2,5,2)$.

Hint: choose two free vectors to start from.

Chapter 8 · Parameters

Lines and planes are of major importance in setting up digital landscapes. In paragraph 4.2 we outlined how to obtain the equation of a straight line, in order to border a landscape or to describe an object's trajectory. Such an equation suits only 2D space (see formula (4.3)). Because 3D applications are taking over, it becomes inevitable for us to know how to describe straight lines in 3D space. Additionally, we also explain the equation of a plane in 3D coordinates.

8.1 Parametric equations

It is straightforward to describe a moving object by cartesian coordinates (x, y) which are themselves functions of the running time:

$$x = x(t) \qquad \text{and} \qquad y = y(t).$$

We call such an approach, by means of a running parameter t, the **parametric equation** of a function (see formula (6.1)). We typeset its parameter as λ or t as it is often referring to a running time. To each value of the parameter λ, there corresponds exactly one point on the curve drawn by $(x(\lambda), y(\lambda))$.

Example: Consider the parametric equation

$$x = \lambda^2 \qquad \text{and} \qquad y = \lambda.$$

Calculate the x- and y-coordinates for a sample of parameter values λ.

λ	-3	-2	-1	0	1	2	3
x	9	4	1	0	1	4	9
y	-3	-2	-1	0	1	2	3

Plotting points on these locations, we obtain the horizontal parabola (see page 67). Eliminating parameter λ via $x = \lambda^2 = y^2$ yields $y^2 = x$, which is indeed the cartesian equation of the horizontal parabola.

Example: From physics we know the parameter equation of the horizontal throw, for launching an object from a certain height giving it a constant horizontal speed v_0. The parametric equation of its parabolic trajectory is

$$x = x_0 + v_0 t \qquad \text{and} \qquad y = y_0 + \frac{1}{2} g t^2.$$

We recall the constant gravitational acceleration $g \approx 9.8066$ meter per square second. Eliminating the parameter t and adopting initial values $(x_0, y_0) = (0,0)$, we obtain the cartesian equation of a parabola. Because due to $t = \frac{x}{v_0}$ we find $y = \frac{1}{2}g\left(\frac{x}{v_0}\right)^2 = \frac{g}{2v_0^2}x^2$.

8.2 Vector equation of a line

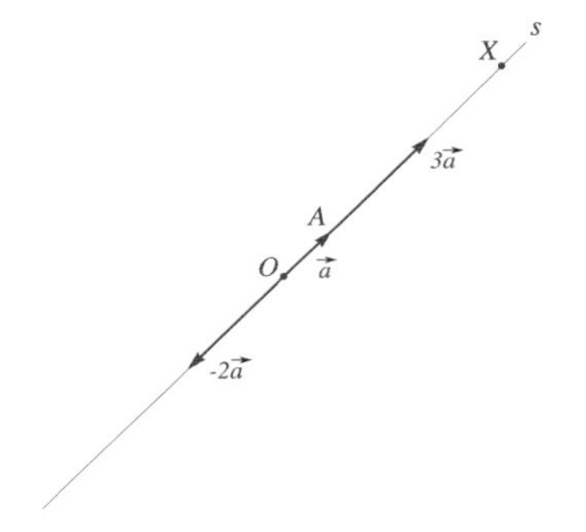

This figure shows several scalar multiples of a location vector $\vec{a} = \overrightarrow{OA}$. The heads of these scalar multiples are all on the same straight line s. The other way around, each point of this straight line running through the origin O and the given point A can be reached by a scalar multiplication of the vector $\vec{a}$ and a real number λ. In other words, for each point X we can express its location vector $\vec{x}$ on the line s as $\vec{x} = \lambda\vec{a}$ given λ a real number. For this reason we call $\vec{x} = \lambda\vec{a}$ a **vector equation** of the line s. In this equation we call $\vec{a}$ a **direction vector** of the straight line s. And the running number λ we call it a free parameter.

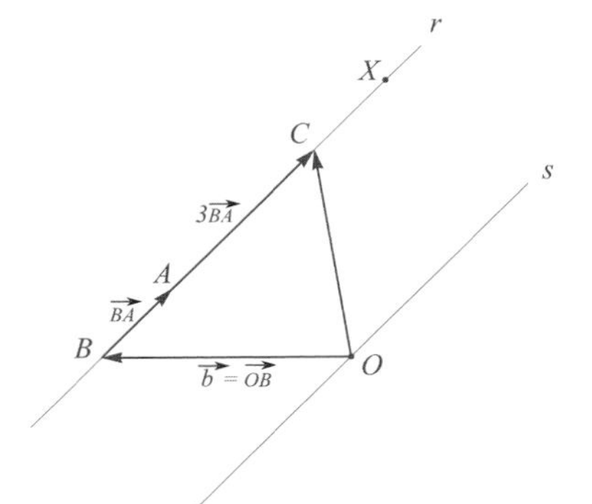

We take a similar approach for straight lines running through any arbitrary **anchor point** but the origin B, along the direction of $\overrightarrow{BA}$. We can express the vector with head C on line r as $\vec{b} + 3\overrightarrow{BA}$. Repeating this for a vector with arbitrary head X on line r we get $\vec{x} = \vec{b} + \lambda\overrightarrow{BA}$ for an arbitrary value of λ. The vector expression $\vec{x} = \vec{b} + \lambda\overrightarrow{BA}$ means that each point on the line r can be reached by taking an appropriate distance from the anchor point B along the direction vector $\overrightarrow{BA}$. In other words, for each parameter value λ the location vector $\vec{x}$ finds his head on the line r. For this reason we call $\vec{x} = \vec{b} + \lambda\overrightarrow{BA}$ a vector equation of the line r.

We may call the location vector $\vec{b} = \overrightarrow{OB}$ a **position vector** of the straight line r. We require the position vector $\vec{b}$ to aim at a chosen anchor point B on the line r and require the **direction vector** $\overrightarrow{BA}$ to be different from the null vector.

We realize that only the *orientation* of the direction vector is essential, not its length nor sense. For instance the direction vectors $\begin{pmatrix} 4 \\ 2 \end{pmatrix}$, $\begin{pmatrix} 2 \\ 1 \end{pmatrix}$ or $\begin{pmatrix} -1 \\ -0.5 \end{pmatrix}$ describe the same line. Secondly, we realize that *each* vector which takes the origin as tail and any point at the line as head, makes a valid position vector. Given the above, we understand we have several ways to describe the same line. We conclude a straight line has no unique vector equation.

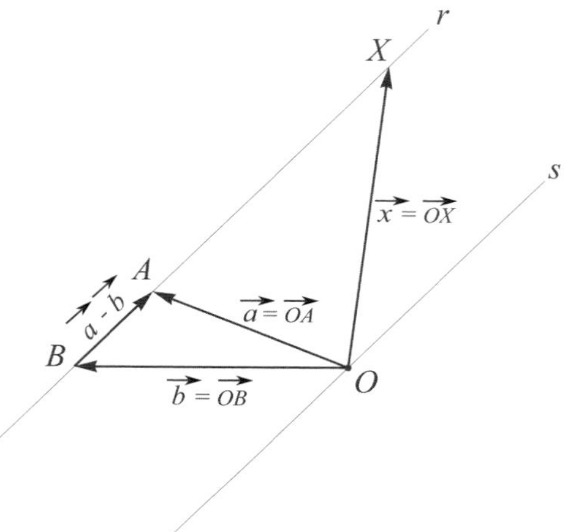

Figure 8.1: A straight line r through two points A and B

In general we consider a straight line r through two points A and B, pinned by their location vectors $\vec{a} = \overrightarrow{OA}$ and $\vec{b} = \overrightarrow{OB}$. To find its vector equation, we just need to choose an appropriate position vector and a direction vector. We again can express the location vector $\vec{x}$ of a variable point X on the line r as the sum vector

$$\vec{x} = \vec{b} + \overrightarrow{BX}.$$

Since $\overrightarrow{BX}$ and $\overrightarrow{BA}$ are parallel vectors, we are able to express $\overrightarrow{BX} = \lambda \overrightarrow{BA} = \lambda(\vec{a} - \vec{b})$. As for a vector equation of a straight line connecting two given points A and B we conclude:

$$\vec{x} = \vec{b} + \lambda(\vec{a} - \vec{b}).$$

We memorize such a vector equation for a straight line more easily in words:

$$\vec{x} = \text{position vector} + \lambda \text{direction vector}. \tag{8.1}$$

Rewriting a vector equation $\vec{x} = \vec{b} + \lambda\vec{c}$ of a straight line on position vector $\vec{b}$ along direction vector $\vec{a}$ by its vector components $\vec{b} = \begin{pmatrix} b_1 \\ b_2 \\ b_3 \end{pmatrix}$ and $\vec{c} = \begin{pmatrix} c_1 \\ c_2 \\ c_3 \end{pmatrix}$ in 3D space, we obtain its **parametric equation**:

$$\begin{cases} x = b_1 + \lambda c_1 \\ y = b_2 + \lambda c_2 \\ z = b_3 + \lambda c_3 \end{cases}.$$

Eliminating the parameter λ from its three equations, we obtain its **cartesian equation**:

$$\frac{x-b_1}{c_1} = \frac{y-b_2}{c_2} = \frac{z-b_3}{c_3} (= \lambda).$$

Example: We aim for an equation of the straight line connecting the points $A(8,2,4)$ and $B(-2,-2,-2)$.

We determine the orientation of this line by

$$\overrightarrow{AB} = \overrightarrow{OB} - \overrightarrow{OA} = \begin{pmatrix} -2-8 \\ -2-2 \\ -2-4 \end{pmatrix} = \begin{pmatrix} -10 \\ -4 \\ -6 \end{pmatrix}.$$

We may take $\begin{pmatrix} -10 \\ -4 \\ -6 \end{pmatrix}$ for direction vector, but since $\begin{pmatrix} -10 \\ -4 \\ -6 \end{pmatrix} = -2 \begin{pmatrix} 5 \\ 2 \\ 3 \end{pmatrix}$, also the vector $\begin{pmatrix} 5 \\ 2 \\ 3 \end{pmatrix}$ is a valid direction vector for the straight line through the points A and B. Any location vector to an arbitrary point of the line makes a valid position vector for the line AB. We may for instance choose $\overrightarrow{OA} = \begin{pmatrix} 8 \\ 2 \\ 4 \end{pmatrix}$ for it. The above decisions yield this vector equation for the line AB:

$$\vec{x} = \begin{pmatrix} 8 \\ 2 \\ 4 \end{pmatrix} + \lambda \begin{pmatrix} 5 \\ 2 \\ 3 \end{pmatrix},$$

or rewritten to its parametric equation:

$$\begin{cases} x = 8 + 5\lambda \\ y = 2 + 2\lambda \\ z = 4 + 3\lambda \end{cases}.$$

For instance, at parameter value $\lambda = -1$ we generate its point $P(3,0,1)$. After elimination of this free parameter λ we find the corresponding cartesian equation for the line AB in 3D space:

$$\frac{x-8}{5} = \frac{y-2}{2} = \frac{z-4}{3}.$$

Alternatively, we could have chosen the direction vector $\overrightarrow{AB}$ and for instance $\overrightarrow{OB}$ as the position vector. Consequently, another valid vector equation describing the same straight line AB would be

$$\vec{x} = \begin{pmatrix} -2 \\ -2 \\ -2 \end{pmatrix} + \mu \begin{pmatrix} -10 \\ -4 \\ -6 \end{pmatrix}.$$

Referring to its point $P(3,0,1)$ again, in this equation it is generated at $\mu = \frac{-1}{2}$.

Example: To further illustrate that a vector equation of a line is not unique, we hereby show some alternatives for the straight line AB through the points $A(2,7)$ and $B(-3,-3)$ in 2D space.

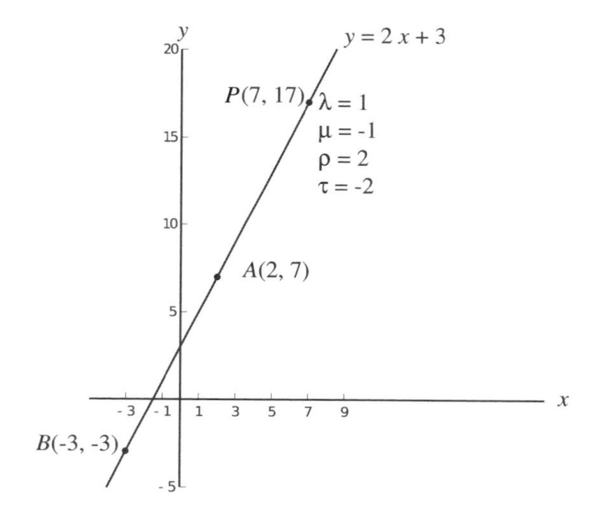

position vector	direction vector	vector equation	cartesian equation
$\vec{a}=\begin{pmatrix}2\\7\end{pmatrix}$	$\overrightarrow{BA}=\begin{pmatrix}5\\10\end{pmatrix}$	$\begin{pmatrix}x\\y\end{pmatrix}=\begin{pmatrix}2\\7\end{pmatrix}+\lambda\begin{pmatrix}5\\10\end{pmatrix}$	$\frac{x-2}{5}=\frac{y-7}{10}$ $\Leftrightarrow y=2x+3$
$\vec{a}=\begin{pmatrix}2\\7\end{pmatrix}$	$\overrightarrow{AB}=\begin{pmatrix}-5\\-10\end{pmatrix}$	$\begin{pmatrix}x\\y\end{pmatrix}=\begin{pmatrix}2\\7\end{pmatrix}+\mu\begin{pmatrix}-5\\-10\end{pmatrix}$	$\frac{x-2}{-5}=\frac{y-7}{-10}$ $\Leftrightarrow y=2x+3$
$\vec{b}=\begin{pmatrix}-3\\-3\end{pmatrix}$	$\overrightarrow{BA}=\begin{pmatrix}5\\10\end{pmatrix}$	$\begin{pmatrix}x\\y\end{pmatrix}=\begin{pmatrix}-3\\-3\end{pmatrix}+\rho\begin{pmatrix}5\\10\end{pmatrix}$	$\frac{x+3}{5}=\frac{y+3}{10}$ $\Leftrightarrow y=2x+3$
$\vec{b}=\begin{pmatrix}-3\\-3\end{pmatrix}$	$\overrightarrow{AB}=\begin{pmatrix}-5\\-10\end{pmatrix}$	$\begin{pmatrix}x\\y\end{pmatrix}=\begin{pmatrix}-3\\-3\end{pmatrix}+\tau\begin{pmatrix}-5\\-10\end{pmatrix}$	$\frac{x+3}{-5}=\frac{y+3}{-10}$ $\Leftrightarrow y=2x+3$

We can reach any point on the line AB using one of these four alternative vector equations. For instance its point $P(7,17)\in AB$ corresponds alternatively to $\lambda=1$, $\mu=-1$, $\rho=2$ or $\tau=-2$.

8.3 Intersecting straight lines

We consider the two straight lines r given by $\vec{x}=\vec{b}+\lambda\vec{c}$ and s given by $\vec{x}=\vec{q}+\mu\vec{p}$. To find their possible intersection point, we have to compare their recipes in one equation which we solve for the parameters λ and μ (see paragraph 4.3).

In case of a solution we will find the intersection point of both lines. In case of an inconsistent system, both lines have no intersection point. This latter case can be caused either by parallel lines (which do not coincide) or by skew lines. We define **intersecting** lines (in 2D and 3D) in case they have exactly one point in common. We define **parallel** lines (in 2D and 3D) in case they are not intersecting and lying in a common plane. We define **skew** lines (in 3D) in case they are not intersecting and *not* lying in a common plane. Notice that coinciding lines are also considered as parallel lines.

Example 1: We aim for the possible intersection point of the straight lines

$$r:\begin{pmatrix}x\\y\\z\end{pmatrix}=\begin{pmatrix}1\\1\\0\end{pmatrix}+\lambda\begin{pmatrix}1\\1\\1\end{pmatrix}\text{ and } s:\begin{pmatrix}x\\y\\z\end{pmatrix}=\begin{pmatrix}2\\0\\2\end{pmatrix}+\mu\begin{pmatrix}1\\-1\\2\end{pmatrix}.$$

Comparing their recipes in one equation and solving it by elimination for the parameters λ and μ, yields a solution to their underdetermined system (see paragraph 2.2).

$$\begin{cases}1+\lambda &= 2+\mu\\ 1+\lambda &= -\mu\\ \lambda &= 2+2\mu\end{cases}\Leftrightarrow\begin{cases}\lambda-\mu &= 1 & |1 & |1\\ \lambda+\mu &= -1 & |-1 & \\ \lambda-2\mu &= 2 & & |-1\end{cases}$$

$$\Leftrightarrow\begin{cases}\lambda-\mu &= 1\\ -2\mu &= 2\\ \mu &= -1\end{cases}$$

$$\Leftrightarrow\begin{cases}\lambda-\mu &= 1\\ \mu &= -1\\ \mu &= -1\end{cases}$$

$$\Leftrightarrow\begin{cases}\lambda &= 0\\ \mu &= -1\\ \mu &= -1\end{cases}$$

Given this solution we generate the intersection point S of the straight lines r and s. Replacing $\lambda = 0$ in the parametric equation of r, leads to

$$\begin{pmatrix}x\\y\\z\end{pmatrix}=\begin{pmatrix}1\\1\\0\end{pmatrix}+0\begin{pmatrix}1\\1\\1\end{pmatrix}=\begin{pmatrix}1\\1\\0\end{pmatrix}=S.$$

Evaluating the parametric equation of s for $\mu = -1$, obviously leads to the same intersection point S:

$$\begin{pmatrix}x\\y\\z\end{pmatrix}=\begin{pmatrix}2\\0\\2\end{pmatrix}-1\begin{pmatrix}1\\-1\\2\end{pmatrix}=\begin{pmatrix}1\\1\\0\end{pmatrix}=S.$$

Example 2: We try to determine the possible intersection point of the straight lines

$$\begin{cases}x &= \lambda\\ y &= -1+\lambda\\ z &= 1+\lambda\end{cases}\quad\text{and}\quad\begin{cases}x &= -1+2\mu\\ y &= 4-\mu\\ z &= 0.\end{cases}$$

After comparing both recipes in one equation, we again apply the elimination method for solving it.

$$\begin{cases} \lambda & = & -1+2\mu \\ -1+\lambda & = & 4-\mu \\ 1+\lambda & = & 0 \end{cases} \Leftrightarrow \begin{cases} \lambda-2\mu & = & -1 & |1 \\ \lambda+\mu & = & 5 & |2 \\ \lambda & = & -1 \end{cases}$$

$$\Leftrightarrow \begin{cases} \lambda-2\mu & = & -1 \\ 3\lambda & = & 9 \\ \lambda & = & -1 \end{cases}$$

$$\Leftrightarrow \begin{cases} \lambda-2\mu & = & -1 \\ \lambda & = & 3 \\ \lambda & = & -1 \end{cases}$$

Because the parameter λ can impossibly hold simultaneously values 3 and -1, we classify the above system as inconsistent. For this reason, we conclude that both straight lines do not intersect at all.

8.4 Vector equation of a plane

From the previous chapter we understand there can be many vector equations describing the same straight line. As a constraint to these equations we can only choose from direction vectors which are equal apart from a nonzero number factor. Every direction vector we can choose is a scalar multiple of another direction vector for that straight line. We call such vectors **linearly dependent**. Two vectors $\vec{a}$ and $\vec{b}$ are linearly dependent when there is a real number k relating both vectors by scalar multiplication $k\vec{a} = \vec{b}$. In case there does not exist such a scalar factor $k \in \mathbb{R}$ to relate them, then we have two **linearly independent** vectors.

Therefore, two linearly independent vectors $\vec{a}$ and $\vec{b}$ can not lie on the same straight line. This allows us to decompose any arbitrary location vector $\vec{x}$ lying in the plane spanned by $\vec{a}$ and $\vec{b}$, in two directions: one along vector $\vec{a}$ and one along vector $\vec{b}$. In other words, we are able to write location vector $\vec{x}$ as a linear combination of the independent vectors $\vec{a}$ and $\vec{b}$, briefly $\vec{x} = \lambda\vec{a} + \mu\vec{b}$.

In 3D space, all heads X of these linear combinations make up the plane v_O through the origin O. In a similar way as for straight lines, we can express such a plane through the origin O by means of a vector equation: $\vec{x} = \lambda\vec{a} + \mu\vec{b}$.

We call those vectors $\vec{a}$ and $\vec{b}$ **direction vectors** of the plane v_O. The plane v_O is spanned by both direction vectors $\vec{a}$ and $\vec{b}$. Any two linearly independent vectors lying in the

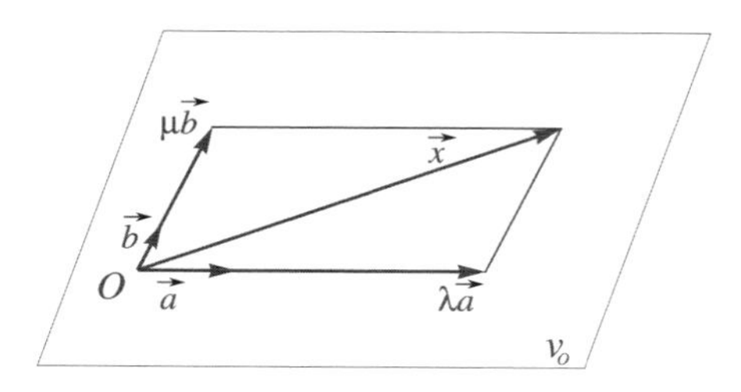

Figure 8.2: The plane ν_O through the origin O

plane ν_O can be chosen as direction vectors of it. Consequently, we have several ways to describe the same plane. We conclude a plane has no unique vector equation.

We take a similar approach for a plane ν_C through an arbitrary **anchor point** different from the origin C. Such a plane ν_C has been shifted over the location vector $\vec{c} = \overrightarrow{OC}$ $(\neq \vec{o})$ with respect to the plane ν_O. This allows us to express any vector $\vec{x}$ of such a plane as a sum of a vector lying in ν_O and a position vector heading from the origin O to the plane ν_C.

Summarized, the vector equation of a general plane ν_C is $\vec{x} = \vec{c} + \lambda\overrightarrow{CA} + \mu\overrightarrow{CB}$. Vector $\vec{c} = \overrightarrow{OC}$ is the **position vector** and vectors $\overrightarrow{CA}$ and $\overrightarrow{CB}$ are both **direction vectors** of the plane ν_C. Just like in case of the plane ν_O, any two linearly independent vectors lying in the plane ν_C can be chosen as direction vectors for it. We are also free to choose an appropriate position vector $\vec{c}$, as long as it is heading from the origin O to an anchor point C lying in the plane ν_C.

We memorize such a vector equation for a plane more easily in words:

$$\vec{x} = \text{position vector} + \lambda \text{direction vector}_1 + \mu \text{direction vector}_2, \tag{8.2}$$

given its position vector running from the origin to an arbitrary anchor point in the plane and both direction vectors linearly independent.

Eliminating parameters λ and μ from the plane's vector equation, we obtain its cartesian equation

$$\nu_C : ax + by + cz + d = 0.$$

Example: We aim for the vector equation of the plane ν_A determined by the points $A(1,2,7), B(-1,0,-3)$ and $C(0,3,8)$. The anchor point of the plane ν_A is, as hinted by its notation, chosen to be $A(1,2,7)$.

We need a position vector and two independent direction vectors for this.

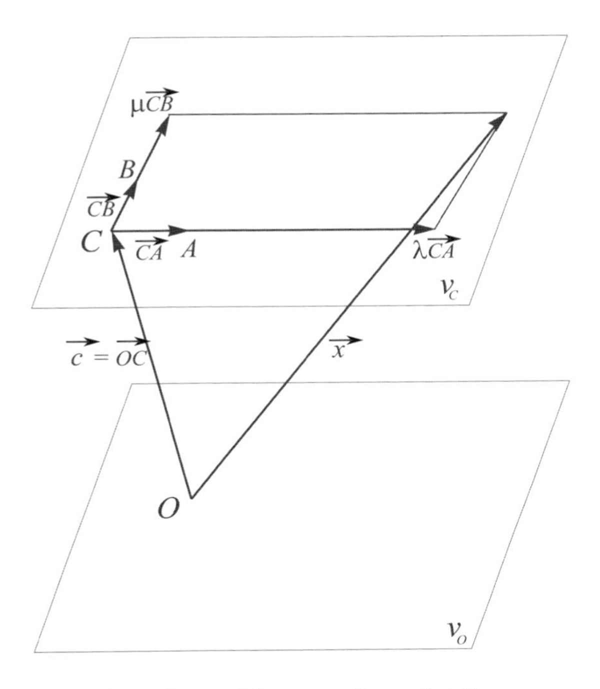

Figure 8.3: The plane v_C through an arbitrary anchor point C

- ▷ position vector: $\overrightarrow{OA} = \begin{pmatrix} 1 \\ 2 \\ 7 \end{pmatrix}$
- ▷ direction vectors: $\overrightarrow{AB} = \begin{pmatrix} -2 \\ -2 \\ -10 \end{pmatrix}$ and $\overrightarrow{AC} = \begin{pmatrix} -1 \\ 1 \\ 1 \end{pmatrix}$

Vectors $\overrightarrow{AB}$ and $\overrightarrow{AC}$ are indeed linearly independent, because there is no $k \in \mathbb{R}$ relating $\overrightarrow{AB} = k\overrightarrow{AC}$. Conclusively the desired vector equation is

$$\begin{pmatrix} x \\ y \\ z \end{pmatrix} = \begin{pmatrix} 1 \\ 2 \\ 7 \end{pmatrix} + \lambda \begin{pmatrix} -2 \\ -2 \\ -10 \end{pmatrix} + \mu \begin{pmatrix} -1 \\ 1 \\ 1 \end{pmatrix}.$$

or separated into its parametric shape

$$\begin{cases} x = 1 - 2\lambda - \mu \\ y = 2 - 2\lambda + \mu \\ z = 7 - 10\lambda + \mu. \end{cases}$$

To obtain the cartesian equation from this parametric equation we need to eliminate its parameters. We can for instance solve the first equation to μ, which results in: $\mu = 1 - 2\lambda - x$. Replacing this in the second equation yields

$$y = 2 - 2\lambda + \mu = 2 - 2\lambda + (1 - 2\lambda - x) = 3 - 4\lambda - x.$$

Subsequently solving the second equation to λ, results in: $\lambda = \frac{1}{4}(3 - x - y)$. Finally we replace the expressions for λ and μ in the third equation.

$$\begin{aligned} z &= 7 - 10\lambda + \mu \\ &= 7 - \frac{10}{4}(3 - x - y) + (1 - 2\lambda - x) \\ &= 7 - \frac{30}{4} + \frac{10}{4}x + \frac{10}{4}y + 1 - \frac{1}{2}(3 - x - y) - x \\ &= 7 - \frac{30}{4} + \frac{10}{4}x + \frac{10}{4}y + 1 - \frac{3}{2} + \frac{1}{2}x + \frac{1}{2}y - x \\ &= 2x + 3y - 1 \end{aligned}$$

We conclude that the cartesian equation of the plane v_A is $2x + 3y - z - 1 = 0$.

8.5 Exercises

Exercise 64 Referring to chapter 6, find the parametric equation of a circle $C(O,r)$.

Exercise 65

1) Find the parametric equation of the straight line running through the point $P(1,0,4)$ and parallel to the straight line AB, given its points $A(6,3,7)$ and $B(2,-1,-1)$.
2) Find the vector equation of the straight line running through the point $P(1,-2,7)$ and parallel to the straight line

$$\frac{x+3}{4} = \frac{y+1}{3} = z-1.$$

Exercise 66

1) Find the vector equation of the straight line r through the point $P(4,2,8)$ and parallel to the vector $\vec{a} = \begin{pmatrix} -1 \\ 4 \\ 3 \end{pmatrix}$. Continue aiming for its parametric and cartesian equation.
2) Find the parametric equation of the straight line r through the points $M(5,8,21)$ and $N(7,10,31)$, culminating in its cartesian equation.
3) Find the possible intersection point of the above straight lines r and s.

Exercise 67

1) Find the cartesian equation of the straight line through the point $P(2,2,3)$ and perpendicular to the plane v_C spanned by the direction vectors $\vec{u} = \begin{pmatrix} 1 \\ 1 \\ 0 \end{pmatrix}$, $\vec{e} = \begin{pmatrix} 0 \\ 0 \\ 1 \end{pmatrix}$ and anchored by the point $C(5,5,5)$.
2) Find a parametric equation of this plane. Prove that the point $P(8,8,4)$ belongs to this plane.

Exercise 68 The three points $P(3,1,0), Q(-4,1,1)$ and $R(5,9,3)$ determine the plane v_P.

1) Find a vector equation of this plane v_P.
2) Find a normal vector $\vec{n}$ on this plane v_P.

Exercise 69

1) Find a parametric equation of the plane v_P anchored in $P(3,6,2)$ and parallel to the plane v_O through the origin O and the points $A(1,2,3)$ and $B(4,2,1)$.

2) Find the cartesian equation of this plane v_P.

Exercise 70 Vectors $\vec{v} = \begin{pmatrix} 1 \\ 2 \\ 2 \end{pmatrix}$ and $\vec{w} = \begin{pmatrix} 5 \\ -1 \\ 1 \end{pmatrix}$ are given. A polygon of a 3D object is pinned by the origin O and the terminal points of vectors $\vec{v}$ and $\vec{w}$. Find a parametric equation and the cartesian equation of this polygon's plane. Find a normal vector $\vec{n}$ on this polygon.

Exercise 71 Are the straight lines

$$l : \frac{x+8}{5} = \frac{y-2}{2} = \frac{z+4}{3}$$

and

$$r : \frac{x-4}{9} = \frac{y+1}{1} = \frac{z-2}{6}$$

1) parallel (be it coinciding),

2) skew, or

3) intersecting?

In case of the latter situation, find their intersection.

Exercise 72 Find the intersection of the straight line $l = AB$ given $A(6,3,7)$ and $B(2,-1,-1)$ and the plane v_P containing the points $P(3,1,0)$, $Q(-4,1,1)$ and $R(5,9,3)$.

Exercise 73 Find and explain the intersection of the three planes

$$u_A : 8x - 7y + 2z = 3,$$

$$v_A : 11x + 5y - 7z = 9,$$

$$w_A : x - 2y + z = 0.$$

Exercise 74 Find and explain the intersection of the straight line

$$l : \frac{x-1}{4} = \frac{y-2}{-2} = \frac{z-3}{3},$$

and the sphere

$$B : (x-1)^2 + (y-2)^2 + (z-3)^2 = 58.$$

Chapter 9 · Collision detection

Programming animations or games, we often meet the need to detect collisions between moving objects. For instance characters are not allowed to walk through walls nor objects.

This chapter outlines some efficient techniques to detect collisions on the screen. Therefore we circumscribe moving 2D objects by a single circle. This obviously is a rough approximation for a targeted object, but its results are satisfactory. Whenever there is any need to, we can refine the approximation using a composite of smaller circles. Similarly, we can circumscribe targeted 3D objects by (a number of) spheres.

9.1 Collision detection and frame rate

We define **frame rate** as the refresh frequency of our runtime screen. The unit for frame rate is Hertz or frames per second, abbreviated to **fps**.

Frame rate is of major importance for a realistic look and feeling of animations and games: this parameter can make or break the experience. One amongst the first 3D computer games ('3D Monster Maze') used a frame rate of only 6 fps. Today's top notch games which challenge their players to interact to fast changing stimuli require minimal frame rates of 30 fps ('Halo 3') or 60 fps ('Call of Duty 4'). Frame rate can be either a runtime variable or a constant parameter (e.g. 30 fps in Flash® applications). As soon as the construction of a frame causes a heavy computational load to CPU or GPU (Central or Graphical Processing Unit), the frame rate will drop. Inversely, a lingering frame rate might accelerate during a less intensive game phase. We will therefore pay some attention to mathematical strategies to overcome this phenomenon.

In this book we distinguish collision detection techniques using circles from those using vectors. However we do not cover collision prediction nor collision handling, for which we recommend the specialized literature.

9.2 Collision detection using circles and spheres

Circles and spheres

Of course, using intersecting circles to detect collisions implies having their equations. For this reason, it is important to realize that a circle pictured on our screen is the locus of all points at equal distance or **radius** to a **center**. In other words, we only need those two elements, the circle's radius r and center $M(m_1, m_2)$, to find its equation.

- ▷ in words: The **circle** $C(M, r)$ given its center $M = (m_1, m_2)$ and radius r, is the locus of all points P at equal distance r to M.
- ▷ in symbols: Given $P(x, y)$ we define the circle by $d(P, M) = r$, which leads to its equation

  $$\sqrt{(x - m_1)^2 + (y - m_2)^2} = r,$$

  or after squaring both sides:

  $$(x - m_1)^2 + (y - m_2)^2 = r^2. \tag{9.1}$$

All points (x, y) satisfying this equation are lying on the circle $C(M, r)$ and the other way around, all points of the circle will satisfy this equation. We hereby emphasize that a circle is not a function since most arguments x produce two return values y.

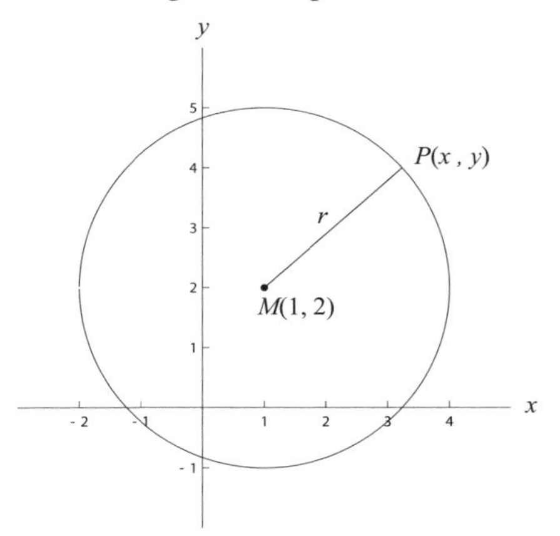

Figure 9.1: Circle $C((1,2),3)$

Example: In figure 9.1 we draw the circle $C(M, 3)$ with radius 3 around center $M(1, 2)$. Its equation is $(x-1)^2 + (y-2)^2 = 9$. Expanding brackets and grouping terms, leads to its **implicit** shape $x^2 + y^2 - 2x - 4y - 4 = 0$.

Generalizing this implicit result, we can write any circle equation as

$$x^2+y^2+ax+by+c=0, \tag{9.2}$$

given $a,b,c \in \mathbb{R}$. Though not every equation of this shape necessarily defines a circle.

Example: We study the implicit equation $x^2+y^2-4x-6y+14=0$. We try two times to rewrite a Perfect Square (see formula (1.2)), step by step:

$$\begin{aligned} x^2+y^2-4x-6y+14=0 & \\ \Leftrightarrow (x^2-4x)+(y^2-6y) &= -14 \\ \Leftrightarrow (x^2-2\cdot 2x+\mathbf{2^2})-\mathbf{2^2}+(y^2-2\cdot 3y+\mathbf{3^2})-\mathbf{3^2} &= -14 \\ \Leftrightarrow (x-2)^2+(y-3)^2-2^2-3^2 &= -14 \\ \Leftrightarrow (x-2)^2+(y-3)^2 &= -1. \end{aligned}$$

Theoretically the right hand side is the square of a radius, which has to be non-negative. Considering the left hand side, as it is a sum of two squares, it should be non-negative too. Conclusively, no single point (x,y) can ever satisfy this equation: it corresponds to the empty set instead of to a circle.

Therefore we recall it is good practice to always check whether an implicit equation $x^2+y^2+ax+by+c=0$ corresponds to a circle, before we start from it.

In 3D space we consequently use a sphere to detect collision between interacting screen objects, whenever such an approximation suits the context. The **sphere** $B(M,r)$ with center $M=(m_1,m_2,m_3)$ and radius r, is the locus of all points $P=(x,y,z)$ at equal distance $d(P,M)=r$ to the sphere center. This definition leads to the final equation of a three-dimensional sphere

$$(x-m_1)^2+(y-m_2)^2+(z-m_3)^2=r^2, \tag{9.3}$$

which is a straightforward extension of the equation of a plane circle.

Example: We consider the sphere with radius $r=8$ around center $M=(-5,2,-3)$. Its equation appears to be $(x+5)^2+(y-2)^2+(z+3)^2=64$. We can rewrite it to its implicit shape $x^2+y^2+z^2+10x-4y+6z-26=0$.

INTERSECTING LINE AND CIRCLE

Imagine we need to prevent a character to walk through a wall. In order to catch this situation in source code, we need to test for the collision between a circumscribing circle (the character) and a line (the wall). We can test a given line t and circle $C(M,r)$ for collision, by comparing the distance between the circle center and the line to the circle's radius. The line and the circle can have zero, one or two intersecting points (see figure 9.2). In case of one intersecting point, the line is tangent to the circle and therefore colliding. A collision we also have in case of two intersecting points.

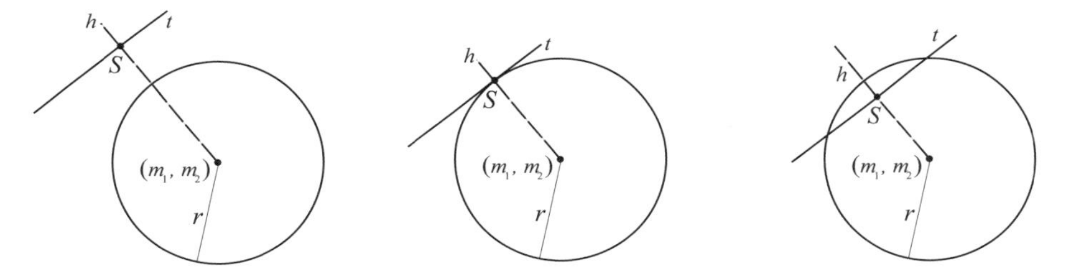

Figure 9.2: Intersecting a line t and a circle $C(M,r)$

Let us design the above outlined test for collision between a straight line and a circle. We therefore determine the altitude h through the circle center M perpendicular to the line t given by its recipe $y = ax + b$. The slopes of perpendicular lines relate as $a \cdot u = -1$ and evaluating $M(m_1, m_2) \in h$ yields an intercept v, completing the recipe for the altitude h as $y = ux + v$. This enables us to find the altitude's foot S on the given line t as the intersecting point of the straight lines h and t by solving their linear system

$$\begin{cases} y = ax + b \\ y = ux + v \end{cases}$$

for x and y, which yields the cartesian coordinates of $S(s_1, s_2)$. Finally, comparing the distance $d(M,S)$ to the radius r reveals or excludes a collision between the given line t and the circle $C(M,r)$. Applying the distance formula (3.3) we need to compare $\sqrt{(m_1 - s_1)^2 + (m_2 - s_2)^2}$ to the radius r. As it is good practice to avoid square roots in computer code because of their computational overhead and precision issues, we square both sides of our collision test for implementation purposes.

We finalize the test for collision between a line and a circle as:

$$\begin{array}{rl} \text{not colliding} & (m_1-s_1)^2+(m_2-s_2)^2 > r^2 \\ \text{tangent line} & (m_1-s_1)^2+(m_2-s_2)^2 = r^2 \\ \text{intersecting} & (m_1-s_1)^2+(m_2-s_2)^2 < r^2 \end{array} \tag{9.4}$$

Example: We want to test a wall on the line $y = 2x - 3$ and an object circumscribed by the circle $C\left((-4,4),\sqrt{5}\right)$ for collision.

We determine the altitude $y = ux + v$ through the circle center perpendicular to the given line. The slopes of both perpendicular lines relate as $2 \cdot u = -1$ which yields $u = -\frac{1}{2}$ for the altitude's slope. Evaluating $M(-4,4) \in h$ means $4 = -\frac{1}{2}(-4) + v$, which yields the altitude's intercept as $v = 2$.

This enables us to find the altitude's foot S solving the linear system

$$\left\{ \begin{array}{rcl} y & = & 2x-3 \\ y & = & -\frac{1}{2}x+2 \end{array} \right. \Longleftrightarrow \left\{ \begin{array}{rl} 2x-y & =3 \\ x+2y & =4 \end{array} \right. \Longleftrightarrow \left\{ \begin{array}{rl} x & =2 \\ y & =1 \end{array} \right.$$

for x and y, which yields the intersecting point $S(2,1)$.

This allows us to evaluate the collision test based on $d(M,S)$:

$$\begin{array}{rcl} (m_1-s_1)^2+(m_2-s_2)^2 & = & \\ (-4-s_1)^2+(4-s_2)^2 & = & \\ (-4-2)^2+(4-1)^2 & = & \\ (-6)^2+3^2 & > & (\sqrt{5})^2 \quad \Longrightarrow \quad 45 > 5 \quad \text{leads to no collision.} \end{array}$$

INTERSECTING CIRCLES AND SPHERES

The collision test for circles is surprisingly easier than the previous collision test (for line and circle). Apart from the obvious simplification a circumscribed circle or a sphere offers, also their test for collision is beyond comparison the most efficient. Of course, circumscribing targeted objects by a circle or sphere can only approximate interactions on the screen, but they offer sufficient 'quick and dirty' collision tests.

To test for collision between two objects, we need to check for an intersection of their circumscribing circles. Two different circles can have zero, one or two intersecting points (see figure 9.3). In case of exactly one intersecting point, both circles are tangent and therefore colliding. In case of two intersecting points, both circles are without any doubt colliding.

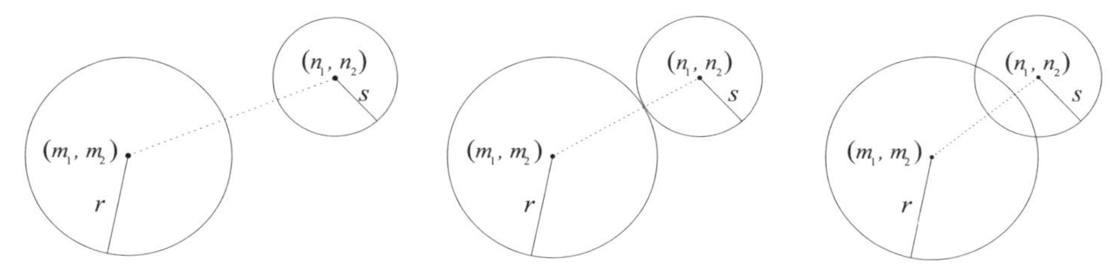

Figure 9.3: Intersecting two circles $C(M,r)$ and $C(N,s)$

Given the frame rate types (see page 161) and the collision prediction techniques (out of the scope of this book), we will not discuss the collision issues outlined in figure 9.4. We choose to focus on the regular collisions as sketched in figure 9.3.

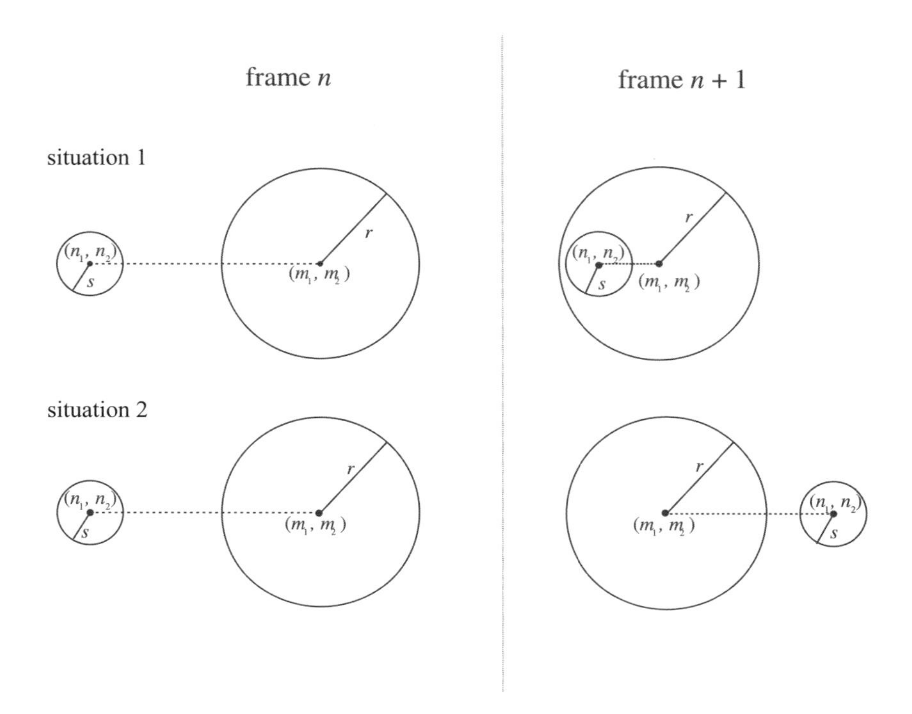

Figure 9.4: Collision detection issues with circles

Two circles do not intersect if the distance between their centers exceeds the sum of their radii. Two circles are tangent if the distance between their centers equals the sum of their radii. Two circles have two intersecting points if the distance between their centers is smaller than the sum of their radii. Applying distance formula (3.3), we need to compare $\sqrt{(m_1-n_1)^2+(m_2-n_2)^2}$ to the sum of the radii $r+s$. Square roots cause computational overhead and precision issues when processed. To overcome those drawbacks, we square both sides of our collision test before implementation. In other words, we finalize the test for collision between two circles as:

$$\begin{array}{rl} \text{not colliding} & (m_1-n_1)^2+(m_2-n_2)^2 > (r+s)^2 \\ \text{tangent circles} & (m_1-n_1)^2+(m_2-n_2)^2 = (r+s)^2 \\ \text{intersecting} & (m_1-n_1)^2+(m_2-n_2)^2 < (r+s)^2 \end{array} \tag{9.5}$$

Conclusively, we only need to implement the latter intersecting test: whether the squared distance between the centers is smaller than or equal to the squared sum of the radii. Each time when this is the case, a collision occurs.

Example: We want to test the circles $C((1,2),3)$ and $C\left((-1,1),\sqrt{2}\right)$ for collision. For a collision to occur, the squared distance between their centers ought to be smaller than or equal to the squared sum of their radii. For that reason we need to calculate $(m_1 - n_1)^2 + (m_2 - n_2)^2 = (1+1)^2 + (2-1)^2 = 5$ and $(r+s)^2 = (3+\sqrt{2})^2 \approx 19,5$. Both circles are colliding because $(m_1 - n_1)^2 + (m_2 - n_2)^2 < (r+s)^2$.

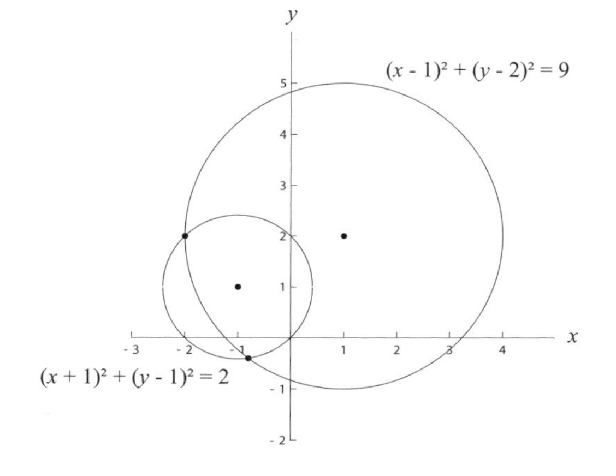

In case we like to inspect their exact intersecting points, we need the circle's equations

$$(x-1)^2 + (y-2)^2 = 9$$

and

$$(x+1)^2 + (y-1)^2 = 2.$$

Each intersecting point (x,y) is ought to satisfy both equations

$$\begin{cases} x^2 + y^2 - 2x - 4y - 4 = 0, \\ x^2 + y^2 + 2x - 2y = 0. \end{cases}$$

To solve this nonlinear system for x and y, we linearly combine the first equation with factor 1 and the second with factor -1 (in other words: we subtract the second equation from the first), which yields the linear equation $-4x - 2y - 4 = 0$, simplified to $y = -2x - 2$. Substituting $y = -2x - 2$ in one of both circles, for instance in $C((1,2),3)$, yields

$$\begin{aligned} (x-1)^2 + (-2x-2-2)^2 = 9 &\Longleftrightarrow (x-1)^2 + (-2x-4)^2 = 9 \\ &\Longleftrightarrow 5x^2 + 14x + 8 = 0. \end{aligned}$$

This is a quadratic equation in x, for which the positive discriminant leads to the solutions $x_1 = -2$ and $x_2 = \frac{-4}{5}$. Because each solution (x,y) needs also to satisfy the former linear substitute $y = -2x - 2$, we calculate $y_1 = -2(-2) - 2 = 2$ and $y_2 = -2\left(\frac{-4}{5}\right) - 2 = \frac{-2}{5}$ respectively. We conclude that these circles intersect in the two points $(-2,2)$ and $\left(\frac{-4}{5}, \frac{-2}{5}\right)$. We realize that a negative discriminant would correspond to no intersection and a zero discriminant to exactly one solution (being the tangent point).

We can straightforward extend the collision test for circles with one dimension, leading to the similar collision test for spheres in 3D space. Again ignoring frame rate issues and collision prediction, we test two spheres: one centered on $M(m_1, m_2, m_3)$ having radius r and one with center $N(n_1, n_2, n_3)$ and radius s for collision via

$$\begin{array}{rl} \text{not colliding} & (m_1-n_1)^2+(m_2-n_2)^2+(m_3-n_3)^2 > (r+s)^2 \\ \text{tangent spheres} & (m_1-n_1)^2+(m_2-n_2)^2+(m_3-n_3)^2 = (r+s)^2 \\ \text{intersecting} & (m_1-n_1)^2+(m_2-n_2)^2+(m_3-n_3)^2 < (r+s)^2 \end{array}$$

9.3 Collision detection using vectors

We recall 3D objects are completely built up by polygons. We define a **polygon** as a convex cut out from a plane by segments $[P_0P_1], [P_1P_2], [P_2P_3], \ldots, [P_{n-1}P_n]$ and $[P_nP_0]$, which connect a sequence of points $P_0, P_1, P_2, \ldots, P_n$, given $n \geqslant 3$.

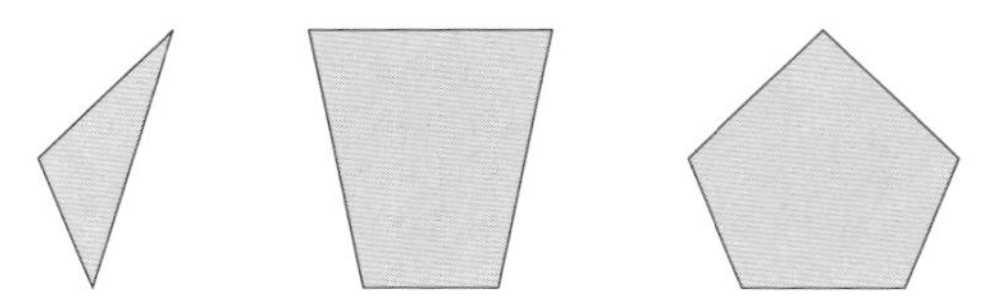

Figure 9.5: Some convex polygons

Most render engines make use of triangular polygons to build up 3D objects. We will further outline how to test whether a sphere collides with a polygon, by the use of vectors. In this book we will typeset a triangular polygon as ABC and the plane where it resides in as ν_A.

Location of a point with respect to other points

We define **collinear** points as points lying on the same straight line.

Given three collinear points P, Q, R, we find the point Q lying in between the two remaining points P and R if

$$(\vec{v}\cdot\vec{p}-\vec{v}\cdot\vec{q})(\vec{v}\cdot\vec{r}-\vec{v}\cdot\vec{q}) \leqslant 0.$$

This test requires the dot products of the location vectors $\vec{p} = \overrightarrow{OP}, \vec{q} = \overrightarrow{OQ}$ and $\vec{r} = \overrightarrow{OR}$ with a freely chosen constant vector $\vec{v} \neq \vec{o}$. This test holds irrespectively the position of

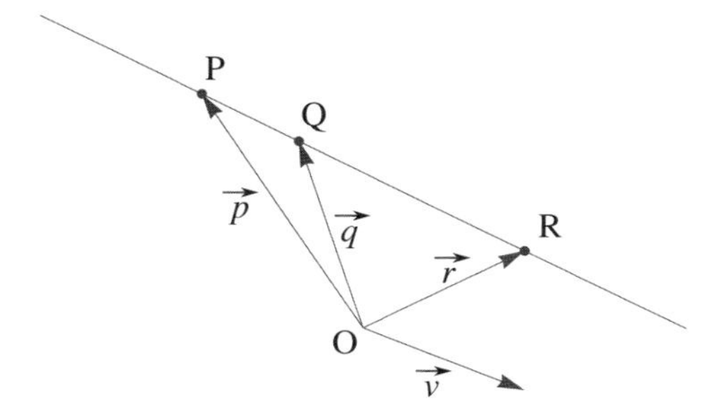

Figure 9.6: Mutual location of collinear points

the straight line PR with respect to the origin O nor the direction of the constant vector $\vec{v} \neq \vec{o}$.

Proof: We assume the three points P, Q and R to be collinear. Consequently, the vectors $\overrightarrow{QP}$ and $\overrightarrow{QR}$ are parallel or antiparallel. In other words, the internal angle between the vectors $\overrightarrow{QP}$ and $\overrightarrow{QR}$ equals $0°$ or $180°$.

We rewrite the test $(\vec{v}\cdot\vec{p}-\vec{v}\cdot\vec{q})(\vec{v}\cdot\vec{r}-\vec{v}\cdot\vec{q}) \leqslant 0$ as

$$\begin{aligned}(\vec{v}\cdot\vec{p}-\vec{v}\cdot\vec{q})(\vec{v}\cdot\vec{r}-\vec{v}\cdot\vec{q}) &= (\vec{v}\cdot\overrightarrow{OP}-\vec{v}\cdot\overrightarrow{OQ})(\vec{v}\cdot\overrightarrow{OR}-\vec{v}\cdot\overrightarrow{OQ})\\ &= \vec{v}\cdot(\overrightarrow{OP}-\overrightarrow{OQ})\cdot\vec{v}\cdot(\overrightarrow{OR}-\overrightarrow{OQ})\\ &= \vec{v}^2(\overrightarrow{OP}-\overrightarrow{OQ})\cdot(\overrightarrow{OR}-\overrightarrow{OQ})\\ &= \vec{v}^2(\overrightarrow{QP}\cdot\overrightarrow{QR}).\end{aligned}$$

In other words, $\vec{v}^2(\overrightarrow{QP}\cdot\overrightarrow{QR}) \leqslant 0$ should hold. Since all squares are non-negative, $\overrightarrow{QP}\cdot\overrightarrow{QR}$ should be negative. Applying the dot product (7.8) yields $||\overrightarrow{QP}||\ ||\overrightarrow{QR}||\cos\theta \leqslant 0$. Consequently, $\cos\theta$ is negative. This can only be achieved by an internal angle between the vectors $\overrightarrow{QP}$ and $\overrightarrow{QR}$ equal to $180°$, which guarantees that the point Q is lying in between the points P and R. ■

Altitude to a straight line

We consider the straight line r running through two points A and B. For this line we can suggest for instance $\overrightarrow{AB}$ as direction vector. The distance from a point P to this straight line r is the distance from the point P to the foot of its altitude perpendicular on r.

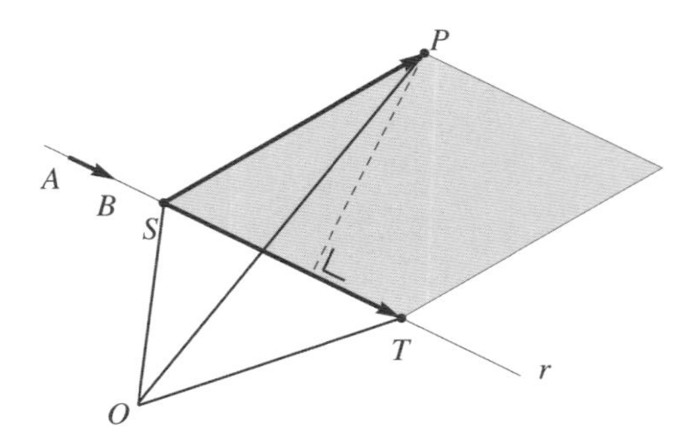

Figure 9.7: Distance from a point to a straight line

We therefore choose two points S and T on the straight line r in such a way that the sense of their vector $\overrightarrow{ST}$ equals the sense of the direction vector $\overrightarrow{AB}$ and with length $||\overrightarrow{ST}|| = 1$. In other words, the free vector $\overrightarrow{ST}$ is a unit vector sharing the same sense with the direction vector $\overrightarrow{AB}$. Consequently, we can express this unit vector as $\overrightarrow{ST} = \vec{e}_{AB} = \frac{\overrightarrow{AB}}{||\overrightarrow{AB}||}$.

Figure 9.7 shows how the inclining vectors $\overrightarrow{ST}$ and $\overrightarrow{SP}$ span the grey parallelogram. The height d of this parallelogram is exactly the distance we are aiming for. The area of a parallelogram equals its base times its height. We set its base as the length of the unit vector $\overrightarrow{ST}$. In other words,

$$\text{area}_{parallelogram} = ||\overrightarrow{ST}|| \cdot d = 1 \cdot d = d.$$

Moreover, we recall that the area of a parallelogram spanned by two vectors, equals the length of their cross product (see page 126). In other words, our parallelogram area equals $||\overrightarrow{ST} \times \overrightarrow{SP}||$. Since we have already found that this area also equals the distance d, this leads us to

$$d = ||\overrightarrow{ST} \times \overrightarrow{SP}|| = \left|\left| \frac{\overrightarrow{AB}}{||\overrightarrow{AB}||} \times \overrightarrow{SP} \right|\right|.$$

We finally calculate the distance d from the point P to the line r, given $\overrightarrow{AB}$ as a direction vector for the line r and an arbitrary point S on the line r by

$$d(P,r) = \frac{||\overrightarrow{AB} \times \overrightarrow{SP}||}{||\overrightarrow{AB}||}. \tag{9.6}$$

We mention that the arbitrary point S can as well be one of the given points A or B.

Example: We calculate the distance from the point $P(5,3,-2)$ to the straight line r, which runs through the points $A(1,0,1)$ and $B(3,5,3)$. The x-, y- and z-axis are put in centimeters.

We need an arbitrary point on the straight line r, for instance the point A. We also need a direction vector of the straight line r, for instance $\overrightarrow{AB} = \begin{pmatrix} 2 \\ 5 \\ 2 \end{pmatrix}$. We then evaluate the above distance formula as

$$d(P,r) = \frac{||\overrightarrow{AB} \times \overrightarrow{AP}||}{||\overrightarrow{AB}||}.$$

In this formula we set $\overrightarrow{AP} = \begin{pmatrix} 5 \\ 3 \\ -2 \end{pmatrix} - \begin{pmatrix} 1 \\ 0 \\ 1 \end{pmatrix} = \begin{pmatrix} 4 \\ 3 \\ -3 \end{pmatrix}$ and as cross product we calculate
$\overrightarrow{AB} \times \overrightarrow{AP} = \begin{pmatrix} 2 \\ 5 \\ 2 \end{pmatrix} \times \begin{pmatrix} 4 \\ 3 \\ -3 \end{pmatrix} = \begin{pmatrix} -21 \\ 14 \\ -14 \end{pmatrix}$. Consequently, this distance formula for point-to-line evaluates as

$$d(P,r) = \frac{\sqrt{(-21)^2 + 14^2 + (-14)^2}}{\sqrt{2^2 + 5^2 + 2^2}} \approx 5.02 \text{ cm}.$$

We mention obtaining the same result when choosing for the arbitrary point $S = B$ and the direction vector $\overrightarrow{AB}$. And of course, we could as well have chosen $\overrightarrow{BA}$ for the direction vector.

Altitude to a plane

Assume that we know the normal vector $\vec{n}$ on the plane v_A. This plane v_A holds a given point A in 3D space. We assume that $\vec{n}$ was normalized into a unit normal on its plane ($\|\vec{n}\| = 1$). We adjust the sense of this unit normal $\vec{n}$ according to our 3D software. We typeset the distance from a point $P\,(p_1, p_2, p_3)$ to a plane v_A, given $A\,(a_1, a_2, a_3)$ simply as $d(P, v_A)$.

We set the vector $\vec{v} = \overrightarrow{AP}$ running from the reference point A of the plane v_A to the head P, from which we want to measure the distance of the point P to the plane v_A. The components of this free vector $\vec{v} = \overrightarrow{AP}$ are $\vec{v} = \begin{pmatrix} p_1 - a_1 \\ p_2 - a_2 \\ p_3 - a_3 \end{pmatrix}$. We now dot product this free vector $\vec{v}$ with the unit normal $\vec{n}$ in order to calculate the distance from the point P to

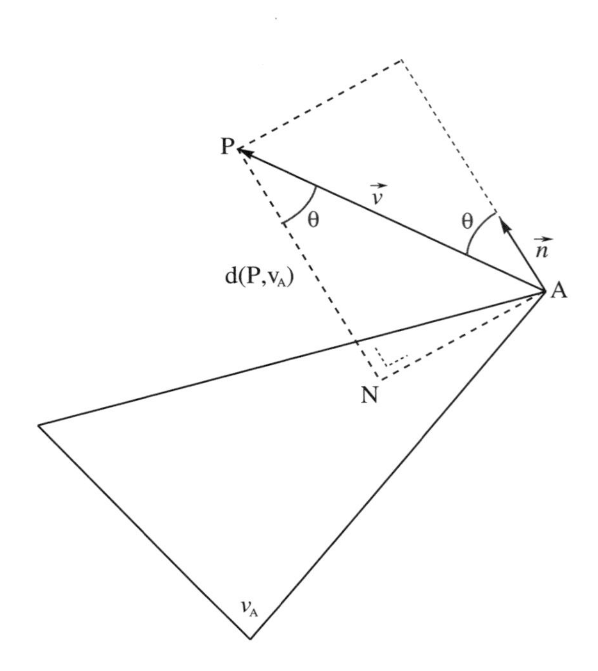

Figure 9.8: Distance from a point to a plane

the plane v_A. We recall the dot product in components as

$$\begin{aligned}\vec{v}\cdot\vec{n} &= v_1n_1 + v_2n_2 + v_3n_3 \\ &= (p_1 - a_1)n_1 + (p_2 - a_2)n_2 + (p_3 - a_3)n_3.\end{aligned}$$

On the other hand, we recall the dot product as such (see formula (7.8)).

$$\vec{v}\cdot\vec{n} = \|\vec{v}\|\ \|\vec{n}\| \cos\theta = \|\vec{v}\| \cos\theta$$

Connecting both above expressions, yields

$$\|\vec{v}\| \cos\theta = (p_1 - a_1)n_1 + (p_2 - a_2)n_2 + (p_3 - a_3)n_3.$$

Applying some trigonometry in the right triangle APN yields $\cos\theta = \frac{d(P,v_A)}{\|\vec{v}\|}$. This leads us to

$$\|\vec{v}\| \frac{d(P,v_A)}{\|\vec{v}\|} = (p_1 - a_1)n_1 + (p_2 - a_2)n_2 + (p_3 - a_3)n_3.$$

In other words, for a plane v_A holding the reference point A and having unit normal $\vec{n}$, we calculate the 'signed' distance from a point P to this given plane v_A by

$$d(P,v_A) = (p_1 - a_1)n_1 + (p_2 - a_2)n_2 + (p_3 - a_3)n_3. \tag{9.7}$$

In case of negative results, we simply apply the absolute value to get the geometrical distance from point the P to the plane v_A (see page 97).

Example: We calculate the distance from the point $P(1,5,-2)$ to the plane determined by the points $A(-2,0,1), B(1,5,1)$ and $C(0,7,2)$. The x-, y- and z-axis are put in centimeters.

We find a unit normal vector $\vec{n} = \begin{pmatrix} -0.40 \\ 0.24 \\ -0.88 \end{pmatrix}$ (see page 129). We calculate the free vector running from the reference point A in the plane v_A to the head P, as $\overrightarrow{AP} = \begin{pmatrix} 3 \\ 5 \\ -3 \end{pmatrix}$. This allows us to calculate the distance from the point P to the plane v_A as

$$d(P, v_A) = 3 \cdot (-0.40) + 5 \cdot 0.24 - 3 \cdot (-0.88) = 2.64 \text{ cm}.$$

Frame rate issues

The naive collision detection performs its test based upon the available coordinate values, frame after frame. This approach risks to miss the collisions which occur between two successive frames, depending on the size of the circumscribed circle or sphere and the actual frame rate. To overcome this issue, the collision detection techniques using vectors may perform better, compared to the similar techniques using circles or spheres. Alternatively, a wide variety of inter frame collision detection techniques is being developed, from which we briefly outline two of them.

▷ **Frame rate enhancement**
A locally applied timestep decreasing along the moving object's trajectory may be efficient to prevent an inter frame collision. Although this approach tends to cause computational overload in case of too cumbersome trajectories.

▷ **Inter frame collision detection**
We develop an inter frame collision detector, fit for plane passages, straightforwardly starting from the altitude to a plane. We reuse our previous vector based distance from a point to a plane (see formula (9.7)) to detect an inter frame passage from a point P through the plane v_A. When we use an invariable unit normal $\vec{n}$ and the 'signed distance' $d(P, v_A)$ switches from positive to negative (or the other way around) between two successive frames, then a point-plane collision occured regardless from the actual frame rate. Indexing any two successive frames n and $n+1$ we can easily implement this inter frame collision test as $d_n(P, v_A) > 0$ followed by $d_{n+1}(P, v_A) \leqslant 0$ (or the other way around).

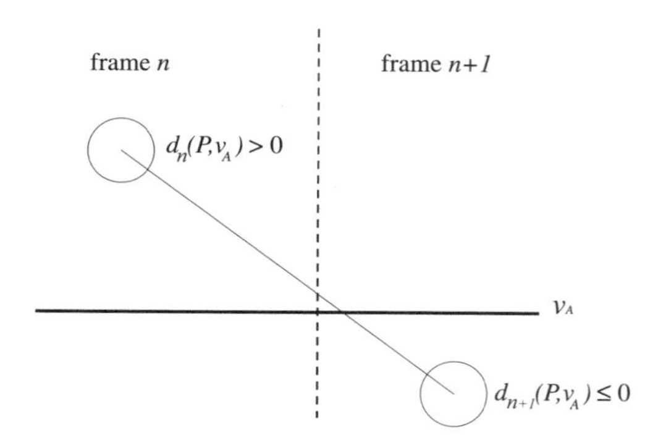

Figure 9.9: Collision detection issue with a plane

LOCATION OF A POINT WITH RESPECT TO A POLYGON

For a more accurate collision detection we need to be sure whether objects really make contact or not. We will develop a test using vectors, for an object to hit a certain polygon ABC. We assume we already measured the distance from a point P to the polygon's plane v_A (see page 159).

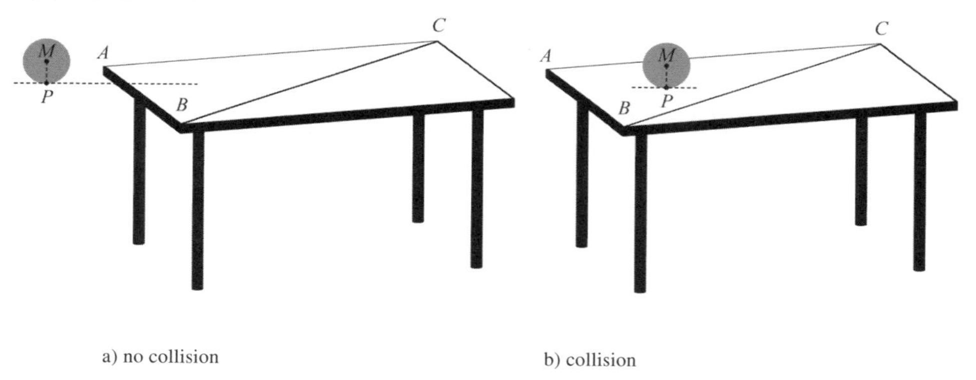

Figure 9.10: Location of a point P within a polygon ABC

When we for instance make use of a sphere, we assume we already know that the distance from its center M to the polygon's plane v_A equals its radius. In figure 9.10 we witness there is no collision in situation a, whereas situation b shows collision. Despite the point P is located in the polygon's plane v_A, it is not located inside the polygon ABC in situation a. This confirms the need for determining where a point is located with respect to a polygon.

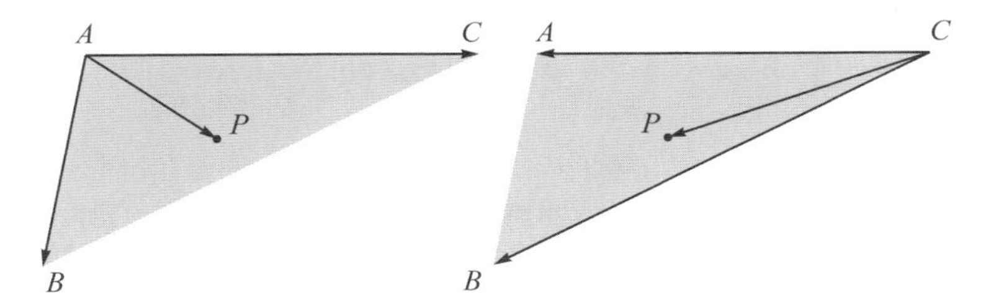

Figure 9.11: The point P is located inside of the polygon ABC

We outline such a location test which involves determining the sense of normal vectors on the polygon. In case of three times equality of senses, then the point P is located inside of the polygon ABC. In all other cases the point P is located outside the polygon ABC.

Applying the right-hand rule or corkscrew-rule, we test for the point P and each polygon edge, whether the sense of two normal vectors equals.

1) Is the sense of $\overrightarrow{AB} \times \overrightarrow{AC}$ equal to the sense of $\overrightarrow{AP} \times \overrightarrow{\mathbf{AC}}$?
2) Is the sense of $\overrightarrow{AC} \times \overrightarrow{AB}$ equal to the sense of $\overrightarrow{AP} \times \overrightarrow{\mathbf{AB}}$?
3) Is the sense of $\overrightarrow{CA} \times \overrightarrow{CB}$ equal to the sense of $\overrightarrow{CP} \times \overrightarrow{\mathbf{CB}}$?

Since we encounter three times an equal sense for the normal vectors, we conclude the point P is located inside the polygon ABC.

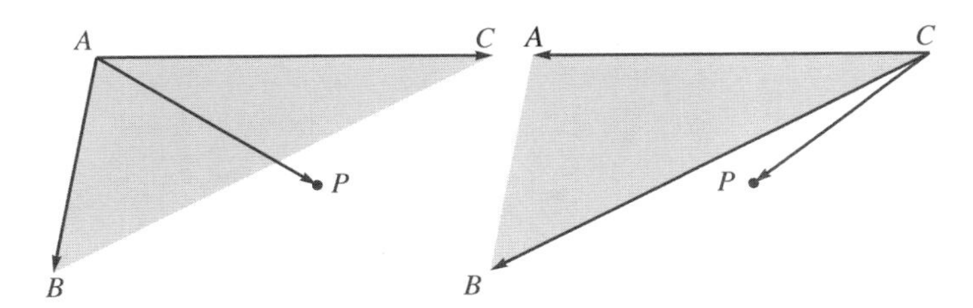

Figure 9.12: The point P is located outside of the polygon ABC

To make it even more clear, we illustrate in figure 9.12 the situation of a point P lying in the plane spanned by A, B and C, but being located outside of the polygon ABC. Initially the test seems to hold: $\overrightarrow{AB} \times \overrightarrow{AC}$ has a sense equal to the sense of $\overrightarrow{AP} \times \overrightarrow{\mathbf{AC}}$. But applying the right-hand rule or corkscrew-rule, we find the normal vector $\overrightarrow{CA} \times \overrightarrow{CB}$ having a sense opposite to the sense of the normal vector $\overrightarrow{CP} \times \overrightarrow{\mathbf{CB}}$. For this reason we conclude that the point P is located outside of the polygon ABC.

As a comment to the above test, we warn that it fails for points collinear with for instance the vertices A and C. Because in this case the cross product leads to $\overrightarrow{AP} \times \overrightarrow{\mathbf{AC}} = \vec{o}$, and the null vector $\vec{o}$ is without direction.

Combining the above 'location-in-polygon' test with the 'point-line' distance formula, we obtain a kind of collision detection model as the one used in the 3D game 'DooM'. Despite 'DooM' (1993) may sound somewhat outdated, combining even the most elementary collision detection techniques already assures cool runtime effects. Astonishingly, usually the most basic collision detections offer the best runtime results. Given these facts, we kindly recommend the professional game programmer to start writing or collecting a personal collision detection library, containing at least the methods outlined in this chapter.

9.4 Exercises

Exercise 75 Verify whether the equations below represent circles or spheres. In case of a circle or a sphere, determine its center and radius.

1) $x^2+y^2+z^2-2x-6y-10z+19=0.$
2) $x^2+y^2+z^2-6x-12y+6z+50=0.$
3) $x^2-12x+y^2-4y+40=0.$

Exercise 76 Determine center and radius of the circle containing the points $(0,0),(2,0)$ and $(0,4)$,

1) by applying algebra.
2) by a ruler-and-compass construction.

Exercise 77 We decided to make use of a circumscribed circle for collision detection in a car game. The car's center is $(20,50)$ and the car's most extreme edge point has cartesian coordinates $(60,80)$. Find the circumscribed circle to this car by its cartesian equation.

Exercise 78 We decided to make use of circumscribed circles for collision detection of players in a 2D baseball game. In the actual frame one player is circumscribed by the circle $(x-50)^2+(y-20)^2=900$ and the other player by $(x+10)^2+(y-10)^2=400$. Do they collide in this frame?

Exercise 79 A car which is circumscribed by a circle with center $(x,-x)$ and radius $\frac{1}{\sqrt{2}}$ moves during seven successive frames on the straight line $y=-x$, for each new frame incrementing x with $\frac{1}{2}$. Does this car collide with the parked lorry circumscribed by the circle with radius $\sqrt{2}$ around center $(2,-2)$, when it takes off in frame 0 from the initial point $(-3,3)$. In case of collision, calculate their intersecting points.

Exercise 80 Calculate the distance from point $P(1,0,1)$ to the straight line a, which is running through the origin $O(0,0,0)$ in the direction of the vector $\vec{a}=(2,-1,3)$.

Exercise 81 Test for collision between the squash ball $S((-1,2,-2),5)$ and the plane wall through the points $A(-2,0,1)$, $B(1,5,2)$ and $C(2,7,0)$.

Exercise 82 Prove that the point $Q(1,0,1)$ is lying on the line PR in between its points $P(-1,0,-1)$ and $R(3,0,3)$.

Exercise 83 A landing strip POQ at the stern of an aircraft carrier in a flight simulator is marked by the origin $O(0,0,0)$ and two remaining vertices $P(-5,0,0)$ and $Q(-3,0,-5)$. We now test two touch down moves from a jet V.

1) Calculate the height above the landing platform of this jet in point $V(10,11,12)$.
2) Is the landing attempt successful at location $V(-1,0,\frac{-5}{2})$?

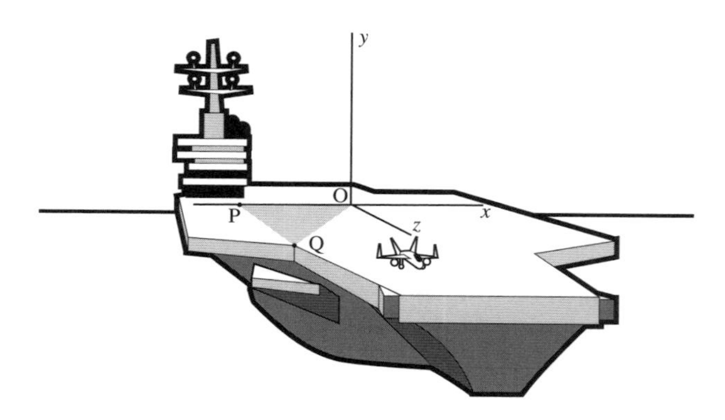

Exercise 84 Use vectors to determine whether the center $S(\frac{17}{3},5,\sqrt{15})$ of a snooker ball ended up in the triangle ABC filled with numbered balls. The plane of the pool table is described by $y=5$ and the vertices of the billiard triangle are respectively having the cartesian coordinates $A(3,5,0)$, $B(0,5,4)$ and $C(6,5,4)$.

Exercise 85 Detect between which frame numbers the projectile P modelled by a variable point, hits the wall modelled by the plane ν_A through the point $A(1,1,1)$ with unit normal $\vec{n}=(-\frac{4}{5},0,\frac{3}{5})$. During four successive frames, the projectile P receives the following actual coordinate values:

1) $P(-2,1,0)$
2) $P(0,4,1)$
3) $P(2,7,2)$
4) $P(4,10,3)$

Exercise 86 Detect between which frame numbers the tank shell modelled by the unit sphere $E((t_1,t_2,t_3),1)$ hits the enemy tank circumscribed by the sphere $B((v_1,v_2,v_3),14)$. During four successive frames, both centers T and V receive the following actual coordinate values:

1) $T(78,61,40)$ and $V(11,8,10)$
2) $T(38,31,20)$ and $V(10,7,9)$
3) $T(-2,1,0)$ and $V(9,6,8)$
4) $T(-42,-29,-20)$ and $V(9,6,8)$

Chapter 10 · Matrices

We often make use of 'rectangles' of numbers, for instance when we declare nested *arrays* during (game) programming (see page 109). Such configurations of numbers or symbols are called matrices. Matrices are extremely useful as they are capable from just holding data, over moving screen pixels to describing Bézier curves. For this reason we study them here in detail, in order to apply matrices optimally in our final chapters.

10.1 The concept of a matrix

The concept of a **matrix** was introduced by the British mathematician **James Joseph Sylvester** (1814–1897) in all its straightforward simplicity. A matrix is a rectangle of numbers

$$A = \begin{pmatrix} a_{11} & a_{12} & \cdots & a_{1n} \\ a_{21} & a_{22} & \cdots & a_{2n} \\ \vdots & \vdots & \vdots & \vdots \\ a_{m1} & a_{m2} & \cdots & a_{mn} \end{pmatrix},$$

of which a_{ij} are called its **matrix elements**. We typeset a matrix using an uppercase latin character and its indexed elements by the same lowercase character.

Horizontally we count m matrix **rows** and vertically n matrix **columns**. This way we find element a_{ij} residing in the ith row and j^{th} column.

$$\begin{array}{c} j^{th}\text{column} \\ A = \begin{pmatrix} & \vdots & \\ \cdots & a_{ij} & \cdots \\ & \vdots & \end{pmatrix} i^{th}\text{row} \end{array}$$

In other words, the first index i denotes the row of the element a_{ij} and the second index j denotes its column. We typeset a matrix of m rows and n columns as $A_{m\times n}$ or for instance $A \in \mathbb{R}^{m\times n}$ in case its elements are real numbers. We call a matrix of this size an 'm by n' matrix.

If the number of rows is 1, in other words for $m = 1$, we have a **row matrix** or **row vector**.

$$B = \begin{pmatrix} 1 & 0 & 1 & 1 & 1 & 0 & 1 & 0 & 0 \end{pmatrix} \text{ is a } 1\times 9 \text{ row vector.}$$

If the number of columns is 1, in other words for $n = 1$, we have a **column matrix** or **column vector**.

$$C = \begin{pmatrix} x \\ y \\ z \end{pmatrix} \text{ is a } 3\times 1 \text{ column vector.}$$

We define a **zero matrix** as a matrix for which all elements equal zero. We typeset such a zero matrix or null matrix, using the uppercase character O for varying matrix sizes $O_{m\times n}$.

If the number of rows equals the number of columns, in other words for $m = n$, we have a **square** matrix.

$$D = \begin{pmatrix} 41 & 32 \\ 10 & 16 \end{pmatrix} \text{ is a square } 2\times 2 \text{ matrix.}$$

We define the sequence of elements $e_{11}, e_{22}, \ldots, e_{nn}$ of a square matrix as its **main diagonal**. The main diagonal runs from the upper left to the lower right element. The antidiagonal runs from the lower left to the upper right element.

$$E = \begin{pmatrix} e_{11} & & & \\ & e_{22} & & \\ & & \ddots & \\ & & & e_{nn} \end{pmatrix}$$

We define an **identity matrix** as a square matrix for which all elements are zero, except for all main diagonal elements which are one. We typeset an identity matrix using its capital first character I for varying square sizes I_n.

$$I_2 = \begin{pmatrix} 1 & 0 \\ 0 & 1 \end{pmatrix} \text{ is a } 2\times 2 \text{ identity matrix.}$$

Two matrices A and B are **equal** when they are of the same size and all of their corresponding elements are equal.

10.2 Determinant of a square matrix

Each square matrix can produce a return value called the **determinant** of the matrix, which we typeset as $\det A$ or $|A|$. We calculate this scalar value based upon the elements of the square matrix. Its determinant determines whether the square matrix has an inverse or not.

The determinant of a 1×1 matrix, in other words one scalar, is simply equal to this scalar.

The determinant of a 2×2 matrix is

$$\begin{vmatrix} a_{11} & a_{12} \\ a_{21} & a_{22} \end{vmatrix} = a_{11}a_{22} - a_{21}a_{12}. \tag{10.1}$$

We can easily memorize this calculation for a determinant of a 2×2 matrix as its 'main diagonal minus antidiagonal'.

When we delete the row and column of a square matrix element a_{ij} and calculate the determinant of its underlying submatrix, we define this return value as the **minor** of the element a_{ij}. We define the **cofactor** of the element a_{ij} as the multiplication of its minor with $(-1)^{i+j}$, given i the row and j the column index of the element. Based upon these definitions, we develop the determinant of an $n \times n$ matrix as the sum of the products of the elements of a single arbitrary row (or column) with their respective cofactors.

The determinant of a 3×3 matrix is, developed to its first row,

$$\begin{aligned}
&\begin{vmatrix} a_{11} & a_{12} & a_{13} \\ a_{21} & a_{22} & a_{23} \\ a_{31} & a_{32} & a_{33} \end{vmatrix} \\
&= (-1)^{1+1} a_{11} \begin{vmatrix} a_{22} & a_{23} \\ a_{32} & a_{33} \end{vmatrix} + (-1)^{1+2} a_{12} \begin{vmatrix} a_{21} & a_{23} \\ a_{31} & a_{33} \end{vmatrix} \\
&\qquad + (-1)^{1+3} a_{13} \begin{vmatrix} a_{21} & a_{22} \\ a_{31} & a_{32} \end{vmatrix} \\
&= a_{11} \begin{vmatrix} a_{22} & a_{23} \\ a_{32} & a_{33} \end{vmatrix} - a_{12} \begin{vmatrix} a_{21} & a_{23} \\ a_{31} & a_{33} \end{vmatrix} + a_{13} \begin{vmatrix} a_{21} & a_{22} \\ a_{31} & a_{32} \end{vmatrix} \qquad (10.2) \\
&= a_{11}(a_{22}a_{33} - a_{32}a_{23}) - a_{12}(a_{21}a_{33} - a_{31}a_{23}) + a_{13}(a_{21}a_{32} - a_{31}a_{22}).
\end{aligned}$$

Example:

$$\begin{aligned}
\begin{vmatrix} 3 & 2 & 1 \\ 7 & 4 & 2 \\ -2 & 0 & 5 \end{vmatrix} &= 3 \begin{vmatrix} 4 & 2 \\ 0 & 5 \end{vmatrix} - 2 \begin{vmatrix} 7 & 2 \\ -2 & 5 \end{vmatrix} + 1 \begin{vmatrix} 7 & 4 \\ -2 & 0 \end{vmatrix} \\
&= 3(4 \cdot 5 - 0 \cdot 2) - 2(7 \cdot 5 - (-2) \cdot 2) + 1(7 \cdot 0 - (-2) \cdot 4) \\
&= -10
\end{aligned}$$

We can even apply the above formula to easily memorize the component formula of the cross product of two vectors (see formula (7.11)). We rewrite this formula as the determinant of a 3×3 matrix, given $\vec{e_1}, \vec{e_2}$ and $\vec{e_3}$ the unit vectors spanning the x-, y- and z-axis respectively.

$$\begin{pmatrix} a_1 \\ a_2 \\ a_3 \end{pmatrix} \times \begin{pmatrix} b_1 \\ b_2 \\ b_3 \end{pmatrix} = \begin{vmatrix} \vec{e_1} & \vec{e_2} & \vec{e_3} \\ a_1 & a_2 & a_3 \\ b_1 & b_2 & b_3 \end{vmatrix}$$

$$= (a_2b_3 - b_2a_3)\vec{e_1} + (-a_1b_3 + b_1a_3)\vec{e_2} + (a_1b_2 - b_1a_2)\vec{e_3}$$

$$= \begin{pmatrix} a_2b_3 - b_2a_3 \\ -a_1b_3 + b_1a_3 \\ a_1b_2 - b_1a_2 \end{pmatrix}$$

We realize we may only use the above determinant to memorize the cross product formula, since it is not a valid determinant due to its first row of vectors. We are aware that the components of the first vector $\vec{a}$ belong to the second determinant row and those of the second vector $\vec{b}$ to the last determinant row, since the cross product is not commutative.

10.3 Addition of matrices

Adding matrices is as straightforward as it is important. We can only add or subtract two matrices A and B of the same size. As a result their **matrix sum** $A+B$ or **difference** $A-B$ inherits this size.

- ▷ in words: We add two matrices by adding their corresponding matrix elements.
- ▷ in symbols: For $A, B \in \mathbb{R}^{m\times n}$ we calculate their sum $A+B \in \mathbb{R}^{m\times n}$ as

$$\begin{pmatrix} a_{11} & a_{12} & \cdots & a_{1n} \\ a_{21} & a_{22} & \cdots & a_{2n} \\ \vdots & \vdots & \vdots & \vdots \\ a_{m1} & a_{m2} & \cdots & a_{mn} \end{pmatrix} + \begin{pmatrix} b_{11} & b_{12} & \cdots & b_{1n} \\ b_{21} & b_{22} & \cdots & b_{2n} \\ \vdots & \vdots & \vdots & \vdots \\ b_{m1} & b_{m2} & \cdots & b_{mn} \end{pmatrix}$$

$$= \begin{pmatrix} a_{11}+b_{11} & a_{12}+b_{12} & \cdots & a_{1n}+b_{1n} \\ a_{21}+b_{21} & a_{22}+b_{22} & \cdots & a_{2n}+b_{2n} \\ \vdots & \vdots & \vdots & \vdots \\ a_{m1}+b_{m1} & a_{m2}+b_{m2} & \cdots & a_{mn}+b_{mn} \end{pmatrix}.$$

We define the difference $A-B$ of two matrices A and B as the sum of A with the opposite matrix of B. We subtract two matrices by subtracting their corresponding matrix elements.

Example: $A=\begin{pmatrix} -1 & 5 & \sqrt{2} \\ 4 & -7 & \sqrt{3} \end{pmatrix}$ $\qquad B=\begin{pmatrix} 3 & 2 & -1 \\ 0 & -1 & -2 \end{pmatrix}$

Since $A\in\mathbb{R}^{2\times 3}$ and $B\in\mathbb{R}^{2\times 3}$ are of the same size, we can calculate $A+B$ and $A-B$.

$$\begin{aligned} A+B &= \begin{pmatrix} -1 & 5 & \sqrt{2} \\ 4 & -7 & \sqrt{3} \end{pmatrix} + \begin{pmatrix} 3 & 2 & -1 \\ 0 & -1 & -2 \end{pmatrix} \\ &= \begin{pmatrix} -1+3 & 5+2 & \sqrt{2}+(-1) \\ 4+0 & -7+(-1) & \sqrt{3}+(-2) \end{pmatrix} \\ &= \begin{pmatrix} 2 & 7 & \sqrt{2}-1 \\ 4 & -8 & \sqrt{3}-2 \end{pmatrix} \end{aligned}$$

$$\begin{aligned} A-B &= \begin{pmatrix} -1 & 5 & \sqrt{2} \\ 4 & -7 & \sqrt{3} \end{pmatrix} - \begin{pmatrix} 3 & 2 & -1 \\ 0 & -1 & -2 \end{pmatrix} \\ &= \begin{pmatrix} -1-3 & 5-2 & \sqrt{2}-(-1) \\ 4-0 & -7-(-1) & \sqrt{3}-(-2) \end{pmatrix} \\ &= \begin{pmatrix} -4 & 3 & \sqrt{2}+1 \\ 4 & -6 & \sqrt{3}+2 \end{pmatrix} \end{aligned}$$

Algebraic structure: Given A, B and $C\in\mathbb{R}^{m\times n}$ matrices of the same size, we summarize all properties of $+$ in $\mathbb{R}^{m\times n}$:

closure	$A+B\in\mathbb{R}^{m\times n}$,
associative property	$A+(B+C)=(A+B)+C$,
zero matrix	$A+O_{m\times n}=A=O_{m\times n}+A$,
opposite matrix	$A+(-A)=O_{m\times n}=-A+A$,
commutative property	$A+B=B+A$.

We conclude $(\mathbb{R}^{m\times n},+)$ is a commutative **group**.

10.4 Scalar multiplication of matrices

We define the **scalar multiplication** of a matrix as a real number times the matrix.

- ▷ in words: We scalar multiply a matrix with a real number λ by multiplying all matrix elements with λ.
- ▷ in symbols: We multiply the matrix $A \in \mathbb{R}^{m\times n}$ with a number $\lambda \in \mathbb{R}$ as

$$\lambda \begin{pmatrix} a_{11} & a_{12} & \cdots & a_{1n} \\ a_{21} & a_{22} & \cdots & a_{2n} \\ \vdots & \vdots & \vdots & \vdots \\ a_{m1} & a_{m2} & \cdots & a_{mn} \end{pmatrix} = \begin{pmatrix} \lambda a_{11} & \lambda a_{12} & \cdots & \lambda a_{1n} \\ \lambda a_{21} & \lambda a_{22} & \cdots & \lambda a_{2n} \\ \vdots & \vdots & \vdots & \vdots \\ \lambda a_{m1} & \lambda a_{m2} & \cdots & \lambda a_{mn} \end{pmatrix}.$$

Example 1: Is $A = \begin{pmatrix} -1 & 5 & \sqrt{2} \\ 4 & -7 & \sqrt{3} \end{pmatrix}$, we can for instance calculate its scalar multiple

$$\begin{aligned} -2A = -2\begin{pmatrix} -1 & 5 & \sqrt{2} \\ 4 & -7 & \sqrt{3} \end{pmatrix} &= \begin{pmatrix} -2\cdot(-1) & -2\cdot 5 & -2\cdot\sqrt{2} \\ -2\cdot 4 & -2\cdot(-7) & -2\cdot\sqrt{3} \end{pmatrix} \\ &= \begin{pmatrix} 2 & -10 & -2\sqrt{2} \\ -8 & 14 & -2\sqrt{3} \end{pmatrix}. \end{aligned}$$

Example 2: For $A = \begin{pmatrix} -1 & 5 & \sqrt{2} \\ 4 & -7 & \sqrt{3} \end{pmatrix}$, we define its **opposite matrix** $-A$ as

$$\begin{aligned} (-1)A = (-1)\begin{pmatrix} -1 & 5 & \sqrt{2} \\ 4 & -7 & \sqrt{3} \end{pmatrix} &= \begin{pmatrix} (-1)\cdot(-1) & (-1)\cdot 5 & (-1)\cdot\sqrt{2} \\ (-1)\cdot 4 & (-1)\cdot(-7) & (-1)\cdot\sqrt{3} \end{pmatrix} \\ &= \begin{pmatrix} 1 & -5 & -\sqrt{2} \\ -4 & 7 & -\sqrt{3} \end{pmatrix}. \end{aligned}$$

Properties: Given $\lambda, \mu \in \mathbb{R}$ and $A, B \in \mathbb{R}^{m\times n}$, we summarize their arithmetic laws.

mixed distributive law	$\lambda(A+B)$	=	$\lambda A + \lambda B$,
	$(\lambda+\mu)A$	=	$\lambda A + \mu A$,
mixed associative law	$(\lambda\cdot\mu)A$	=	$\lambda(\mu A)$.

10.5 Transpose of a matrix

To transpose a matrix is to tilt its rectangular shape, which we define in detail below.

- in words: We define the **transpose of an** $m \times n$ **matrix** A as the $n \times m$ matrix obtained by rewriting all rows of A into columns, and consequently the columns of A into rows. We typeset this tilted version of the matrix A as A^T, realizing this is just a notation and not a matrix power. In other words, the first row of A corresponds to the first column of A^T, the second row of A corresponds to the second column of A^T, ...
- in symbols: We tilt a matrix $A \in \mathbb{R}^{m\times n}$ into its transposed matrix A^T as

$$\begin{pmatrix} a_{11} & a_{12} & \cdots & a_{1n} \\ a_{21} & a_{22} & \cdots & a_{2n} \\ \vdots & \vdots & \vdots & \vdots \\ a_{m1} & a_{m2} & \cdots & a_{mn} \end{pmatrix}^T = \begin{pmatrix} a_{11} & a_{21} & \cdots & a_{m1} \\ a_{12} & a_{22} & \cdots & a_{m2} \\ \vdots & \vdots & \vdots & \vdots \\ a_{1n} & a_{2n} & \cdots & a_{mn} \end{pmatrix} \in \mathbb{R}^{n\times m}.$$

Example: Given a matrix $A = \begin{pmatrix} -1 & 5 & \sqrt{2} \\ 4 & -7 & \sqrt{3} \end{pmatrix} \in \mathbb{R}^{2\times 3}$, we transpose it as

$$A^T = \begin{pmatrix} -1 & 4 \\ 5 & -7 \\ \sqrt{2} & \sqrt{3} \end{pmatrix} \in \mathbb{R}^{3\times 2}.$$

Properties: Given $A, B \in \mathbb{R}^{m\times n}$, we mention already

$$\begin{aligned} (A^T)^T &= A, \\ (A+B)^T &= A^T + B^T. \end{aligned}$$

We define a **symmetric matrix** as a square matrix A for which $A^T = A$.

10.6 Dot product of matrices

Introduction

A computer store sells three different models of laptops: model 1, model 2 and model 3. A matrix L is holding a double order, meant for two different destinations, of a major customer.

$$L = \begin{pmatrix} 60 & 10 & 0 \\ 100 & 50 & 150 \end{pmatrix} \begin{matrix} \leftarrow 1^{\text{st}} \text{ order} \\ \leftarrow 2^{\text{nd}} \text{ order} \end{matrix}$$
$$\begin{matrix} \uparrow & \uparrow & \uparrow \\ \text{model 1} & \text{model 2} & \text{model 3} \end{matrix}$$

A column matrix S is meanwhile holding all model sale prices in euro.

$$S = \begin{pmatrix} 900 \\ 1200 \\ 1500 \end{pmatrix} \begin{matrix} \leftarrow \text{model 1} \\ \leftarrow \text{model 2} \\ \leftarrow \text{model 3} \end{matrix}$$

The customer pays for his first order the full amount of

$$60 \cdot 900 \quad + \quad 10 \cdot 1200 \quad + \quad 0 \cdot 1500 \quad = 66000$$

and for his second order

$$100 \cdot 900 \quad + \quad 50 \cdot 1200 \quad + \quad 150 \cdot 1500 \quad = 375000.$$

Putting both subtotals in a column matrix

$$P = \begin{pmatrix} 66000 \\ 375000 \end{pmatrix},$$

we can interpret P as a product of the matrices L and S. We realize how the elements of P were calculated: for the first order

$$\begin{aligned} \begin{pmatrix} 60 & 10 & 0 \\ 100 & 50 & 150 \end{pmatrix} \cdot \begin{pmatrix} 900 \\ 1200 \\ 1500 \end{pmatrix} &= \begin{pmatrix} 60 \cdot 900 + 10 \cdot 1200 + 0 \cdot 1500 \\ \cdots \end{pmatrix} \\ &= \begin{pmatrix} 66000 \\ \cdots \end{pmatrix} \end{aligned}$$

and for the second order

$$\begin{aligned} \begin{pmatrix} 60 & 10 & 0 \\ 100 & 50 & 150 \end{pmatrix} \cdot \begin{pmatrix} 900 \\ 1200 \\ 1500 \end{pmatrix} &= \begin{pmatrix} \cdots \\ 100 \cdot 900 + 50 \cdot 1200 + 150 \cdot 1500 \end{pmatrix} \\ &= \begin{pmatrix} \cdots \\ 375000 \end{pmatrix}. \end{aligned}$$

We obtain the element on the first row of P by multiplying the elements of the first row of L with the corresponding elements of S and then adding all these products. We realize how multiplying a $\mathbf{2} \times 3$ matrix with a $3 \times \mathit{1}$ matrix results into a $\mathbf{2} \times \mathit{1}$ product matrix.

Condition

We multiply two matrices by repeatedly multiplying a row of the first matrix with a column of the second matrix.

$$\begin{pmatrix} 1 & 2 \\ 3 & 4 \\ 5 & 6 \end{pmatrix} \cdot \begin{pmatrix} 1 & 2 & 3 \\ 4 & 5 & 6 \end{pmatrix} = \begin{pmatrix} 9 & 12 & 15 \\ 19 & 26 & 33 \\ 29 & 40 & 51 \end{pmatrix}$$

For instance the element 33 in the product matrix residing on the second row and third column, is the sum of the products of the elements of the second row of the first matrix with the corresponding elements of the third column of the second matrix: $3 \cdot 3 + 4 \cdot 6 = 33$. Consequently, in order to multiply two matrices, the number of columns of the first matrix needs to equal the number of rows of the second matrix. The multiplication of a 3×2 matrix with a compatible 2×5 matrix results into a 3×5 product matrix. We can highlight this condition to easily memorize it by matrix sizes: $\mathbf{3} \times 2$ and $2 \times \mathbf{5}$ yields $\mathbf{3} \times \mathbf{5}$.

Definition

We can multiply two matrices A and B *in this order* if the number of columns of A equals the number of rows of B. The number of rows of their product matrix $A \cdot B$ corresponds to the number of rows of A; the number of columns of $A \cdot B$ corresponds to the number of columns of B.

We typeset the **dot** or **matrix product** of $A \in \mathbb{R}^{m \times n}$ and $B \in \mathbb{R}^{n \times p}$ *in this order* as $A \cdot B = C$. Consequently their **product matrix** C will belong to $\mathbb{R}^{m \times p}$. We hereby define this dot product for matrices

- in words: Product matrix $A \cdot B$ holds at its ith *row and its* j^{th} *column* the result of multiplying all elements of the ith *row of matrix* A with the corresponding elements of the jth *column of matrix* B and then adding these products.
- in symbols:

$$\begin{pmatrix} a_{i1} & a_{i2} & a_{i3} & \cdots & a_{in} \end{pmatrix} \cdot \begin{pmatrix} b_{1j} \\ b_{2j} \\ b_{3j} \\ \vdots \\ b_{nj} \end{pmatrix} = \begin{pmatrix} & \vdots & \\ \cdots & c_{ij} & \cdots \\ & \vdots & \end{pmatrix}$$

expressing

$$c_{ij} = a_{i1}b_{1j} + a_{i2}b_{2j} + a_{i3}b_{3j} + \cdots + a_{in}b_{nj} = \sum_{k=1}^{n} a_{ik}b_{kj}$$

for each $1 \leqslant i \leqslant m$ and $1 \leqslant j \leqslant p$.

Example: We declare two matrices

$$A = \begin{pmatrix} -1 & 5 & \sqrt{2} \\ 4 & -7 & \sqrt{3} \end{pmatrix} \in \mathbb{R}^{2\times 3} \text{ and } B = \begin{pmatrix} 2 & 8 & 12 \\ -3 & -1 & 5 \\ 0 & 4 & 10 \end{pmatrix} \in \mathbb{R}^{3\times 3}.$$

We are able to calculate their dot product $A \cdot B$ because the matrix sizes $\mathbf{2} \times 3$ and $3 \times \mathbf{3}$ are compatible and the size of the product matrix will be $\mathbf{2} \times \mathbf{3}$.

$$\begin{pmatrix} -1 & 5 & \sqrt{2} \\ 4 & -7 & \sqrt{3} \end{pmatrix} \cdot \begin{pmatrix} 2 & 8 & 12 \\ -3 & -1 & 5 \\ 0 & 4 & 10 \end{pmatrix}$$

$$= \begin{pmatrix} -1\cdot 2+5\cdot(-3)+\sqrt{2}\cdot 0 & -1\cdot 8+5\cdot(-1)+\sqrt{2}\cdot 4 & -1\cdot 12+5\cdot 5+\sqrt{2}\cdot 10 \\ 4\cdot 2-7\cdot(-3)+\sqrt{3}\cdot 0 & 4\cdot 8-7\cdot(-1)+\sqrt{3}\cdot 4 & 4\cdot 12-7\cdot 5+\sqrt{3}\cdot 10 \end{pmatrix}$$

$$= \begin{pmatrix} -17 & -13+4\sqrt{2} & 13+10\sqrt{2} \\ 29 & 39+4\sqrt{3} & 13+10\sqrt{3} \end{pmatrix}$$

It is impossible to calculate the dot product $B \cdot A$ as the number of columns of B does not equal the number of rows of A. Their matrix sizes $\mathbf{3} \times 3$ and $2 \times \mathbf{3}$ are incompatible for a dot product.

Properties

1) A product matrix $A \cdot B$ is in general not equal to the product matrix $B \cdot A$. We already experienced that one of them can even not exist. In other words, for the dot product of matrices we have ***no** commutativity*. As a consequence, respecting the order of the matrices is of major importance to multiply them.

2) Assuming the matrix sizes of A, B and C allow for all their dot products, we have:

associative property	$(A \cdot B) \cdot C = A \cdot (B \cdot C)$,
right distributive law	$(A + B) \cdot C = A \cdot C + B \cdot C$,
left distributive law	$C \cdot (A + B) = C \cdot A + C \cdot B$.

3) Zero divisors

Given $A = \begin{pmatrix} 1 & -2 \\ -1 & 2 \end{pmatrix}$ and $B = \begin{pmatrix} 2 & -10 \\ 1 & -5 \end{pmatrix}$, their dot product $A \cdot B$ is somehow astonishing

$$A \cdot B = \begin{pmatrix} 1 & -2 \\ -1 & 2 \end{pmatrix} \cdot \begin{pmatrix} 2 & -10 \\ 1 & -5 \end{pmatrix} = \begin{pmatrix} 0 & 0 \\ 0 & 0 \end{pmatrix}.$$

We have to conclude that the dot product of two matrices can result into the zero matrix, without any of the initial matrices being the zero matrix. We define a **zero divisor** as a matrix $A \neq O$ for which $A \cdot B = O$ or $B \cdot A = O$, for some matrix B different from the zero matrix.

While for scalar numbers $a, b \in \mathbb{R}$ we can deduct from their product $a \cdot b = 0$ either $a = 0$ (and)or $b = 0$, it does not apply at all to matrices. A dot product $A \cdot B = O_{m \times n}$ offers no conclusion about A (and)or B being zero matrices.

4) We can only apply a **matrix power** to square matrices.

$$\begin{pmatrix} 1 & -2 & 3 \\ -1 & 8 & 6 \\ 0 & -5 & 2 \end{pmatrix}^2 = \begin{pmatrix} 1 & -2 & 3 \\ -1 & 8 & 6 \\ 0 & -5 & 2 \end{pmatrix} \cdot \begin{pmatrix} 1 & -2 & 3 \\ -1 & 8 & 6 \\ 0 & -5 & 2 \end{pmatrix}$$
$$= \begin{pmatrix} 3 & -33 & -3 \\ -9 & 36 & 57 \\ 5 & -50 & -26 \end{pmatrix}$$

5) An identity matrix I_n is the matrix equivalent of the number $1 \in \mathbb{R}$, which means $I_n \cdot A = A \cdot I_n = A$, for any $n \times n$ matrix A.

$$\begin{pmatrix} 1 & 0 & 0 \\ 0 & 1 & 0 \\ 0 & 0 & 1 \end{pmatrix} \cdot \begin{pmatrix} 1 & -2 & 3 \\ -1 & 8 & 6 \\ 0 & -5 & 2 \end{pmatrix} = \begin{pmatrix} 1 & -2 & 3 \\ -1 & 8 & 6 \\ 0 & -5 & 2 \end{pmatrix} \cdot \begin{pmatrix} 1 & 0 & 0 \\ 0 & 1 & 0 \\ 0 & 0 & 1 \end{pmatrix}$$
$$= \begin{pmatrix} 1 & -2 & 3 \\ -1 & 8 & 6 \\ 0 & -5 & 2 \end{pmatrix}$$

6) The transpose of a matrix product reverses its order:

$$(S \cdot B)^T = B^T \cdot S^T.$$

We can easily memorize this effect as the 'Boots-and-Socks' rule. In the morning we first put on our Socks, then we put on our Boots. We 'Transpose' this in the evening by first Taking off our Boots, then Taking off our Socks.

7) We have the mixed associative property for the scalar multiplication and the matrix product. For a scalar $\lambda \in \mathbb{R}$ and assuming the matrix sizes of A and B allow for their dot product, we express this property as

$$\lambda(A \cdot B) = \lambda A \cdot B = A \cdot (\lambda B).$$

10.7 Inverse of a matrix

INTRODUCTION

We first consider the inverse of a real number:

$$\sqrt{3} \cdot \frac{1}{\sqrt{3}} = 1.$$

We say b is the inverse number of a if and only if $a \cdot b = 1$, and we typeset $b = \frac{1}{a} = a^{-1}$.

Next, we consider two 3×3 matrices

$$A = \begin{pmatrix} 1 & 5 & 2 \\ 1 & 1 & 7 \\ 0 & -3 & 4 \end{pmatrix} \quad \text{and} \quad B = \begin{pmatrix} -25 & 26 & -33 \\ 4 & -4 & 5 \\ 3 & -3 & 4 \end{pmatrix}.$$

At a first glance we do not spot a special relation between the matrix A and the matrix B, until we look at their dot product $A \cdot B$

$$A \cdot B = \begin{pmatrix} 1 & 0 & 0 \\ 0 & 1 & 0 \\ 0 & 0 & 1 \end{pmatrix} = I_3.$$

Similarly to numbers, we typeset $B = A^{-1}$, calling it the inverse matrix of the matrix A.

DEFINITION

Consider an $n \times n$ matrix A. We define the unique $n \times n$ matrix B for which

$$A \cdot B = B \cdot A = I_n$$

as the **inverse matrix** of the matrix A, typeset as A^{-1} if this matrix B exists. In this respect, we call the matrix A **invertible**. In other words, if the matrix $A \in \mathbb{R}^{n \times n}$ is invertible, then the unique matrix A^{-1} exists for which

$$A \cdot A^{-1} = A^{-1} \cdot A = I_n.$$

Conditions

- ▷ The inverse matrix can only exist for square matrices.
- ▷ Many square matrices do not have an inverse matrix. (In the real numbers only zero does not have an inverse number. Instead it leads to the concept of *infinity* which is not a number.)
  - ► In case the determinant of an $n \times n$ matrix A equals zero, there is no inverse matrix of A. In this case we call the matrix A **singular**.
  - ► In case the determinant of an $n \times n$ matrix A is not zero, there is a unique inverse matrix of A. In this case we call the matrix A **invertible**.
- ▷ For an invertible square matrix A, its inverse matrix is unique. In other words, there is exactly one matrix A^{-1} for which $A \cdot A^{-1} = A^{-1} \cdot A = I$.
- ▷ For real numbers we equivalently typeset $a^{-1} = \frac{1}{a}$. The former notation is used but the latter is invalid for matrices. We realize division of matrices is not defined.
- ▷ $(A^{-1})^{-1} = A$, given A^{-1} exists.

Row reduction

Three **elementary row operations** can be used on matrices (see also page 35). We perform these row operations for matrix inversion, as the process of finding the inverse of a square matrix.

1) Swap the positions of two rows.

   This row operation simply reorders the rows of a given matrix. Take for instance the matrix

$$\begin{pmatrix} 0 & 1 & 0 \\ 1 & 0 & 0 \\ 0 & 0 & 1 \end{pmatrix}.$$

   Swap its first two rows in order to acquire the identity matrix

$$\begin{pmatrix} 0 & 1 & 0 \\ 1 & 0 & 0 \\ 0 & 0 & 1 \end{pmatrix} \overset{R_1 \leftrightarrow R_2}{\sim} \begin{pmatrix} 1 & 0 & 0 \\ 0 & 1 & 0 \\ 0 & 0 & 1 \end{pmatrix}.$$

2) Multiply a row by a nonzero scalar $\lambda \neq 0$.

   We sometimes need or like to multiply each element of a certain matrix row by a nonzero number.

We for instance multiply the first row of this matrix

$$\begin{pmatrix} \frac{3}{2} & 0 & 0 \\ 0 & 1 & 0 \\ 0 & 0 & 1 \end{pmatrix}$$

by $\frac{2}{3}$ in order to acquire the identity matrix.

$$\begin{pmatrix} \frac{3}{2} & 0 & 0 \\ 0 & 1 & 0 \\ 0 & 0 & 1 \end{pmatrix} \overset{R_1 \to \frac{2}{3} R_1}{\sim} \begin{pmatrix} 1 & 0 & 0 \\ 0 & 1 & 0 \\ 0 & 0 & 1 \end{pmatrix}$$

3) Adding two rows.

   This row operation adds a certain row to another row, which we overwrite by the result. Let us for instance consider this matrix

$$\begin{pmatrix} 1 & -1 & 0 \\ 0 & 1 & 0 \\ 0 & 0 & 1 \end{pmatrix}.$$

   We add the second row to the first row, overwriting the first row by their sum, in order to acquire once more the identity matrix.

$$\begin{pmatrix} 1 & -1 & 0 \\ 0 & 1 & 0 \\ 0 & 0 & 1 \end{pmatrix} \overset{R_1 \to R_1 + R_2}{\sim} \begin{pmatrix} 1 & 0 & 0 \\ 0 & 1 & 0 \\ 0 & 0 & 1 \end{pmatrix}$$

Matrix inversion involves applying a strategic sequence of those elementary row operations.

Matrix inversion

There are different ways to calculate the inverse matrix of a given square matrix. In case of an invertible square matrix, we can for instance choose to calculate its inverse matrix by **gaussian elimination**.

1) We augment the given square matrix $A \in \mathbb{R}^{n \times n}$ to the right by the identity matrix I_n, which results in an $n \times 2n$ **block matrix** $(A|I_n)$.

2) We then perform elementary row operations to reduce the block matrix $(A|I_n)$ until the initial matrix A in its left block turns into the identity matrix.

3) When this identity matrix can be achieved in the left block, the matrix A is invertible and its inverse matrix appears in the right block of the augmented matrix. We have reduced the initial augmented matrix $(A|I_n)$ to the final block matrix $(I_n|A^{-1})$.

4) Alternatively, a row full of zeros appearing in the left block means the matrix A is not invertible but singular.

Example: We calculate the inverse of this matrix

$$A = \begin{pmatrix} 1 & 5 & 2 \\ 1 & 1 & 7 \\ 0 & -3 & 4 \end{pmatrix}.$$

Its inverse matrix A^{-1} exists, because of

$$\det A = 1(1\cdot 4-(-3)\cdot 7)-5(1\cdot 4-0\cdot 7)+2(1\cdot(-3)-0\cdot 1) = -1.$$

We now perform row operations on the augmented matrix $(A|I_3)$ until its left block displays the identity matrix I_3. As there is a multitude of routes to achieve this, we like to demonstrate two of them. We kick off with outlining a first approach along three stages.

Step 1: we aim for zeros in the column of the first element on the first row.

$$\left(\begin{array}{ccc|ccc} 1 & 5 & 2 & 1 & 0 & 0 \\ 1 & 1 & 7 & 0 & 1 & 0 \\ 0 & -3 & 4 & 0 & 0 & 1 \end{array}\right)$$

$$\sim \left(\begin{array}{ccc|ccc} 1 & 5 & 2 & 1 & 0 & 0 \\ 0 & -4 & 5 & -1 & 1 & 0 \\ 0 & -3 & 4 & 0 & 0 & 1 \end{array}\right) \qquad (R_2 \to R_2 - R_1)$$

Step 2: we tune the element on the second row, second column, to 1 and turn the remaining elements of that column to zeros.

$$\sim \left(\begin{array}{ccc|ccc} 1 & 5 & 2 & 1 & 0 & 0 \\ 0 & 1 & \frac{-5}{4} & \frac{1}{4} & \frac{-1}{4} & 0 \\ 0 & -3 & 4 & 0 & 0 & 1 \end{array}\right) \qquad (R_2 \to \frac{-1}{4} R_2)$$

$$\sim \left(\begin{array}{ccc|ccc} 1 & 0 & \frac{33}{4} & \frac{-1}{4} & \frac{5}{4} & 0 \\ 0 & 1 & \frac{-5}{4} & \frac{1}{4} & \frac{-1}{4} & 0 \\ 0 & -3 & 4 & 0 & 0 & 1 \end{array}\right) \qquad (R_1 \to R_1 - 5R_2)$$

$$\sim \left(\begin{array}{ccc|ccc} 1 & 0 & \frac{33}{4} & \frac{-1}{4} & \frac{5}{4} & 0 \\ 0 & 1 & \frac{-5}{4} & \frac{1}{4} & \frac{-1}{4} & 0 \\ 0 & 0 & \frac{1}{4} & \frac{3}{4} & \frac{-3}{4} & 1 \end{array}\right) \qquad (R_3 \to R_3 + 3R_2)$$

Step 3: we tune the element on the third row, third column, to 1 and turn the remaining elements of that column to zeros.

$$\sim \left(\begin{array}{ccc|ccc} 1 & 0 & \frac{33}{4} & \frac{-1}{4} & \frac{5}{4} & 0 \\ 0 & 1 & \frac{-5}{4} & \frac{1}{4} & \frac{-1}{4} & 0 \\ 0 & 0 & 1 & 3 & -3 & 4 \end{array}\right) \qquad (R_3 \to 4R_3)$$

$$\sim \left(\begin{array}{ccc|ccc} 1 & 0 & 0 & -25 & 26 & -33 \\ 0 & 1 & \frac{-5}{4} & \frac{1}{4} & \frac{-1}{4} & 0 \\ 0 & 0 & 1 & 3 & -3 & 4 \end{array}\right) \qquad (R_1 \to R_1 - \frac{33}{4}R_3)$$

$$\sim \left(\begin{array}{ccc|ccc} 1 & 0 & 0 & -25 & 26 & -33 \\ 0 & 1 & 0 & 4 & -4 & 5 \\ 0 & 0 & 1 & 3 & -3 & 4 \end{array}\right) \qquad (R_2 \to R_2 + \frac{5}{4}R_3)$$

Finally, we may conclude

$$\begin{pmatrix} 1 & 5 & 2 \\ 1 & 1 & 7 \\ 0 & -3 & 4 \end{pmatrix}^{-1} = \begin{pmatrix} -25 & 26 & -33 \\ 4 & -4 & 5 \\ 3 & -3 & 4 \end{pmatrix}$$

as it is confirmed by the dot product

$$\begin{pmatrix} 1 & 5 & 2 \\ 1 & 1 & 7 \\ 0 & -3 & 4 \end{pmatrix} \cdot \begin{pmatrix} -25 & 26 & -33 \\ 4 & -4 & 5 \\ 3 & -3 & 4 \end{pmatrix} = \begin{pmatrix} 1 & 0 & 0 \\ 0 & 1 & 0 \\ 0 & 0 & 1 \end{pmatrix}.$$

An alternative route, avoiding fractions, leads of course to the same inverse matrix.

$$\left(\begin{array}{ccc|ccc} 1 & 5 & 2 & 1 & 0 & 0 \\ 1 & 1 & 7 & 0 & 1 & 0 \\ 0 & -3 & 4 & 0 & 0 & 1 \end{array}\right)$$

step 1

$$\sim \left(\begin{array}{ccc|ccc} 1 & 5 & 2 & 1 & 0 & 0 \\ 0 & -4 & 5 & -1 & 1 & 0 \\ 0 & -3 & 4 & 0 & 0 & 1 \end{array}\right) \qquad (R_2 \to R_2 - R_1)$$

$$\sim \left(\begin{array}{ccc|ccc} 1 & 5 & 2 & 1 & 0 & 0 \\ 0 & -1 & 1 & -1 & 1 & -1 \\ 0 & -3 & 4 & 0 & 0 & 1 \end{array}\right) \qquad (R_2 \to R_2 - R_3)$$

step 2

$$\sim \left(\begin{array}{ccc|ccc} 1 & 5 & 2 & 1 & 0 & 0 \\ 0 & 1 & -1 & 1 & -1 & 1 \\ 0 & -3 & 4 & 0 & 0 & 1 \end{array}\right) \qquad (R_2 \to -R_2)$$

$$\sim \left(\begin{array}{ccc|ccc} 1 & 5 & 2 & 1 & 0 & 0 \\ 0 & 1 & -1 & 1 & -1 & 1 \\ 0 & 0 & 1 & 3 & -3 & 4 \end{array}\right) \qquad (R_3 \to R_3 + 3R_2)$$

$$\sim \left(\begin{array}{ccc|ccc} 1 & 0 & 7 & -4 & 5 & -5 \\ 0 & 1 & -1 & 1 & -1 & 1 \\ 0 & 0 & 1 & 3 & -3 & 4 \end{array}\right) \qquad (R_1 \to R_1 - 5R_2)$$

$$\sim \left(\begin{array}{ccc|ccc} 1 & 0 & 0 & -25 & 26 & -33 \\ 0 & 1 & -1 & 1 & -1 & 1 \\ 0 & 0 & 1 & 3 & -3 & 4 \end{array}\right) \qquad (R_1 \to R_1 - 7R_3)$$

step 3

$$\sim \left(\begin{array}{ccc|ccc} 1 & 0 & 0 & -25 & 26 & -33 \\ 0 & 1 & 0 & 4 & -4 & 5 \\ 0 & 0 & 1 & 3 & -3 & 4 \end{array}\right) \qquad (R_2 \to R_2 + R_3)$$

Inverse of a product

The inverse of a matrix product reverses its order:

$$(S \cdot B)^{-1} = B^{-1} \cdot S^{-1}.$$

We recall this effect as the 'Boots-and-Socks' rule, as we already encountered it at the transpose of a matrix product.

Example: Given the matrices $A = \begin{pmatrix} 3 & 2 \\ 4 & 3 \end{pmatrix}$ and $B = \begin{pmatrix} 1 & -2 \\ 1 & 0 \end{pmatrix}$.

Since $\det A = 1$ and $\det B = 2$ we are able to calculate their inverse matrices:

$$A \cdot B = \begin{pmatrix} 5 & -6 \\ 7 & -8 \end{pmatrix}$$

$$A^{-1} = \begin{pmatrix} 3 & -2 \\ -4 & 3 \end{pmatrix}$$

$$B^{-1} = \begin{pmatrix} 0 & 1 \\ -\frac{1}{2} & \frac{1}{2} \end{pmatrix}$$

$$(A \cdot B)^{-1} = \begin{pmatrix} -4 & 3 \\ -\frac{7}{2} & \frac{5}{2} \end{pmatrix}$$

$$A^{-1} \cdot B^{-1} = \begin{pmatrix} 1 & 2 \\ -\frac{3}{2} & -\frac{5}{2} \end{pmatrix}$$

$$B^{-1} \cdot A^{-1} = \begin{pmatrix} -4 & 3 \\ -\frac{7}{2} & \frac{5}{2} \end{pmatrix}$$

Solving systems of linear equations

We are able to express linear systems as matrix products. Expanding the dot product and applying the equality of matrices yields the two equations of the given 2×2 system.

$$\begin{cases} -2x + y & = & -2 \\ x + y & = & 10 \end{cases} \iff \begin{pmatrix} -2 & 1 \\ 1 & 1 \end{pmatrix} \cdot \begin{pmatrix} x \\ y \end{pmatrix} = \begin{pmatrix} -2 \\ 10 \end{pmatrix}$$

After trimming a given $m \times n$ system into its default shape, we perceive three matrices to express it as a matrix product.

- ▷ We define the **coefficient matrix** as the $m \times n$ matrix A taking the coefficients a_{ij} as its elements.
- ▷ The $n \times 1$ column matrix X containing the unknown quantities x_i at the left hand side.
- ▷ The $m \times 1$ column matrix T containing the constant terms t_i at the right hand side.

This allows us to express a linear system

$$\begin{cases} a_{11}x_1 + a_{12}x_2 + \cdots + a_{1n}x_n & = & t_1 \\ a_{21}x_1 + a_{22}x_2 + \cdots + a_{2n}x_n & = & t_2 \\ & \vdots & \\ a_{m1}x_1 + a_{m2}x_2 + \cdots + a_{mn}x_n & = & t_m \end{cases}$$

as the matrix product

$$\iff \begin{pmatrix} a_{11} & a_{12} & \dots & a_{1n} \\ a_{21} & a_{22} & \dots & a_{2n} \\ \vdots & \vdots & & \vdots \\ a_{m1} & a_{m2} & \dots & a_{mn} \end{pmatrix} \cdot \begin{pmatrix} x_1 \\ x_2 \\ \vdots \\ x_n \end{pmatrix} = \begin{pmatrix} t_1 \\ t_2 \\ \vdots \\ t_m \end{pmatrix},$$

or even more efficient as

$$A \cdot X = T.$$

We call the above shape the **matrix form** of the given $m \times n$ system. In case its coefficient matrix A is an invertible square matrix, we can choose to solve the $n \times n$ system using its inverse matrix A^{-1}.

$$\begin{aligned} A \cdot X = T &\Leftrightarrow A^{-1} \cdot (A \cdot X) = A^{-1} \cdot T \\ &\Leftrightarrow I_n \cdot X = A^{-1} \cdot T \\ &\Leftrightarrow X = A^{-1} \cdot T \end{aligned}$$

Example: Let us solve the linear system

$$\begin{cases} -2x + y &= -2 \\ x + y &= 10 \end{cases}.$$

First of all, we express the system in its matrix form.

$$\begin{pmatrix} -2 & 1 \\ 1 & 1 \end{pmatrix} \cdot \begin{pmatrix} x \\ y \end{pmatrix} = \begin{pmatrix} -2 \\ 10 \end{pmatrix}$$

The determinant of this coefficient matrix A equalling -3 assures the inverse coefficient matrix to exist.

$$A^{-1} = \begin{pmatrix} -2 & 1 \\ 1 & 1 \end{pmatrix}^{-1} = \begin{pmatrix} -\frac{1}{3} & \frac{1}{3} \\ \frac{1}{3} & \frac{2}{3} \end{pmatrix}$$

Consequently, we find the solution to this system as

$$\begin{pmatrix} x \\ y \end{pmatrix} = \begin{pmatrix} -\frac{1}{3} & \frac{1}{3} \\ \frac{1}{3} & \frac{2}{3} \end{pmatrix} \cdot \begin{pmatrix} -2 \\ 10 \end{pmatrix} = \begin{pmatrix} 4 \\ 6 \end{pmatrix}.$$

10.8 The Fibonacci operator

We can generate the Fibonacci sequence (see paragraph 5.3) using matrices. Defining the **Fibonacci operator** as the 2×2 matrix $F = \begin{pmatrix} 0 & 1 \\ 1 & 1 \end{pmatrix}$ and its initial Fibonacci vector as the column matrix $\vec{f_0} = \begin{pmatrix} 0 \\ 1 \end{pmatrix}$, we calculate subsequently these dot products:

$$\vec{f_1} = F \cdot \vec{f_0} = \begin{pmatrix} 0 & 1 \\ 1 & 1 \end{pmatrix} \cdot \begin{pmatrix} 0 \\ 1 \end{pmatrix} = \begin{pmatrix} 1 \\ 1 \end{pmatrix},$$

$$\vec{f_2} = F \cdot \vec{f_1} = \begin{pmatrix} 0 & 1 \\ 1 & 1 \end{pmatrix} \cdot \begin{pmatrix} 1 \\ 1 \end{pmatrix} = \begin{pmatrix} 1 \\ 2 \end{pmatrix},$$

$$\vec{f_3} = F \cdot \vec{f_2} = \begin{pmatrix} 0 & 1 \\ 1 & 1 \end{pmatrix} \cdot \begin{pmatrix} 1 \\ 2 \end{pmatrix} = \begin{pmatrix} 2 \\ 3 \end{pmatrix},$$

$$\vec{f_4} = F \cdot \vec{f_3} = \begin{pmatrix} 0 & 1 \\ 1 & 1 \end{pmatrix} \cdot \begin{pmatrix} 2 \\ 3 \end{pmatrix} = \begin{pmatrix} 3 \\ 5 \end{pmatrix}.$$

We rewrite this fourth Fibonacci vector into its matrix form as

$$\vec{f_4} = F \cdot \vec{f_3} = F \cdot F \cdot \vec{f_2} = F \cdot F \cdot F \cdot \vec{f_1} = F \cdot F \cdot F \cdot F \cdot \vec{f_0}$$

$$= F^4 \cdot \vec{f_0} = \begin{pmatrix} 0 & 1 \\ 1 & 1 \end{pmatrix}^4 \cdot \begin{pmatrix} 0 \\ 1 \end{pmatrix} = \begin{pmatrix} 3 \\ 5 \end{pmatrix}.$$

Next, we apply this matrix form to all natural matrix powers

$$\vec{f_k} = F^k \cdot \vec{f_0},$$

given their exponents $k \in \mathbb{N}$.

There is an amazing relation between the matrix powers of the Fibonacci operator F and the golden number Φ (see paragraph 5.1), which we outline without proof:

$$F^k = \begin{pmatrix} 1 & 1 \\ \Phi & \Phi' \end{pmatrix} \cdot \begin{pmatrix} \Phi^k & 0 \\ 0 & \Phi'^k \end{pmatrix} \cdot \begin{pmatrix} 1 & 1 \\ \Phi & \Phi' \end{pmatrix}^{-1}, \tag{10.3}$$

given $\Phi = \frac{1+\sqrt{5}}{2}$ (see definition (5.1)) and $\Phi' = \frac{1-\sqrt{5}}{2}$ (see definition (5.3)) as both roots of the quadratic equation describing the golden number.

Matrix algebra simplifies the matrix power F^k into a single 2×2 product matrix.

$$F^k = \begin{pmatrix} 1 & 1 \\ \Phi & \Phi' \end{pmatrix} \cdot \begin{pmatrix} \Phi^k & 0 \\ 0 & \Phi'^k \end{pmatrix} \cdot \begin{pmatrix} \frac{-1+\sqrt{5}}{2\sqrt{5}} & \frac{1}{\sqrt{5}} \\ \frac{1+\sqrt{5}}{2\sqrt{5}} & -\frac{1}{\sqrt{5}} \end{pmatrix}$$

$$= \begin{pmatrix} \Phi^k & \Phi'^k \\ \Phi^{k+1} & \Phi'^{k+1} \end{pmatrix} \cdot \begin{pmatrix} \frac{-1+\sqrt{5}}{2\sqrt{5}} & \frac{1}{\sqrt{5}} \\ \frac{1+\sqrt{5}}{2\sqrt{5}} & -\frac{1}{\sqrt{5}} \end{pmatrix}$$

$$= \begin{pmatrix} \Phi^k & \Phi'^k \\ \Phi^{k+1} & \Phi'^{k+1} \end{pmatrix} \cdot \frac{1}{\sqrt{5}} \cdot \begin{pmatrix} -\Phi' & 1 \\ \Phi & -1 \end{pmatrix}$$

$$= \frac{1}{\sqrt{5}} \cdot \begin{pmatrix} -\Phi^k \cdot \Phi' + \Phi \cdot \Phi'^k & \Phi^k - \Phi'^k \\ -\Phi^{k+1} \cdot \Phi' + \Phi \cdot \Phi'^{k+1} & \Phi^{k+1} - \Phi'^{k+1} \end{pmatrix}$$

Applying this single matrix expression of F^k for calculating the k-th Fibonacci vector $\vec{f}_k$, yields

$$\begin{aligned} \vec{f}_k &= F^k \cdot \vec{f}_0 \\ &= \frac{1}{\sqrt{5}} \cdot \begin{pmatrix} -\Phi^k \cdot \Phi' + \Phi \cdot \Phi'^k & \Phi^k - \Phi'^k \\ -\Phi^{k+1} \cdot \Phi' + \Phi \cdot \Phi'^{k+1} & \Phi^{k+1} - \Phi'^{k+1} \end{pmatrix} \cdot \begin{pmatrix} 0 \\ 1 \end{pmatrix} \\ &= \frac{1}{\sqrt{5}} \cdot \begin{pmatrix} \Phi^k - \Phi'^k \\ \Phi^{k+1} - \Phi'^{k+1} \end{pmatrix}. \end{aligned}$$

This reformulation of the link between Fibonacci numbers and the golden number by means of matrices, reveals **Binet's formula**. The first component of the Fibonacci vector $\vec{f}_k$ is the k-th Fibonacci number, for which holds

$$f_k = \frac{(\Phi)^k - (\Phi')^k}{\sqrt{5}}. \tag{10.4}$$

In this formula, Φ is the golden number and $\Phi' = \frac{1-\sqrt{5}}{2}$, given natural exponents $k \in \mathbb{N}$.

This formula was already discovered in 1730 by the French mathematician **Abraham de Moivre** (1667–1754) but only to be proved in 1843 by his compatriot **Jacques Binet** (1786–1856).

When we for instance determine the first Fibonacci number f_1 applying Binet's formula, we evaluate it as $f_1 = \frac{1}{\sqrt{5}} \cdot ((\frac{1+\sqrt{5}}{2})^1 - (\frac{1-\sqrt{5}}{2})^1)$ which simplifies to $f_1 = 1$.

10.9 Exercises

Exercise 87 Given $A = \begin{pmatrix} 5 & 10 & 0 & 1 \\ 10 & 200 & 10 & 20 \\ 20 & 10 & 2 & 1 \end{pmatrix}$ and $B = \begin{pmatrix} 4 & 10 & 3 & 0 \\ 10 & 80 & 40 & 50 \\ 10 & 30 & 2 & 0 \end{pmatrix}$,

calculate the following matrices:

1) $A+B$
2) $2B-3A$
3) A^T
4) B^T
5) $(B-A)^T$
6) B^T-A^T

Exercise 88 Calculate, whenever possible, given the matrices

$$A = \begin{pmatrix} 2 & 1 & 3 \\ -1 & 0 & 1 \end{pmatrix}, B = \begin{pmatrix} 1 & -1 \\ 0 & 1 \\ 2 & 3 \end{pmatrix}, C = \begin{pmatrix} 1 & 0 & -2 & 1 \\ 0 & 3 & 1 & 4 \end{pmatrix},$$

the following dot products:

1) $A \cdot B$
2) $B \cdot A$
3) $B \cdot C$
4) $C \cdot B$
5) $(A \cdot B) \cdot C$
6) $A \cdot (B \cdot C)$
7) $(A \cdot B)^T$
8) $B^T \cdot A^T$

Exercise 89 Calculate, whenever possible, given the matrices

$$A = \begin{pmatrix} 8 & 2 \\ 3 & 2 \end{pmatrix}, B = \begin{pmatrix} 3 & 4 \\ 2 & 3 \end{pmatrix}, C = \begin{pmatrix} 1 & 2 & 3 \\ 2 & 5 & 3 \\ 1 & 0 & 8 \end{pmatrix}, D = \begin{pmatrix} 1 & 2 & 3 \\ 2 & 5 & 3 \\ 1 & 0 & 9 \end{pmatrix},$$

the following matrix expressions:

1) A^2
2) A^{-1}
3) B^{-1}
4) C^{-1}
5) D^{-1}
6) $A^{-1}B^{-1}$
7) $B^{-1}A^{-1}$
8) $(A \cdot B)^{-1}$

Exercise 90 Solve the following linear system using the inverse matrix method.

$$\begin{cases} -x & + & y & - & 4z & = 5 \\ 2x & + & 2y & & & = 4 \\ 3x & + & 3y & + & 2z & = 2 \end{cases}$$

Exercise 91 Prove that for all natural exponents $n \in \mathbb{N}$ the dot product $F^n \cdot \vec{f}_0$ generates all Fibonacci vectors $\vec{f}_n$. Hint: apply 'mathematical induction' in $\mathbb{N}$.

1) Verify the conjecture for a small value of n, for instance $n = 1$.
2) Accept the conjecture for an arbitrary value of n, given $n \in \mathbb{N}$.
3) Prove the 'inheritance' of it to the next natural index $n+1$, only based upon both previous induction steps 1 and 2.

Exercise 92 Verify formula (10.3), which links the Fibonacci operator F to the golden number Φ, for exponent values $k = 1$ and $k = 2$.

Exercise 93 Apply Binet's formula (10.4) to generate the first four Fibonacci numbers f_0, f_1, f_2 and f_3.

Exercise 94 We define **idempotent** matrices P as square matrices reproducing themselves by applying the dot product, i.e. $P^2 = P$. We can find more amazing examples of idempotent matrixes than just the identity matrices: $I_n \cdot I_n = I_n$. Verify for instance that all of the following real matrices are idempotent matrices:

$$K = \begin{pmatrix} 1 & 0 \\ 0 & 0 \end{pmatrix}, L = \begin{pmatrix} 1-a & \frac{(1-a)a}{b} \\ b & a \end{pmatrix}, M = \begin{pmatrix} 1 & a & 0 \\ 0 & 0 & 0 \\ 0 & 0 & 1 \end{pmatrix},$$

given $a \in \mathbb{R}, b \in \mathbb{R} \setminus \{0\}$.

Exercise 95 We define **nilpotent** matrices N as square zero divisors, producing the zero matrix O by matrix exponentiating themselves, i.e. $N^k = O$. Verify that all of the following real matrices are nilpotent matrices, and seek their corresponding exponent k:

$$N = \begin{pmatrix} 0 & 1 \\ 0 & 0 \end{pmatrix}, P = \begin{pmatrix} 12 & -18 \\ 8 & -12 \end{pmatrix}, R = \begin{pmatrix} 1 & 2 & 5 \\ 2 & 4 & 10 \\ -1 & -2 & -5 \end{pmatrix}.$$

Chapter 11 · Linear transformations

Animating computer graphics is all about displaying transformations of line segments into line segments. **Linear transformations** are capable of mapping accordingly, line segments onto line segments. Wide spread examples of linear transformations are translations, scalings, reflections, rotations and shearings. We apply these transformations using their matrix operator.

We define such a transformation or its operator L as **linear** when it features both properties

$$L(\vec{a}+\vec{b}) = L(\vec{a}) + L(\vec{b}),$$
$$L(\lambda\vec{a}) = \lambda L(\vec{a}),$$

given the vectors $\vec{a}, \vec{b} \in \mathbb{R}^n$ and a scalar $\lambda \in \mathbb{R}$.

Linear transformations can be combined: a sequence of linear transformations can be replaced by one linear transformation. And the other way around: a linear transformation can be decomposed into a sequence of linear transformations. This decomposition is a powerful practical property, given the fact that a linear transformation and its decomposition yield the same effect.

Since linear transformations are key essential for all 2D and 3D applications, we outline them in detail in this chapter.

11.1 Translation

We start with the basic screen effect of moving in a certain direction along a straight line, from A to B. We call these geometrical shifts **translations** which we typeset as $T_{\overrightarrow{AB}}$. We can realize such a translation from point A to point B either by a vector sum or by a matrix product.

Imagine an object located in a cartesian frame in the point $P(3,2)$, which we want to move one unit to the right and four units upwards. We achieve this effect simply by adding 1 to its x-label and 4 to its y-label. This adjusts its location from point the $P(3,2)$ to the point $P'(3+1, 2+4) = P'(4,6)$.

This approach is straightforward in case of one point or just a handful of points. In case of objects built up by a lot of points, we like to move them more systematically.

A translation is an effect which shifts each point of the plane (or space) by the same free vector $\overrightarrow{AB}$. The previous example translates one point horizontally over distance 1 and vertically over distance 4, which we typeset as

$$\begin{pmatrix} 3 \\ 2 \end{pmatrix} + \begin{pmatrix} 1 \\ 4 \end{pmatrix} = \begin{pmatrix} 4 \\ 6 \end{pmatrix}.$$

If $P(x,y)$ is an original point and $P'(x',y')$ the resulting image point reached by the displacement vector $\overrightarrow{AB} = (\Delta x, \Delta y)$, then we generalize the **two-dimensional translation** using vectors as

$$\begin{pmatrix} x' \\ y' \end{pmatrix} = \begin{pmatrix} x \\ y \end{pmatrix} + \begin{pmatrix} \Delta x \\ \Delta y \end{pmatrix}. \qquad (11.1)$$

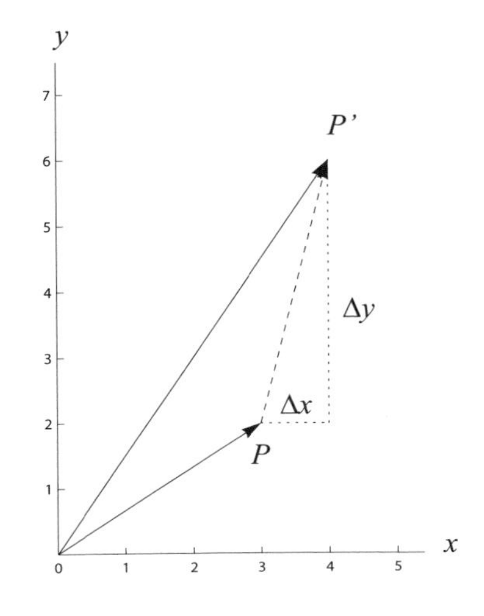

The values of Δx and of Δy do not need to be positive. A negative Δx causes a shift to the left. A negative Δy causes a shift downwards.

By taking the z-label into account, we can translate objects in 3D space as well. We express a **three-dimensional translation** using vectors as

$$\begin{pmatrix} x' \\ y' \\ z' \end{pmatrix} = \begin{pmatrix} x \\ y \\ z \end{pmatrix} + \begin{pmatrix} \Delta x \\ \Delta y \\ \Delta z \end{pmatrix}, \qquad (11.2)$$

with offsets $\Delta x, \Delta y$ and Δz describing the 3D translation vector $\overrightarrow{AB}$.

Example: 2D translation.

We give the general matrix expression to translate flat objects 20 pixels to the right and 10 pixels downwards on the screen, which we will use to shift the golden rectangle bordered by the vertices $A(10,10)$ and $B(20,\ 16.18)$. Assuming that the edges of this rectangle are parallel to the cartesian axes, then both vertices, A at the lower left and B at the upper right corner, determine the complete rectangle.

In order to translate 20 units to the right and 10 units downwards, we set $\Delta x = 20$ and $\Delta y = -10$ (negative because of the downwards effect).

$$\begin{pmatrix} x' \\ y' \end{pmatrix} = \begin{pmatrix} x \\ y \end{pmatrix} + \begin{pmatrix} 20 \\ -10 \end{pmatrix}$$

Evaluating each original vertex, yields its corresponding image vertex. Vertex A translates to the image vertex A' via

$$\begin{pmatrix} x' \\ y' \end{pmatrix} = \begin{pmatrix} 10 \\ 10 \end{pmatrix} + \begin{pmatrix} 20 \\ -10 \end{pmatrix} = \begin{pmatrix} 30 \\ 0 \end{pmatrix}.$$

Similarly, vertex B translates to the image vertex $B'(40,\ 6.18)$. Translating both corner vertices, translates the complete golden rectangle.

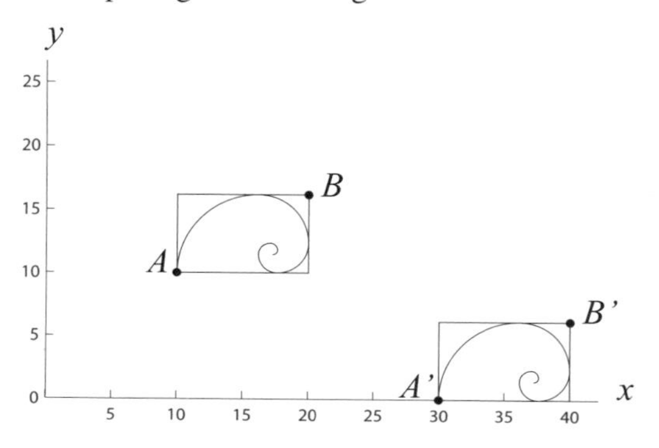

Figure 11.1: Translation of a golden rectangle

If we just need a single translation, we should perform it by vector addition, as it is most simple and fast. If we plan to scale or rotate after translation, we should choose to translate by matrix multiplication (see paragraph 11.6). In this latter case we augment each vector in the translation's matrix expression with an extra component 1 for technical reasons. We call the components of these augmented vectors **homogeneous** and their corresponding points P having **homogeneous coordinates** $P(x,y,z,1)$.

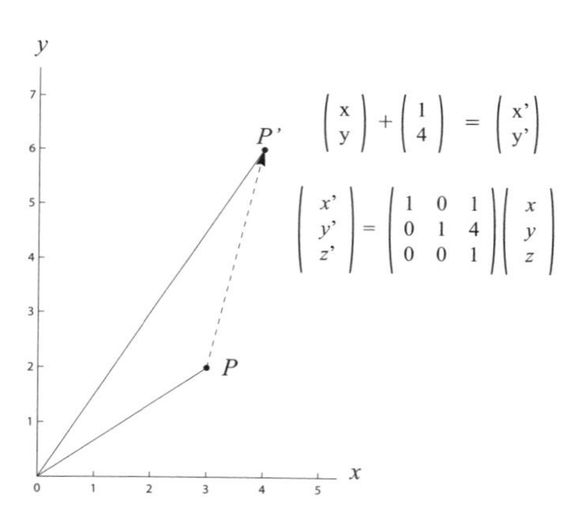

We express a **two-dimensional translation** using matrices by applying a dot product on homogeneous components

$$\begin{pmatrix} x' \\ y' \\ 1 \end{pmatrix} = \begin{pmatrix} 1 & 0 & \Delta x \\ 0 & 1 & \Delta y \\ 0 & 0 & 1 \end{pmatrix} \cdot \begin{pmatrix} x \\ y \\ 1 \end{pmatrix}. \tag{11.3}$$

We define the 3×3 matrix as the translation **operator** $T_{\overrightarrow{AB}}$ taking an original point P as argument to return its image point P', as shifted by the displacement vector $\overrightarrow{AB}$. Its matrix element Δx is the horizontal displacement along the x-axis and the element Δy is the vertical shift along the y-axis.

We emphasize that the extra component 1 and the extra operator row 0 0 1 are without physical meaning. These extra homogeneous components just allow us technically to perform translations by matrix multiplication. We calculate the image value x' by taking its original value x (multiplied by 1), dropping the original value y (multiplied by 0) and adding offset Δx (multiplied by 1). We calculate the image value y' similarly by dropping the original value x, taking its original value y and adding offset Δy to it. In other words, this dot product achieves the former vector addition.

$$\begin{pmatrix} x' \\ y' \\ 1 \end{pmatrix} = \begin{pmatrix} 1 \cdot x + 0 \cdot y + \Delta x \cdot 1 \\ 0 \cdot x + 1 \cdot y + \Delta y \cdot 1 \\ 0 \cdot x + 0 \cdot y + 1 \cdot 1 \end{pmatrix} = \begin{pmatrix} x + \Delta x \\ y + \Delta y \\ 1 \end{pmatrix}$$

We straightforwardly extend this matrix product for translating 3D objects, which involves a 4×4 translation matrix operator.

We express a **three-dimensional translation** by the matrix product

$$\begin{pmatrix} x' \\ y' \\ z' \\ 1 \end{pmatrix} = \begin{pmatrix} 1 & 0 & 0 & \Delta x \\ 0 & 1 & 0 & \Delta y \\ 0 & 0 & 1 & \Delta z \\ 0 & 0 & 0 & 1 \end{pmatrix} \cdot \begin{pmatrix} x \\ y \\ z \\ 1 \end{pmatrix}. \tag{11.4}$$

Calculating the inverse translation matrix $T^{-1}_{\overrightarrow{AB}}$ (see exercise 105), we express the inverse operation as

$$\begin{pmatrix} x \\ y \\ z \\ 1 \end{pmatrix} = \begin{pmatrix} 1 & 0 & 0 & -\Delta x \\ 0 & 1 & 0 & -\Delta y \\ 0 & 0 & 1 & -\Delta z \\ 0 & 0 & 0 & 1 \end{pmatrix} \cdot \begin{pmatrix} x' \\ y' \\ z' \\ 1 \end{pmatrix}. \tag{11.5}$$

We recall Δx as the horizontal and Δy as the vertical displacement, and we interpret Δz logically as the displacement along the z-axis.

Example: 3D translation via matrices

We use the matrix operator for the translation of 2 pixels to the left, 3 pixels upwards and 5 pixels to the front, to translate the cubic Bezier segment (see chapter 14) defined by its control points $A(0,0,0)$, $B(1,2,3)$, $C(2,-1,1)$ and $D(3,1,2)$.

To realize the above translation we set $\Delta x = -2$, $\Delta y = 3$ and $\Delta z = 5$ in its matrix operator.

$$\begin{pmatrix} x' \\ y' \\ z' \\ 1 \end{pmatrix} = \begin{pmatrix} 1 & 0 & 0 & -2 \\ 0 & 1 & 0 & 3 \\ 0 & 0 & 1 & 5 \\ 0 & 0 & 0 & 1 \end{pmatrix} \cdot \begin{pmatrix} x \\ y \\ z \\ 1 \end{pmatrix}$$

Evaluating each original point yields its accordingly shifted image point. Control point $A(0,0,0)$ translates to the image point A' via

$$\begin{pmatrix} x' \\ y' \\ z' \\ 1 \end{pmatrix} = \begin{pmatrix} 1 & 0 & 0 & -2 \\ 0 & 1 & 0 & 3 \\ 0 & 0 & 1 & 5 \\ 0 & 0 & 0 & 1 \end{pmatrix} \cdot \begin{pmatrix} 0 \\ 0 \\ 0 \\ 1 \end{pmatrix} = \begin{pmatrix} -2 \\ 3 \\ 5 \\ 1 \end{pmatrix}.$$

Similarly translating the control points B, C and D yields $B'(-1,5,8)$, $C'(0,2,6)$ and $D'(1,4,7)$. Translating all four control points, translates the complete Bezier segment.

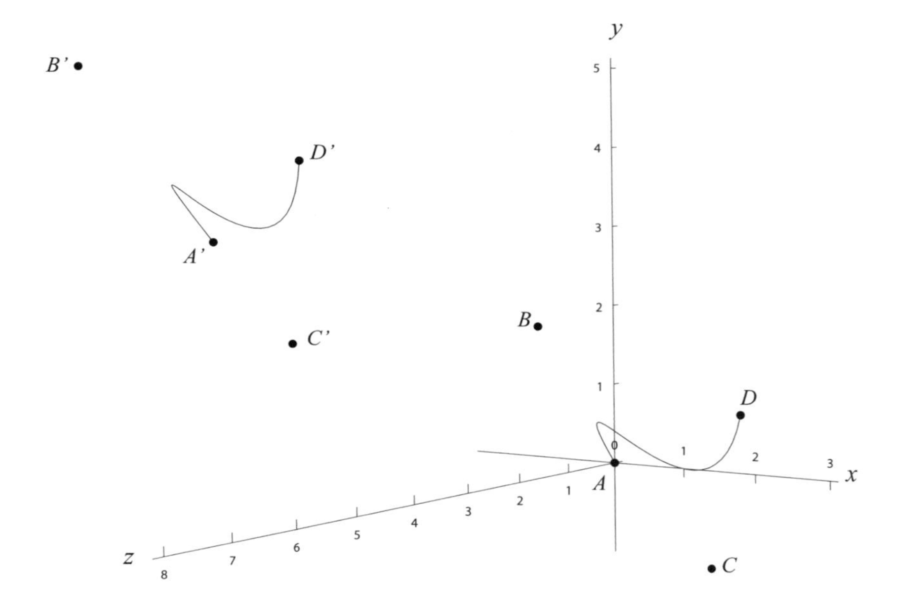

Figure 11.2: Translation of a cubic Bezier segment

11.2 Scaling

The dot product on matrices also suits the scaling of objects. Let us take the same approach as we took for translating objects: we scale a complete object by transforming each of its anchor points separately.

We express the **two-dimensional scaling** S_O in matrix form as

$$\begin{pmatrix} x' \\ y' \\ 1 \end{pmatrix} = \begin{pmatrix} s_x & 0 & 0 \\ 0 & s_y & 0 \\ 0 & 0 & 1 \end{pmatrix} \cdot \begin{pmatrix} x \\ y \\ 1 \end{pmatrix}, \qquad (11.6)$$

given $s_x > 0$ as the scale factor along the x-axis and $s_y > 0$ as the one along the y-axis.

For uniform scaling we set $s_x = s_y$. When we apply different nonzero values $s_x \neq s_y$, the matrix operator causes a non-uniform scaling. Positive scale factors s_x and s_y smaller than 1 will shrink objects, whereas larger than 1 they will enlarge objects.

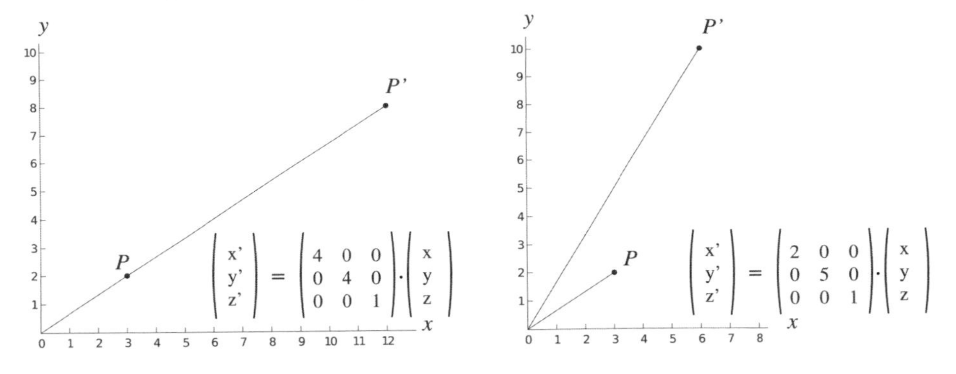

Figure 11.3: Uniform and non-uniform scaling

Example: 2D basic scaling

We give the general matrix expression to enlarge objects three times, which we will use to scale the golden rectangle between the vertices $A(10,10)$ and $B(20,\ 16.18)$.

For this uniform scaling we set $s_x = s_y = 3$, leading to

$$\begin{pmatrix} x' \\ y' \\ 1 \end{pmatrix} = \begin{pmatrix} 3 & 0 & 0 \\ 0 & 3 & 0 \\ 0 & 0 & 1 \end{pmatrix} \cdot \begin{pmatrix} x \\ y \\ 1 \end{pmatrix}.$$

Evaluating each original vertex returns its corresponding scaled vertex. For instance vertex A scales to A' via

$$\begin{pmatrix} 3 & 0 & 0 \\ 0 & 3 & 0 \\ 0 & 0 & 1 \end{pmatrix} \cdot \begin{pmatrix} 10 \\ 10 \\ 1 \end{pmatrix} = \begin{pmatrix} 30 \\ 30 \\ 1 \end{pmatrix}.$$

Indeed, figure 11.4 confirms an image golden rectangle of three times its original size. But as a side effect, our enlarged rectangle also tripled its distance to the origin. This is due to the fact that the **basic transformation** S_O is constructed with respect to the origin O, which pushes enlarged objects away from the origin and pulls shrinked objects towards the origin. To keep a scaled object at its location (for instance anchored at its center point), requires a composed transformation (see paragraph 11.6). We straightforwardly extend

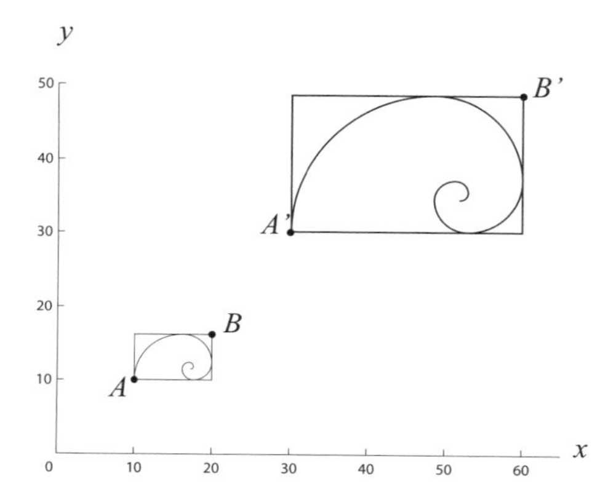

Figure 11.4: 2D uniform scaling

this matrix product for scaling 3D objects, which requires just one dimension extra.

We perform a **three-dimensional basic scaling** by applying the matrix product

$$\begin{pmatrix} x' \\ y' \\ z' \\ 1 \end{pmatrix} = \begin{pmatrix} s_x & 0 & 0 & 0 \\ 0 & s_y & 0 & 0 \\ 0 & 0 & s_z & 0 \\ 0 & 0 & 0 & 1 \end{pmatrix} \cdot \begin{pmatrix} x \\ y \\ z \\ 1 \end{pmatrix}. \tag{11.7}$$

Calculating the inverse basic scale operator S_O^{-1} (see exercise 105), we express the inverse operator as

$$\begin{pmatrix} x \\ y \\ z \\ 1 \end{pmatrix} = \begin{pmatrix} \frac{1}{s_x} & 0 & 0 & 0 \\ 0 & \frac{1}{s_y} & 0 & 0 \\ 0 & 0 & \frac{1}{s_z} & 0 \\ 0 & 0 & 0 & 1 \end{pmatrix} \cdot \begin{pmatrix} x' \\ y' \\ z' \\ 1 \end{pmatrix} \tag{11.8}$$

We recall the positive scale factors s_x along the x-axis, s_y along the y-axis and s_z along the z-axis.

Example: 3D non-uniform scaling.

Let us find the scale operator which doubles the height of 3D objects and shrinks their depth to half the size, in order to rescale the cubic Bezier segment determined by its four control points $A(0,0,0)$, $B(1,2,3)$, $C(2,-1,1)$ and $D(3,1,2)$.

To double an objects' vertical size, we set scale factors $s_x = 1$ (no change) and $s_y = 2$. To shrink an object with 50% along the z-axis, we set $s_z = 0.5$.

$$\begin{pmatrix} x' \\ y' \\ z' \\ 1 \end{pmatrix} = \begin{pmatrix} 1 & 0 & 0 & 0 \\ 0 & 2 & 0 & 0 \\ 0 & 0 & 0.5 & 0 \\ 0 & 0 & 0 & 1 \end{pmatrix} \cdot \begin{pmatrix} x \\ y \\ z \\ 1 \end{pmatrix}$$

Making all four control points of this Bezier segment subject to this scale operator yields $A'(0,0,0)$, $B'(1,\ 4,\ 1.5)$, $C'(2,\ -2,\ 0.5)$ and $D'(3,2,1)$.

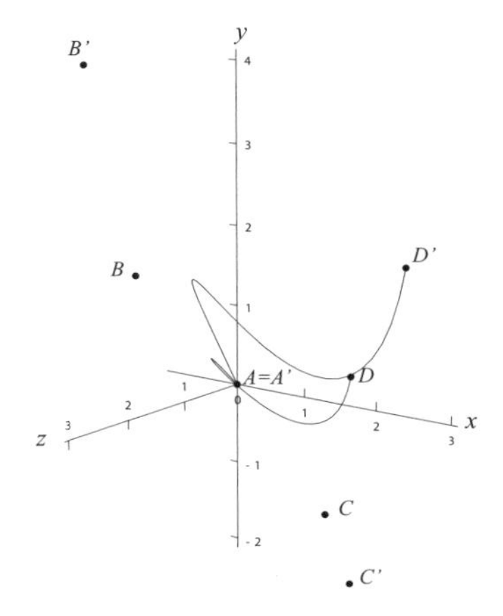

Figure 11.5: Non-uniform scaling of a cubic Bezier segment

11.3 Rotation

There are several ways to rotate objects. There is for instance a method based on (hyper)complex numbers, which we will meet in another chapter (see paragraph 12.5). Alternatively, there is also the method we already used for translation $T_{\overrightarrow{AB}}$ and for basic scaling S_O, based on the matrix product.

Rotation in 2D

Rotating a point P on the unit circle, having P labeled as $(x,y) = (\cos\alpha, \sin\alpha)$, over a positive angle θ returns its image point P' labeled as

$$(x', y') = (\cos(\alpha+\theta), \sin(\alpha+\theta)).$$

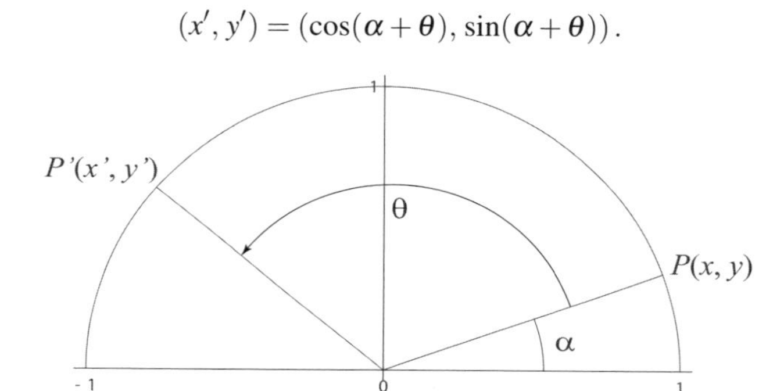

Figure 11.6: Rotation over a positive angle θ

We reverse engineer the well known trigonometric Sum Identities of sine and cosine:

$$\begin{cases} \cos(\alpha+\theta) = \cos\alpha\cos\theta - \sin\alpha\sin\theta \\ \sin(\alpha+\theta) = \sin\alpha\cos\theta + \cos\alpha\sin\theta \end{cases}$$

$$\Leftrightarrow \begin{cases} \cos(\alpha+\theta) = x\cos\theta - y\sin\theta \\ \sin(\alpha+\theta) = y\cos\theta + x\sin\theta \end{cases}$$

$$\Leftrightarrow \begin{cases} x' = x\cos\theta - y\sin\theta \\ y' = y\cos\theta + x\sin\theta \end{cases}$$

$$\Leftrightarrow \begin{pmatrix} x' \\ y' \end{pmatrix} = \underbrace{\begin{pmatrix} \cos\theta & -\sin\theta \\ \sin\theta & \cos\theta \end{pmatrix}}_{\text{rotation operator}} \cdot \begin{pmatrix} x \\ y \end{pmatrix}.$$

In order to combine screen effects in the future, we augment this rotation operator to a matrix compatible to the translation and scale matrices. For this reason we again use homogeneous coordinates, which lead to the matrix operator of the **two-dimensional rotation** $R_{O,\theta}$ around center O and over an angle θ. Since its rotation center lies in the origin O, also this rotator is a basic transformation.

$$\begin{pmatrix} x' \\ y' \\ 1 \end{pmatrix} = \begin{pmatrix} \cos\theta & -\sin\theta & 0 \\ \sin\theta & \cos\theta & 0 \\ 0 & 0 & 1 \end{pmatrix} \cdot \begin{pmatrix} x \\ y \\ 1 \end{pmatrix} \tag{11.9}$$

If we rotate over an angle θ, then the inverse transformation rotates over the angle $-\theta$. We recall that for oppositely signed angles, their cosines are equal and their sines are opposite, which leads us to the inverse basic rotation operator $R_{O,\theta}^{-1}$ (see exercise 105) as

$$\begin{pmatrix} x \\ y \\ 1 \end{pmatrix} = \begin{pmatrix} \cos\theta & \sin\theta & 0 \\ -\sin\theta & \cos\theta & 0 \\ 0 & 0 & 1 \end{pmatrix} \cdot \begin{pmatrix} x' \\ y' \\ 1 \end{pmatrix}.$$

Example: basic 2D rotation.

Let us find the basic matrix operator to rotate flat objects over a right angle, in order to apply it to the golden rectangle determined by its corner vertices $A(10,10)$ and $B(20,\ 16.18)$.

$$\begin{pmatrix} x' \\ y' \\ 1 \end{pmatrix} = \begin{pmatrix} \cos 90^\circ & -\sin 90^\circ & 0 \\ \sin 90^\circ & \cos 90^\circ & 0 \\ 0 & 0 & 1 \end{pmatrix} \cdot \begin{pmatrix} x \\ y \\ 1 \end{pmatrix} = \begin{pmatrix} 0 & -1 & 0 \\ 1 & 0 & 0 \\ 0 & 0 & 1 \end{pmatrix} \cdot \begin{pmatrix} x \\ y \\ 1 \end{pmatrix}$$

Evaluating both vertices A and B returns their rotated image points as $A'(-10,10)$ and $B'(-16.18,\ 20)$.

There again is a side effect to note, as our rotated rectangle also shifted to another location. This is due to the fact that the basic rotation operator $R_{O,\theta}$ is constructed with respect to the origin O, just like the basic scale operator S_O was. The major consequence of it is the rotation of objects around the origin. As soon as we want to rotate objects around their center or around an arbitrary point, we will need to compose standard transformations (see paragraph 11.6).

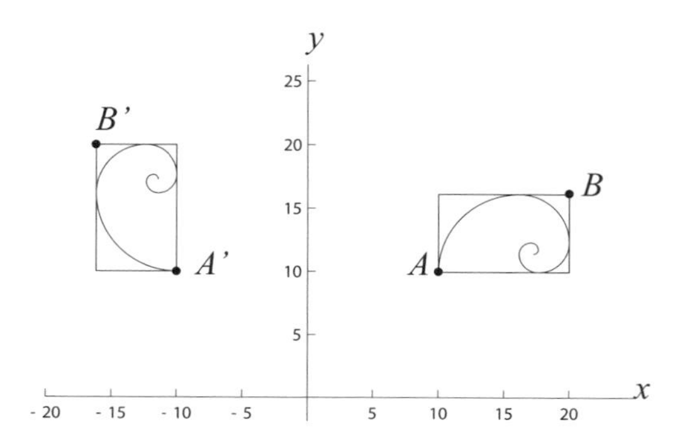

Figure 11.7: Basic 2D rotation

Rotation in 3D

Let us study a 3D rotation in detail. As rotations in 2D operate in a plane (for instance in a screen), rotations in 3D benefit some more degrees of freedom: they operate in three planes.

These are their three planes to rotate in:

- a vertical plane xy in which the 'roll' rotates around the z-axis,
- a longitudinal plane yz in which the 'pitch' rotates around the x-axis,
- a horizontal plane xz in which the 'yaw' rotates around the y-axis.

Let us for a start study the 'roll' around the z-axis, since it is the equivalent of the rotation in 2D. The z-axis is perpendicular to the (flat) screen, so a rotation around it is actually the rotation around the origin O in the xy-plane. Hence its 3D rotation operator $R_{z,\theta}$ is similar to the 2D rotator $R_{O,\theta}$ as it takes just one extra row and column. We call this rotation a **roll**, the term commonly used for it in aviation.

As the three-dimensional rotation around the z-axis $R_{z,\theta}$ only affects the x- and y-labels of points, we deduct its operator matrix as

$$\begin{pmatrix} x' \\ y' \\ z' \\ 1 \end{pmatrix} = \begin{pmatrix} \cos\theta & -\sin\theta & 0 & 0 \\ \sin\theta & \cos\theta & 0 & 0 \\ 0 & 0 & 1 & 0 \\ 0 & 0 & 0 & 1 \end{pmatrix} \cdot \begin{pmatrix} x \\ y \\ z \\ 1 \end{pmatrix}. \qquad (11.10)$$

A second possible 3D rotation is the one around the x-axis, which we typeset as $R_{x,\theta}$ and commonly denote as **pitch**. Rotating around the x-axis does not affect the x-labels, so the first row and column of its operator matrix conserve the x-values.

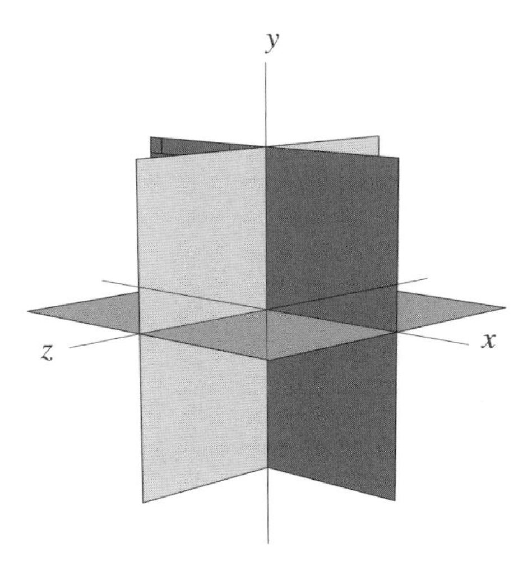

Figure 11.8: Basic 3D rotation planes

Consequently we deduct the three-dimensional rotation $R_{x,\theta}$ around the x-axis over the angle θ as

$$\begin{pmatrix} x' \\ y' \\ z' \\ 1 \end{pmatrix} = \begin{pmatrix} 1 & 0 & 0 & 0 \\ 0 & \cos\theta & -\sin\theta & 0 \\ 0 & \sin\theta & \cos\theta & 0 \\ 0 & 0 & 0 & 1 \end{pmatrix} \cdot \begin{pmatrix} x \\ y \\ z \\ 1 \end{pmatrix}. \tag{11.11}$$

As a third and final type of 3D rotation we commonly call **yaw** the rotation around the y-axis, typeset as $R_{y,\theta}$. Rotating around the y-axis does not affect the y-labels, so the second row and column of its operator matrix conserve the y-values. Consequently we express the three-dimensional rotation $R_{y,\theta}$ around the y-axis over the angle θ as

$$\begin{pmatrix} x' \\ y' \\ z' \\ 1 \end{pmatrix} = \begin{pmatrix} \cos\theta & 0 & \sin\theta & 0 \\ 0 & 1 & 0 & 0 \\ -\sin\theta & 0 & \cos\theta & 0 \\ 0 & 0 & 0 & 1 \end{pmatrix} \cdot \begin{pmatrix} x \\ y \\ z \\ 1 \end{pmatrix} \tag{11.12}$$

We warn for a technical detail: the minus sign in $R_{y,\theta}$ resides at its lower left element. All three types of 3D rotation are basic transformations since they each take a cartesian axis to rotate around. In a future paragraph, we will discuss how to rotate around an arbitrary center and how to combine basic 3D rotations to rotate simultaneously around the x-, y- and z-axis.

Example: Basic 3D rotation around the y-axis.

We construct the matrix operator to rotate 3D objects around the y-axis over a straight angle, in order to rotate the triangle with vertices $A(4,0,1), B(0,3,2)$ and $C(-1,2,-1)$.

$$\begin{pmatrix} x' \\ y' \\ z' \\ 1 \end{pmatrix} = \begin{pmatrix} \cos\pi & 0 & \sin\pi & 0 \\ 0 & 1 & 0 & 0 \\ -\sin\pi & 0 & \cos\pi & 0 \\ 0 & 0 & 0 & 1 \end{pmatrix} \cdot \begin{pmatrix} x \\ y \\ z \\ 1 \end{pmatrix}$$

$$= \begin{pmatrix} -1 & 0 & 0 & 0 \\ 0 & 1 & 0 & 0 \\ 0 & 0 & -1 & 0 \\ 0 & 0 & 0 & 1 \end{pmatrix} \cdot \begin{pmatrix} x \\ y \\ z \\ 1 \end{pmatrix}$$

Evaluating its vertices A, B and C returns the rotated triangle as $A'(-4,0,-1), B'(0,3,-2)$ and $C'(1,2,1)$.

11.4 Reflection

Also reflections can be described using matrices. Two-dimensionally we can basically reflect over the x-axis, over the y-axis or over the origin O. We can describe alternatively the reflection over the origin as the basic rotation over the straight angle, or as the combined reflection over the x-axis and the y-axis simultaneously.

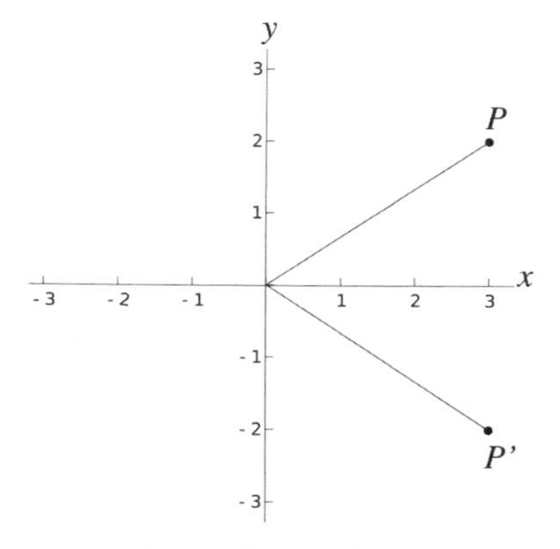

(a) reflection over the x-axis

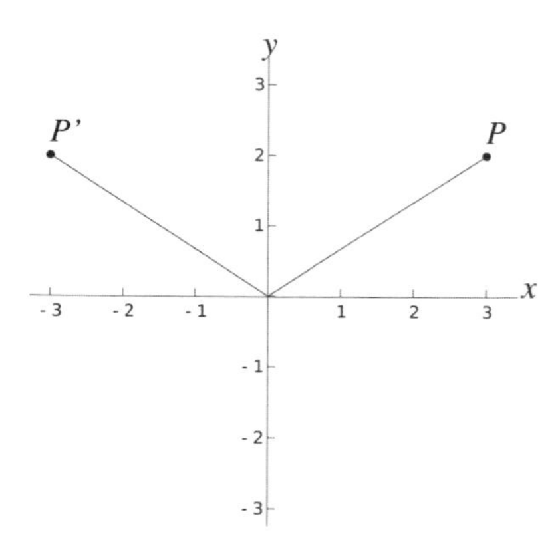

(b) reflection over the y-axis

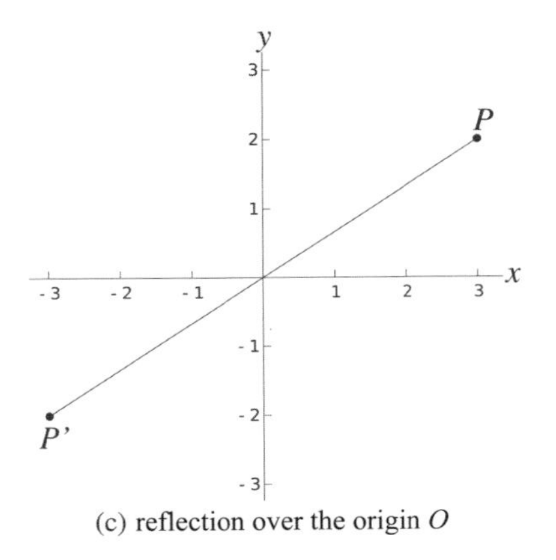

(c) reflection over the origin O

(a) $\begin{vmatrix} 1 & 0 & 0 \\ 0 & -1 & 0 \\ 0 & 0 & 1 \end{vmatrix} \cdot \begin{vmatrix} x \\ y \\ 1 \end{vmatrix} = \begin{vmatrix} x \\ -y \\ 1 \end{vmatrix}$

(b) $\begin{vmatrix} -1 & 0 & 0 \\ 0 & 1 & 0 \\ 0 & 0 & 1 \end{vmatrix} \cdot \begin{vmatrix} x \\ y \\ 1 \end{vmatrix} = \begin{vmatrix} -x \\ y \\ 1 \end{vmatrix}$

(c) $\begin{vmatrix} -1 & 0 & 0 \\ 0 & -1 & 0 \\ 0 & 0 & 1 \end{vmatrix} \cdot \begin{vmatrix} x \\ y \\ 1 \end{vmatrix} = \begin{vmatrix} -x \\ -y \\ 1 \end{vmatrix}$

The **two-dimensional reflection over the x-axis** leaves the x-labels invariant and affects just the sign of the y-labels. We typeset it as M_x and express its matrix operator as

$$\begin{pmatrix} x' \\ y' \\ 1 \end{pmatrix} = \begin{pmatrix} 1 & 0 & 0 \\ 0 & -1 & 0 \\ 0 & 0 & 1 \end{pmatrix} \cdot \begin{pmatrix} x \\ y \\ 1 \end{pmatrix}. \tag{11.13}$$

The **two-dimensional reflection over the y-axis** leaves the y-labels invariant and affects just the sign of the x-labels. We typeset it as M_y and express its matrix operator as

$$\begin{pmatrix} x' \\ y' \\ 1 \end{pmatrix} = \begin{pmatrix} -1 & 0 & 0 \\ 0 & 1 & 0 \\ 0 & 0 & 1 \end{pmatrix} \cdot \begin{pmatrix} x \\ y \\ 1 \end{pmatrix}. \tag{11.14}$$

The **two-dimensional reflection over the origin** affects the sign of both the x- and the y-labels. We typeset it as M_O and calculate its matrix operator via $R_{O,180°}$ as

$$\begin{pmatrix} x' \\ y' \\ 1 \end{pmatrix} = \begin{pmatrix} -1 & 0 & 0 \\ 0 & -1 & 0 \\ 0 & 0 & 1 \end{pmatrix} \cdot \begin{pmatrix} x \\ y \\ 1 \end{pmatrix}. \tag{11.15}$$

11.5 Shearing

A **basic shearing** deforms an object in only one direction, proportionally to the distance to an invariable axis. We illustrate this effect via the '*false italic*' typesetting font of which the bottom pixels remain invariable and all other pixels are pushed to the right, the more up the more pushed to the right.

The **basic shearing along the x-axis** by an angle called the **shear strain** σ_x, conserves the y-labels and affects the x-labels, but conserves the x-axis itself. We typeset this horizontal shearing as S_{σ_x} and express its matrix operator as

$$\begin{pmatrix} x' \\ y' \\ 1 \end{pmatrix} = \begin{pmatrix} 1 & \tan\sigma_x & 0 \\ 0 & 1 & 0 \\ 0 & 0 & 1 \end{pmatrix} \cdot \begin{pmatrix} x \\ y \\ 1 \end{pmatrix}.$$

Proof:

The basic shearing of an original point $P(x,y)$ by S_{σ_x} conserves its y-label and adds to its x-label a proportional length $|PP'|$, typeset as

$$\begin{pmatrix} x' \\ y' \end{pmatrix} = \begin{pmatrix} x + |PP'| \\ y \end{pmatrix}.$$

We retrieve this proportional length $|PP'|$ from the right triangle OPP' (see figure 11.9) as

$$\tan\sigma_x = \frac{|PP'|}{|OP|} = \frac{|PP'|}{y} \quad \Longrightarrow \quad |PP'| = y\tan\sigma_x.$$

Replacing $|PP'|$ by $y\tan\sigma_x$, we reverse engineer the column matrix to a matrix product

$$\begin{pmatrix} x' \\ y' \end{pmatrix} = \begin{pmatrix} x + y\tan\sigma_x \\ y \end{pmatrix} = \begin{pmatrix} 1 & \tan\sigma_x \\ 0 & 1 \end{pmatrix} \cdot \begin{pmatrix} x \\ y \end{pmatrix}.$$

Adding the homogeneous component finalizes the proof. ∎

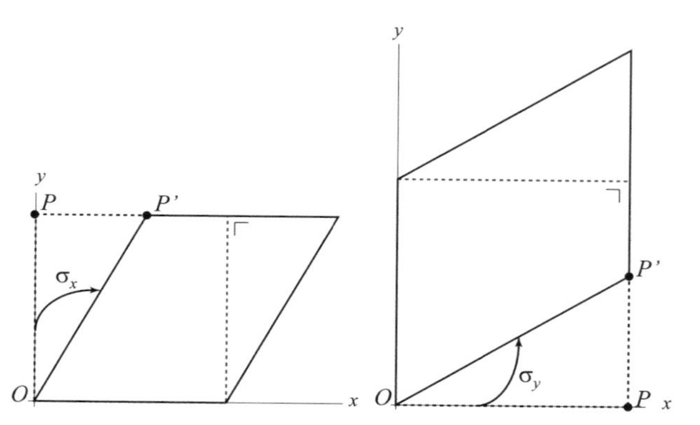

Figure 11.9: Basic shearing along the x-axis and along the y-axis

The **basic shearing along the y-axis** by an angle called the **shear strain** σ_y conserves the x-labels and affects the y-labels, but conserves the y-axis itself. We typeset this vertical shearing as S_{σ_y} and express its matrix operator as

$$\begin{pmatrix} x' \\ y' \\ 1 \end{pmatrix} = \begin{pmatrix} 1 & 0 & 0 \\ \tan\sigma_y & 1 & 0 \\ 0 & 0 & 1 \end{pmatrix} \cdot \begin{pmatrix} x \\ y \\ 1 \end{pmatrix}.$$

We prove this matrix operator completely similar to the previous proof for the S_{σ_x}-operator.

Eventually we are able to shear an object *simultaneously* in both the direction of the x-axis and the y-axis. We realize such a **combined basic shearing** by both shear strains σ_x and σ_y, by applying its matrix operator

$$S_{\sigma_x,\sigma_y} = \begin{pmatrix} 1 & \tan\sigma_x & 0 \\ \tan\sigma_y & 1 & 0 \\ 0 & 0 & 1 \end{pmatrix}.$$

Example: We construct the matrix operator to shear flat objects simultaneously by $\sigma_x = 50^\circ$ along the x-axis and by $\sigma_y = 20^\circ$ along the y-axis. We then apply this operator to transform the golden rectangle bordered by the vertices $A(0,0)$ and $B(10,\ 6.18)$.

$$\begin{pmatrix} x' \\ y' \\ 1 \end{pmatrix} = \begin{pmatrix} 1 & \tan 50^\circ & 0 \\ \tan 20^\circ & 1 & 0 \\ 0 & 0 & 1 \end{pmatrix} \begin{pmatrix} x \\ y \\ 1 \end{pmatrix}.$$

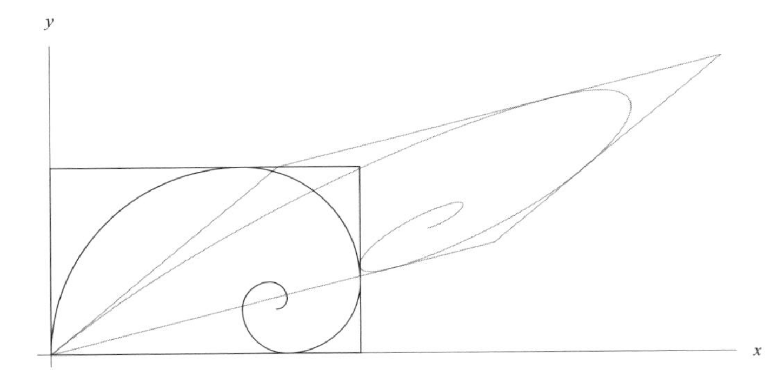

Figure 11.10: The combined basic shearing $S_{50^\circ,20^\circ}$ along both cartesian axes

Properties:

- ▷ The plane basic shearing conserves area.
- ▷ The inverse basic shearing takes the opposite angle:

$$S_{\sigma_x}^{-1} = S_{-\sigma_x} \qquad \text{and} \qquad S_{\sigma_y}^{-1} = S_{-\sigma_y}.$$

▷ The inverse combined basic shearing does not take the opposite angles:

$$S_{\sigma_x,\sigma_y}^{-1} \neq S_{-\sigma_x,-\sigma_y}.$$

▷ The combined basic shearing is *not* a combination of basic shearings:

$$S_{\sigma_x,\sigma_y} \neq S_{\sigma_x} \cdot S_{\sigma_y} \qquad \text{and} \qquad S_{\sigma_x,\sigma_y} \neq S_{\sigma_y} \cdot S_{\sigma_x}.$$

11.6 Composing transformations

Let us now compose translations and the basic scaling, rotation, reflection and shearing into one transformation. This will allow us to tackle some former challenges such as scaling with respect to the object center, rotation around an arbitrary point and combining 3D rotations.

We are aware that the matrix product is not commutative: altering the order of the matrices we multiply, changes the result in general. We visualize this non-commutative property by composing a translation (10 pixels upwards and 20 pixels to the right) and a basic reflection over the origin O. We apply this composed transformation to the golden rectangle bordered by the vertices $A(10,10)$ and $B(20,\ 16.18)$. Firstly, we reflect after translation. Secondly, we will translate after reflection.

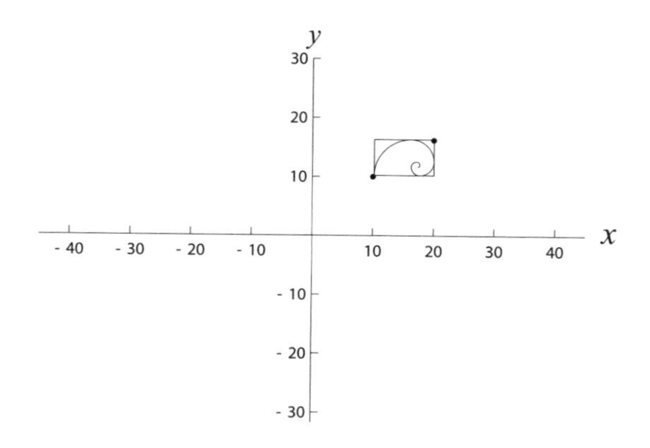

1. first translate

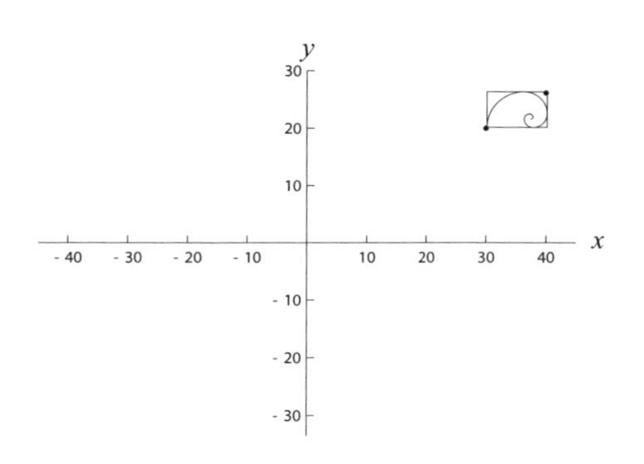

then reflect

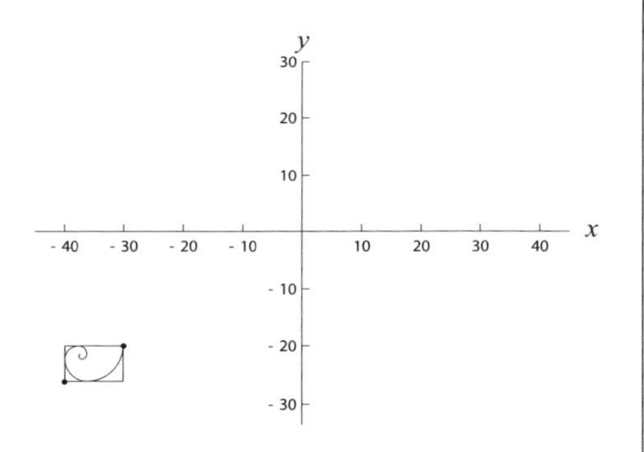

2. first reflect

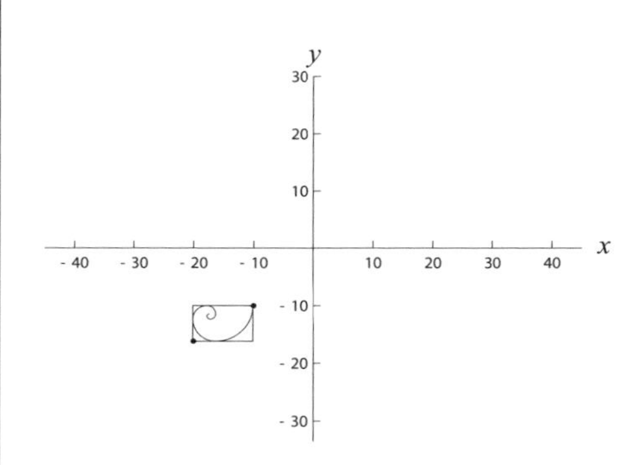

then translate

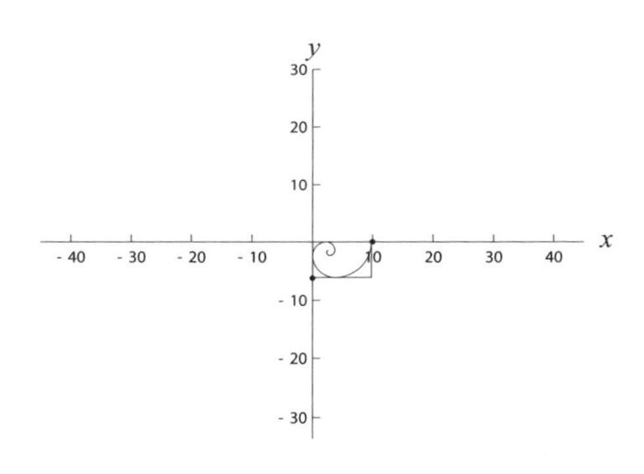

We set the translation operator T and the basic reflection operator M_O as

$$T = \begin{pmatrix} 1 & 0 & 20 \\ 0 & 1 & 10 \\ 0 & 0 & 1 \end{pmatrix} \text{ and } M_O = \begin{pmatrix} -1 & 0 & 0 \\ 0 & -1 & 0 \\ 0 & 0 & 1 \end{pmatrix}.$$

1. first translate

$$\begin{pmatrix} 1 & 0 & 20 \\ 0 & 1 & 10 \\ 0 & 0 & 1 \end{pmatrix} \cdot \begin{pmatrix} 10 \\ 10 \\ 1 \end{pmatrix} = \begin{pmatrix} 30 \\ 20 \\ 1 \end{pmatrix}$$

hen reflect

$$\begin{pmatrix} -1 & 0 & 0 \\ 0 & -1 & 0 \\ 0 & 0 & 1 \end{pmatrix} \cdot \begin{pmatrix} 30 \\ 20 \\ 1 \end{pmatrix} = \begin{pmatrix} -30 \\ -20 \\ 1 \end{pmatrix}$$

2. first reflect

$$\begin{pmatrix} -1 & 0 & 0 \\ 0 & -1 & 0 \\ 0 & 0 & 1 \end{pmatrix} \cdot \begin{pmatrix} 10 \\ 10 \\ 1 \end{pmatrix} = \begin{pmatrix} -10 \\ -10 \\ 1 \end{pmatrix}$$

then translate

$$\begin{pmatrix} 1 & 0 & 20 \\ 0 & 1 & 10 \\ 0 & 0 & 1 \end{pmatrix} \cdot \begin{pmatrix} -10 \\ -10 \\ 1 \end{pmatrix} = \begin{pmatrix} 10 \\ 0 \\ 1 \end{pmatrix}$$

composed

$$\begin{pmatrix} -1 & 0 & 0 \\ 0 & -1 & 0 \\ 0 & 0 & 1 \end{pmatrix} \cdot \begin{pmatrix} 1 & 0 & 20 \\ 0 & 1 & 10 \\ 0 & 0 & 1 \end{pmatrix} \cdot \begin{pmatrix} 10 \\ 10 \\ 1 \end{pmatrix}$$

$$= \begin{pmatrix} -1 & 0 & -20 \\ 0 & -1 & -10 \\ 0 & 0 & 1 \end{pmatrix} \cdot \begin{pmatrix} 10 \\ 10 \\ 1 \end{pmatrix}$$

$$= \begin{pmatrix} -30 \\ -20 \\ 1 \end{pmatrix}$$

composed

$$\begin{pmatrix} 1 & 0 & 20 \\ 0 & 1 & 10 \\ 0 & 0 & 1 \end{pmatrix} \cdot \begin{pmatrix} -1 & 0 & 0 \\ 0 & -1 & 0 \\ 0 & 0 & 1 \end{pmatrix} \cdot \begin{pmatrix} 10 \\ 10 \\ 1 \end{pmatrix}$$

$$= \begin{pmatrix} -1 & 0 & 20 \\ 0 & -1 & 10 \\ 0 & 0 & 1 \end{pmatrix} \cdot \begin{pmatrix} 10 \\ 10 \\ 1 \end{pmatrix}$$

$$= \begin{pmatrix} 10 \\ 0 \\ 1 \end{pmatrix}$$

Altering the product order obviously changes the composed operator matrix. Aiming for a composed transformation, once we know its successive steps and in which order they have to be taken, we are able to express its composed operator matrix. Due to the non-commutativity of the matrix product, ordering its factors correctly is of major importance. The homogeneous column matrix containing the original location has to be placed as the outer right factor, then stacked with each successive operator, from right to left building up a matrix product.

We study a few examples of composed transformations to trigger our understanding of how to compose a translation and basic scaling, rotation, reflection or shearing.

2D ROTATION AROUND AN ARBITRARY CENTER

We refer to the former example of the basic 2D rotation (see page 201) in which we rotated the golden rectangle over a straight angle around the origin. If we aim to rotate this golden rectangle around its bottom left vertex $A(10, 10)$, we need to take three steps instead of one.

step 0

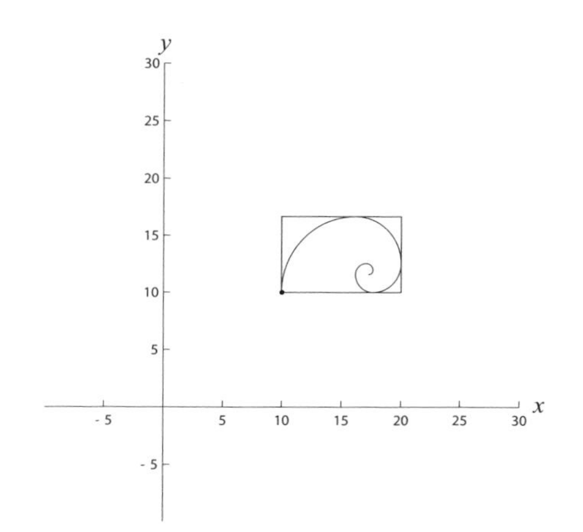

step 1: translate by $T_{\overrightarrow{AO}}$

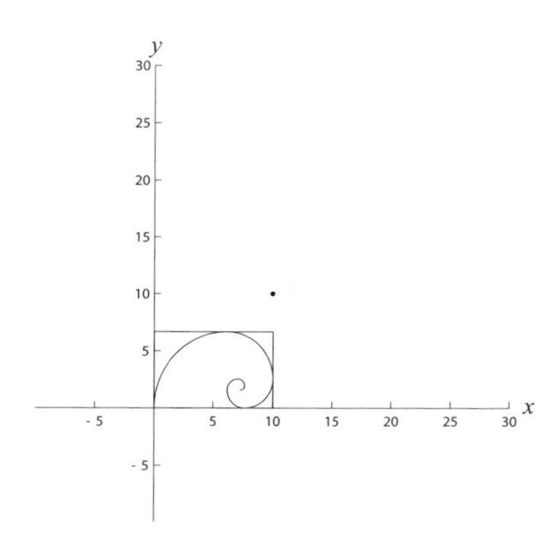

step 2: rotate by $R_{O,90^\circ}$

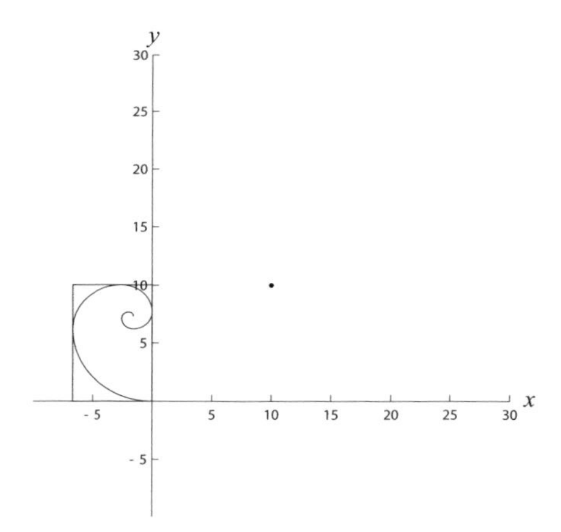

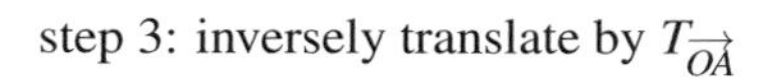

step 3: inversely translate by $T_{\overrightarrow{OA}}$

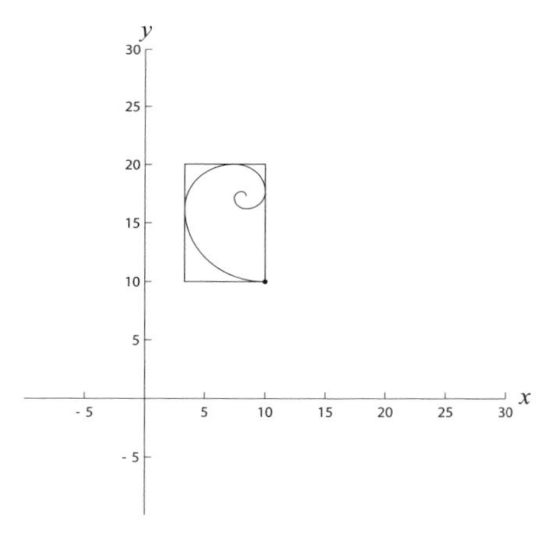

1) A translation by $T_{\overrightarrow{AO}}$ in order to get the bottom left vertex $A(10,10)$ in the origin $O(0,0)$. This implies a shift of 10 pixels to the left and 10 pixels downwards, as dictated by $\overrightarrow{AO} = \vec{o} - \vec{a}$ (see formula (7.5)).

$$\begin{pmatrix} 1 & 0 & -10 \\ 0 & 1 & -10 \\ 0 & 0 & 1 \end{pmatrix}$$

2) A basic rotation over 90° around O.

$$\begin{pmatrix} \cos 90^\circ & -\sin 90^\circ & 0 \\ \sin 90^\circ & \cos 90^\circ & 0 \\ 0 & 0 & 1 \end{pmatrix}$$

3) The inverse translation by $T_{\overrightarrow{OA}}$ to restore the position of the bottom left vertex A. This implies a shift of 10 pixels to the right and 10 pixels upwards.

$$\begin{pmatrix} 1 & 0 & 10 \\ 0 & 1 & 10 \\ 0 & 0 & 1 \end{pmatrix}$$

We summarize the matrix expression holding the composed operator to calculate all image points for this example as

$$\begin{pmatrix} x' \\ y' \\ 1 \end{pmatrix} = \begin{pmatrix} 1 & 0 & 10 \\ 0 & 1 & 10 \\ 0 & 0 & 1 \end{pmatrix} \cdot \begin{pmatrix} \cos 90^\circ & -\sin 90^\circ & 0 \\ \sin 90^\circ & \cos 90^\circ & 0 \\ 0 & 0 & 1 \end{pmatrix} \cdot \begin{pmatrix} 1 & 0 & -10 \\ 0 & 1 & -10 \\ 0 & 0 & 1 \end{pmatrix} \cdot \begin{pmatrix} x \\ y \\ 1 \end{pmatrix}$$

$$= \begin{pmatrix} 0 & -1 & 20 \\ 1 & 0 & 0 \\ 0 & 0 & 1 \end{pmatrix} \cdot \begin{pmatrix} x \\ y \\ 1 \end{pmatrix}.$$

Evaluating both bordering vertices A and B calculates the golden rectangle's new position as $A'(10,10)$ and $B'(3.82,\ 20)$.

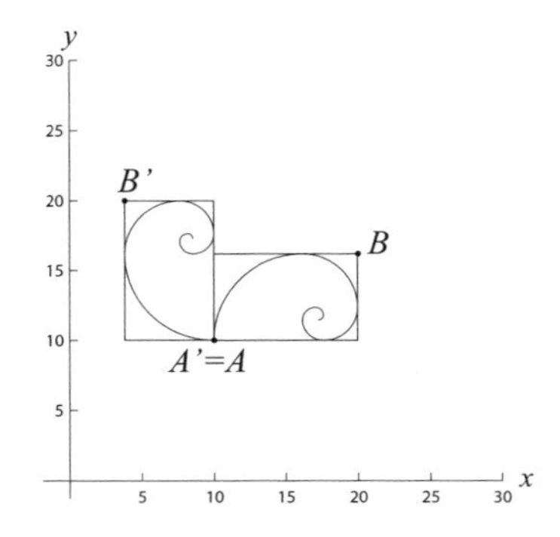

Figure 11.11: 2D rotation around an arbitrary point

3D SCALING ABOUT AN ARBITRARY CENTER

Imagine we like to scale a 3D object, keeping it steady at its original location. Let us find the matrix operator to realize this, applying positive scale factors s_x, s_y and s_z with respect to its own center $C(x_c, y_c, z_c)$. This transformation requires three steps:

1) translation by $T_{\overrightarrow{CO}}$ in order to get the object's center $C(x_c, y_c, z_c)$ in the origin $O(0,0,0)$,

$$T_{\overrightarrow{CO}} = \begin{pmatrix} 1 & 0 & 0 & -x_c \\ 0 & 1 & 0 & -y_c \\ 0 & 0 & 1 & -z_c \\ 0 & 0 & 0 & 1 \end{pmatrix}$$

2) basic scaling S_O applying positive scale factors s_x, s_y and s_z,

$$S_O = \begin{pmatrix} s_x & 0 & 0 & 0 \\ 0 & s_y & 0 & 0 \\ 0 & 0 & s_z & 0 \\ 0 & 0 & 0 & 1 \end{pmatrix}$$

3) inverse translation by $T_{\overrightarrow{OC}}$ to restore the position of the object's center C

$$T_{\overrightarrow{OC}} = \begin{pmatrix} 1 & 0 & 0 & x_c \\ 0 & 1 & 0 & y_c \\ 0 & 0 & 1 & z_c \\ 0 & 0 & 0 & 1 \end{pmatrix}.$$

We summarize the matrix expression holding the composed operator as

$$\begin{pmatrix} x' \\ y' \\ z' \\ 1 \end{pmatrix} = \begin{pmatrix} 1 & 0 & 0 & x_c \\ 0 & 1 & 0 & y_c \\ 0 & 0 & 1 & z_c \\ 0 & 0 & 0 & 1 \end{pmatrix} \cdot \begin{pmatrix} s_x & 0 & 0 & 0 \\ 0 & s_y & 0 & 0 \\ 0 & 0 & s_z & 0 \\ 0 & 0 & 0 & 1 \end{pmatrix} \cdot \begin{pmatrix} 1 & 0 & 0 & -x_c \\ 0 & 1 & 0 & -y_c \\ 0 & 0 & 1 & -z_c \\ 0 & 0 & 0 & 1 \end{pmatrix} \cdot \begin{pmatrix} x \\ y \\ z \\ 1 \end{pmatrix}$$

$$= \begin{pmatrix} s_x & 0 & 0 & x_c - s_x x_c \\ 0 & s_y & 0 & y_c - s_y y_c \\ 0 & 0 & s_z & z_c - s_z z_c \\ 0 & 0 & 0 & 1 \end{pmatrix} \cdot \begin{pmatrix} x \\ y \\ z \\ 1 \end{pmatrix}.$$

2D REFLECTION OVER AN AXIS THROUGH THE ORIGIN

We determine the general matrix expression to reflect flat objects over a straight line r described by $y = ax$, inclined to the x-axis by an angle φ. We recall to determine this angle φ via the slope of r as $\tan\varphi = a$ (see formula (4.2) in chapter 4).

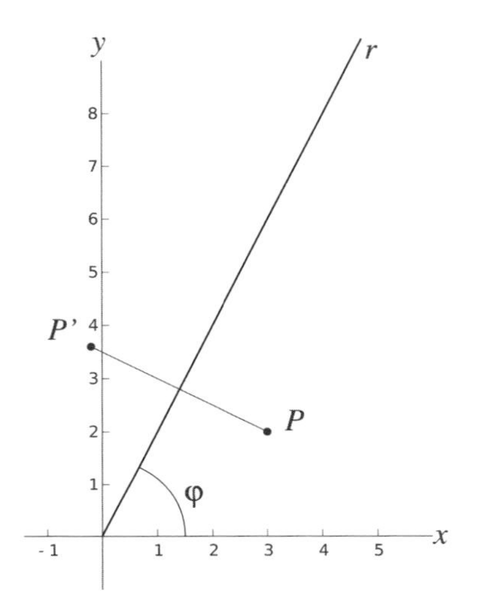

Figure 11.12: 2D reflection line r through O

Achieving this reflection requires composing three transformations in their appropriate order. Firstly, we need to rotate the line r (as well as all points P of the plane) clockwise around O over an angle φ, to get the line r onto the x-axis. Secondly, we perform the basic reflection over the x-axis. Finally, we inversely rotate the line r (and the complete plane) counter clockwise over the angle φ, to restore the original orientation of the line r.

1) basic clockwise rotation over the angle φ by

$$R_{O,-\varphi} = \begin{pmatrix} \cos(-\varphi) & -\sin(-\varphi) & 0 \\ \sin(-\varphi) & \cos(-\varphi) & 0 \\ 0 & 0 & 1 \end{pmatrix} = \begin{pmatrix} \cos\varphi & \sin\varphi & 0 \\ -\sin\varphi & \cos\varphi & 0 \\ 0 & 0 & 1 \end{pmatrix}$$

2) basic reflection over the x-axis by

$$M_x = \begin{pmatrix} 1 & 0 & 0 \\ 0 & -1 & 0 \\ 0 & 0 & 1 \end{pmatrix}$$

3) basic counter clockwise rotation over the angle φ by

$$R_{O,\varphi} = \begin{pmatrix} \cos\varphi & -\sin\varphi & 0 \\ \sin\varphi & \cos\varphi & 0 \\ 0 & 0 & 1 \end{pmatrix}.$$

This leads to a composed matrix operator which is dependent of the line's inclination φ.

$$\begin{pmatrix} x' \\ y' \\ 1 \end{pmatrix} = \begin{pmatrix} \cos\varphi & -\sin\varphi & 0 \\ \sin\varphi & \cos\varphi & 0 \\ 0 & 0 & 1 \end{pmatrix} \cdot \begin{pmatrix} 1 & 0 & 0 \\ 0 & -1 & 0 \\ 0 & 0 & 1 \end{pmatrix} \cdot \begin{pmatrix} \cos\varphi & \sin\varphi & 0 \\ -\sin\varphi & \cos\varphi & 0 \\ 0 & 0 & 1 \end{pmatrix} \cdot \begin{pmatrix} x \\ y \\ 1 \end{pmatrix}$$

$$= \begin{pmatrix} \cos(2\varphi) & \sin(2\varphi) & 0 \\ \sin(2\varphi) & -\cos(2\varphi) & 0 \\ 0 & 0 & 1 \end{pmatrix} \cdot \begin{pmatrix} x \\ y \\ 1 \end{pmatrix}$$

2D REFLECTION OVER AN ARBITRARY AXIS

We extend the previous reflection in order to reflect over an arbitrary line not necessarily through the origin O. We assume the line r described by $y = ax + b$ to intersect the y-axis at intercept $B(0,b)$ and to incline to the x-axis at an angle φ. We achieve reflecting a point P over such a line r by again composing basic transformations. This specific reflection we aim for, requires five basic transformations. Firstly, we shift all points vertically over a distance b until the line r runs through the origin. Secondly, we rotate around O over an angle $-\varphi$, putting the line r onto the x-axis. Essentially, we then reflect over the x-axis. Step four inversely rotates line r over the angle φ. Finally, we inversely shift over the distance b, restoring the plane in its original position.

1) vertical translation over the distance b to the origin O by

$$T_{\overrightarrow{BO}} = \begin{pmatrix} 1 & 0 & 0 \\ 0 & 1 & -b \\ 0 & 0 & 1 \end{pmatrix}$$

2) basic clockwise rotation by

$$R_{O,-\varphi} = \begin{pmatrix} \cos(-\varphi) & -\sin(-\varphi) & 0 \\ \sin(-\varphi) & \cos(-\varphi) & 0 \\ 0 & 0 & 1 \end{pmatrix} = \begin{pmatrix} \cos\varphi & \sin\varphi & 0 \\ -\sin\varphi & \cos\varphi & 0 \\ 0 & 0 & 1 \end{pmatrix}$$

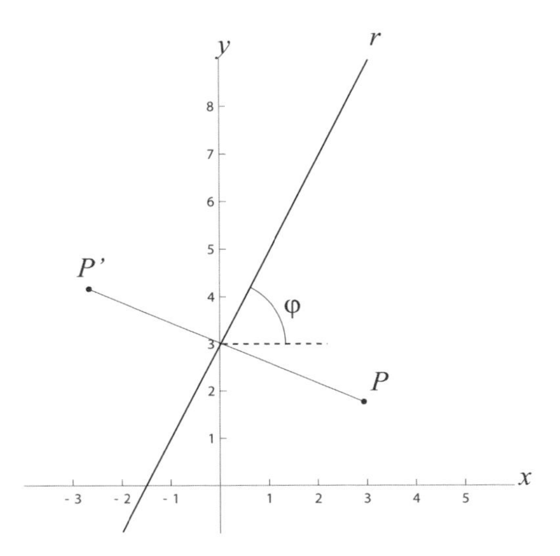

Figure 11.13: 2D arbitrary reflection line

3) basic reflection over the x-axis by

$$M_x = \begin{pmatrix} 1 & 0 & 0 \\ 0 & -1 & 0 \\ 0 & 0 & 1 \end{pmatrix}$$

4) basic counter clockwise rotation by

$$R_{O,\varphi} = \begin{pmatrix} \cos\varphi & -\sin\varphi & 0 \\ \sin\varphi & \cos\varphi & 0 \\ 0 & 0 & 1 \end{pmatrix}$$

5) vertical translation over the distance b to the point B by

$$T_{\overrightarrow{OB}} = \begin{pmatrix} 1 & 0 & 0 \\ 0 & 1 & b \\ 0 & 0 & 1 \end{pmatrix}.$$

This leads to a composed matrix operator which is dependent on the line's inclination φ and its intercept b.

$$\begin{pmatrix} x' \\ y' \\ 1 \end{pmatrix} = \begin{pmatrix} \cos(2\varphi) & \sin(2\varphi) & -b\sin(2\varphi) \\ \sin(2\varphi) & -\cos(2\varphi) & b+b\cos(2\varphi) \\ 0 & 0 & 1 \end{pmatrix} \begin{pmatrix} x \\ y \\ 1 \end{pmatrix}.$$

Example: We construct the matrix operator to reflect objects over the straight line $y = \frac{1}{2}x + 20$. We apply this transformation to reflect the golden rectangle with corner vertices $A(10,10)$ en $B(20,\ 16.18)$ over the above reflection line.

Constructing its matrix operator requires the reflection line's intercept b and its inclination angle φ.

- ▷ The line $y = \frac{1}{2}x + 20$ intersects the y-axis in the point $B(0,20)$.
- ▷ We interpret the line's slope $a = \frac{1}{2}$ trigonometrically as $\frac{1}{2} = \tan\varphi$ which yields $\varphi = \arctan\frac{1}{2} \approx 26.57°$. Consequently we calculate $\cos(2\varphi) \approx 0.6$ and $\sin(2\varphi) \approx 0.8$.

Evaluating both $b = 20$ and $\varphi \approx 26.57°$ leads to the composed matrix operator which reflects points over the line $y = \frac{1}{2}x + 20$.

$$\begin{pmatrix} x' \\ y' \\ 1 \end{pmatrix} = \begin{pmatrix} 0.6 & 0.8 & -16 \\ 0.8 & -0.6 & 32 \\ 0 & 0 & 1 \end{pmatrix} \cdot \begin{pmatrix} x \\ y \\ 1 \end{pmatrix}$$

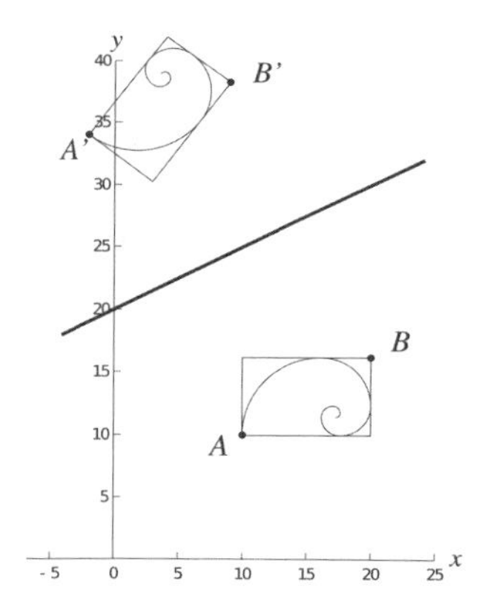

Figure 11.14: Reflection over the line $y = \frac{1}{2}x + 20$

We calculate both image vertices of the reflected golden rectangle as

$$\begin{pmatrix} 0.6 & 0.8 & -16 \\ 0.8 & -0.6 & 32 \\ 0 & 0 & 1 \end{pmatrix} \cdot \begin{pmatrix} 10 \\ 10 \\ 1 \end{pmatrix} = \begin{pmatrix} -2 \\ 34 \\ 1 \end{pmatrix},$$

$$\begin{pmatrix} 0.6 & 0.8 & -16 \\ 0.8 & -0.6 & 32 \\ 0 & 0 & 1 \end{pmatrix} \cdot \begin{pmatrix} 20 \\ 16.18 \\ 1 \end{pmatrix} = \begin{pmatrix} 8.9 \\ 38.29 \\ 1 \end{pmatrix}.$$

3D COMBINED ROTATION

Finally we aim for the composed matrix operator to rotate 3D objects, in this order by: a roll over 30°, a pitch over 180° and a yaw over 90°. We apply this composed transformation to rotate the triangle ABC with vertices $A(220,0,-30), B(0,50,-150)$ and $C(40,20,-100)$.

We need to compose three basic 3D rotations in their given order.

1) the basic rotation

$$R_{z,30^\circ} = \begin{pmatrix} \cos 30^\circ & -\sin 30^\circ & 0 & 0 \\ \sin 30^\circ & \cos 30^\circ & 0 & 0 \\ 0 & 0 & 1 & 0 \\ 0 & 0 & 0 & 1 \end{pmatrix} = \begin{pmatrix} \frac{\sqrt{3}}{2} & -\frac{1}{2} & 0 & 0 \\ \frac{1}{2} & \frac{\sqrt{3}}{2} & 0 & 0 \\ 0 & 0 & 1 & 0 \\ 0 & 0 & 0 & 1 \end{pmatrix}$$

2) the basic rotation

$$R_{x,180^\circ} = \begin{pmatrix} 1 & 0 & 0 & 0 \\ 0 & \cos 180^\circ & -\sin 180^\circ & 0 \\ 0 & \sin 180^\circ & \cos 180^\circ & 0 \\ 0 & 0 & 0 & 1 \end{pmatrix} = \begin{pmatrix} 1 & 0 & 0 & 0 \\ 0 & -1 & 0 & 0 \\ 0 & 0 & -1 & 0 \\ 0 & 0 & 0 & 1 \end{pmatrix}$$

3) the basic rotation

$$R_{y,90^\circ} = \begin{pmatrix} \cos 90^\circ & 0 & \sin 90^\circ & 0 \\ 0 & 1 & 0 & 0 \\ -\sin 90^\circ & 0 & \cos 90^\circ & 0 \\ 0 & 0 & 0 & 1 \end{pmatrix} = \begin{pmatrix} 0 & 0 & 1 & 0 \\ 0 & 1 & 0 & 0 \\ -1 & 0 & 0 & 0 \\ 0 & 0 & 0 & 1 \end{pmatrix}$$

Composing these basic rotations yields the combined rotation matrix operator.

$$\begin{pmatrix} x' \\ y' \\ z' \\ 1 \end{pmatrix} = \begin{pmatrix} 0 & 0 & 1 & 0 \\ 0 & 1 & 0 & 0 \\ -1 & 0 & 0 & 0 \\ 0 & 0 & 0 & 1 \end{pmatrix} \cdot \begin{pmatrix} 1 & 0 & 0 & 0 \\ 0 & -1 & 0 & 0 \\ 0 & 0 & -1 & 0 \\ 0 & 0 & 0 & 1 \end{pmatrix} \cdot \begin{pmatrix} \frac{\sqrt{3}}{2} & -\frac{1}{2} & 0 & 0 \\ \frac{1}{2} & \frac{\sqrt{3}}{2} & 0 & 0 \\ 0 & 0 & 1 & 0 \\ 0 & 0 & 0 & 1 \end{pmatrix} \cdot \begin{pmatrix} x \\ y \\ z \\ 1 \end{pmatrix}$$

$$= \begin{pmatrix} 0 & 0 & -1 & 0 \\ \frac{-1}{2} & \frac{-\sqrt{3}}{2} & 0 & 0 \\ \frac{-\sqrt{3}}{2} & \frac{1}{2} & 0 & 0 \\ 0 & 0 & 0 & 1 \end{pmatrix} \cdot \begin{pmatrix} x \\ y \\ z \\ 1 \end{pmatrix}$$

Rotating the triangle ABC with the vertices $A(220,0,-30), B(0,50,-150)$ and $C(40,20,-100)$ returns their corresponding image vertices as $A'(30,\ -100,\ 173.2)$, $B'(150,\ -43.3,\ 25)$ and $C'(100,\ -37.32,\ -24.64)$.

11.7 Conventions

Depending on the context we express points P either as column matrices or as row matrices. For instance in the context of this book we typeset dot products $A \cdot P$ of the operator matrix A and column matrices. The same convention is used in OpenGL, whereas DirectX® adopts row matrices to define points.

Transposing the dot product via

$$(A \cdot P)^T = P^T \cdot A^T,$$

implies we need to transpose also the operator matrix to A^T in the latter case (conventionally using row matrices) and to reverse the product order.

All this leads for general two-dimensional matrix operators to

$$\left(\begin{pmatrix} a_{11} & a_{12} & \Delta x \\ a_{21} & a_{22} & \Delta y \\ 0 & 0 & 1 \end{pmatrix} \cdot \begin{pmatrix} x \\ y \\ 1 \end{pmatrix} \right)^T = \begin{pmatrix} x \\ y \\ 1 \end{pmatrix}^T \cdot \begin{pmatrix} a_{11} & a_{12} & \Delta x \\ a_{21} & a_{22} & \Delta y \\ 0 & 0 & 1 \end{pmatrix}^T$$

$$= \begin{pmatrix} x & y & 1 \end{pmatrix} \cdot \begin{pmatrix} a_{11} & a_{21} & 0 \\ a_{12} & a_{22} & 0 \\ \Delta x & \Delta y & 1 \end{pmatrix}.$$

11.8 Exercises

Exercise 96 Determine the matrix operator to double the size of 3D objects along the x-axis and to halve their size along the z-axis. Scale the triangle ABC spanned by the vertices $A(0,30,-100), B(-50,100,-20)$ and $C(-20,0,-300)$ by this operator.

Exercise 97 Draw the flat triangle ABC defined by the vertices $A(0,0), B(2,1)$ and $C(0,1)$. Rotate this triangle basically and counter clockwise over an angle of $45°$ and calculate all image vertices A', B' and C'. Draw this image triangle $A'B'C'$ in the same cartesian frame.

Exercise 98 Determine the 2D transformation matrix to rotate objects counter clockwise around center $(3,1)$ over an angle of $\frac{\pi}{4}$ radians. Apply this operator to rotate the triangle ABC defined by the vertices $A(6,1)$, $B(8,2)$ and $C(7,3)$.

Exercise 99 Prove all four properties of the basic shearing as outlined on page 207.

Exercise 100 Consider the 3D translation matrix given by $\Delta x = -1$, $\Delta y = -1$ and $\Delta z = -1$ for a start. Let this translation be followed by the 3D rotation around the x-axis over $+30°$. Another 3D rotation, around the y-axis over $+45°$, finalizes the transformation. Compose all steps to transform the tetrahedron defined by the vertices $(0,0,0)$, $(1,0,0)$, $(0,1,0)$ and $(0,0,1)$.

Exercise 101

- ▷ Given the triangle ABC defined by the vertices $A(-1,5), B(9,0)$ and $C(-5,10)$. Find its centroid Z as the intersection point of its medians.
- ▷ Find the transformation operator to rotate this triangle clockwise around Z over the right angle of $90°$.
- ▷ Calculate all returned image vertices A', B', C' and draw both triangles, ABC and its image $A'B'C'$, in a different colour.

Exercise 102 We want to resize the cube defined by the vertices $A(2,2,1)$, $B(5,1,2)$, $C(5,1,-1)$, $D(2,2,-1)$, $E(2,5,1)$, $F(5,4,2)$, $G(5,1,4)$ and $H(2,5,4)$ into a cuboid by applying scale factor 2 along the x-axis, factor 4 along the y-axis and factor 3 along the z-axis, with respect to its corner vertex A. Determine the appropriate matrix operator and use it to calculate all image vertices of the returned cuboid.

Exercise 103 Given the hexagon defined by the vertices $A(2,2)$, $B(4,2)$, $C(5,3)$, $D(4,4)$, $E(2,4)$ and $F(1,3)$, determine the transformation matrix to rotate it clockwise around its center Z over an angle of $270°$. Calculate all returned image vertices and draw both the original and the image hexagon in a different colour.

Hint: the center Z of the hexagon is the midpoint of the line segment $[AD]$.

Exercise 104 Reflect the trapezium $ABCD$ defined by the vertices $A(4,1)$, $B(5,2)$, $C(6,4)$ and $D(\frac{9}{2},3)$ over the line described by the parametric equation

$$\begin{cases} x = 2\lambda \\ y = -1 + 4\lambda . \end{cases}$$

Determine the basic matrix operators required at each step and in the correct order, to dot product them to the composed matrix operator. Calculate all returned image vertices and draw both the original and the image trapezium in a different colour.

Exercise 105 Apply row reduction (see page 180) to calculate the 3D inverse of the

- ▷ translation matrix $T^{-1}_{\overrightarrow{AB}}$,
- ▷ basic scale operator S^{-1}_{O},
- ▷ basic rotation operator $R^{-1}_{O,\theta}$.

Exercise 106

1) Outline the required steps in their correct order to reflect over the line described by $y = x$.
2) Dot product the above matrices to calculate the composed matrix operator which reflects over the line described by $y = x$.
3) Apply this composed matrix operator to reflect the point $P(-3,1)$ over the line $y = x$.
4) Draw the reflection line, the original point P and its returned image point P' in the same cartesian plane.

Exercise 107

1) Outline the required steps in their correct order to reflect over the line $y = \frac{1}{\sqrt{3}}x$.
2) Dot product the above matrices to calculate the composed matrix operator which reflects over the line $y = \frac{1}{\sqrt{3}}x$.
3) Apply this composed matrix operator to reflect the circle centered on $M(3,4)$ and containing the point $P(4,5)$.

Chapter 12 · Hypercomplex numbers

Rotator matrices are able to perform any desired rotation in 3D space but may suffer numerical instabilities. As an alternative to matrices, we can realize 3D rotations using quaternions. This chapter initiates us stepwise in the fascinating world of these extended complex numbers.

12.1 Complex numbers

Complex numbers have been discovered already a few centuries ago. Their renaming from 'impossible' over 'imaginary' to 'complex' numbers, illustrates to what extent they gave their users an uneasy feeling. Complex numbers were accidentally discovered in the 16th century as a side-effect of solving cubic equations. Sometimes square roots of negative numbers appeared, which could be used temporarily in an algebraic way. By the end of the 18th century the Swiss mathematician **Leonard Euler** (1707-1783) concluded the quadratic equation $x^2 = -1$ is impossible to solve for real numbers. Hence, 'impossible' numbers were born.

All natural, integer, rational and real numbers are represented on the **number line**. In other words, each number corresponds to a point on this straight line and vice versa. This allows us to interpret number addition and multiplication as linear transformations. Addition of numbers corresponds to translation of points (see paragraph 11.1). For instance, adding 3 to a number corresponds to a shift of 3 to the right on the number line. Similarly, subtracting 5 corresponds to a translation of 5 to the left. Multiplication of a number with a positive factor corresponds to a basic scaling of a point (see paragraph 11.2). For instance the multiplication with factor 2 of the point labeled 1 on the line returns the scaled point to the right labeled 2. Multiplication of a number with a negative factor corresponds to the basic scaling composed with the basic reflection over the origin (see paragraph 11.4). For instance the multiplication with factor -2 of the point labeled 1 on the line returns the point to the left labeled -2, as the image of the basic scaling S_O with the scale factor $s_x = 2$ composed with the basic reflection M_O. Repeating this transformation for a second time, returns the image point labeled 4. We confirm both successive multiplications arithmetically as $(-2) \cdot (-2) = +4$. Multiplication of a number with factor -1 corresponds to the basic reflection M_O, reflecting points labeled x, over the origin to their image point $-x$. Repeating this transformation for a second time, returns the initial point labeled x as image. We confirm both successive multiplications arithmetically as $(-1) \cdot (-1) = +1$ or $(-1)^2 = +1$.

We realize there is no transformation on the number line which returns -1 after two successive operations on $+1$. In other words, there is no real number that squares to -1.

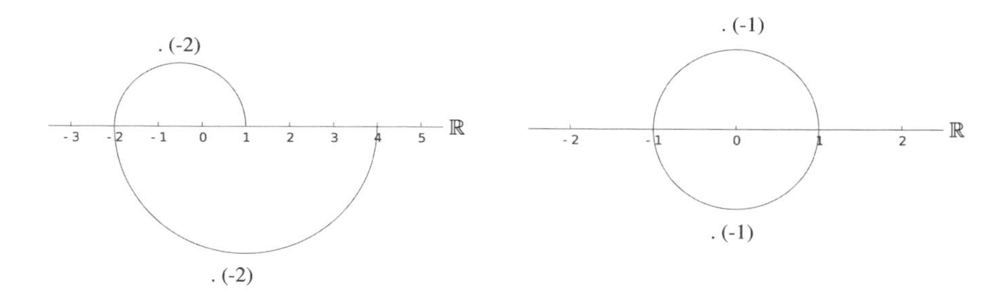

Figure 12.1: The real number line

As a consequence, it is impossible to take the square root of the number -1 in $\mathbb{R}$.

Seeking for an 'impossible number' i which squares to -1, corresponds to seeking for a transformation which returns -1 after two successive operations on the point labeled $+1$. The basic reflection M_O equals the basic rotation around the origin over the straight angle $R_{O,180^\circ}$ (see paragraph 11.3). Two successive basic rotations over the right angle $R_{O,90^\circ}$ yield the reflection over the origin M_O:

$$R_{O,90^\circ} \cdot R_{O,90^\circ} = R_{O,180^\circ}.$$

Hence, the Swiss mathematical politician **Jean-Robert Argand** (1768–1822) interpreted the 'impossible number' i as a basic rotation over 90°.

As a consequence, its image point after one operation on the point labeled $+1$ lies no longer on the (horizontal) number line. Applying the basic rotation $R_{O,90^\circ}$ on the point labeled $+1$ yields its image point labeled i on the vertical line, perpendicular to the real number line. The horizontal 'real' number line inspired Argand to name his vertical number line, containing the unity i, the 'imaginary' number line. We define i as the imaginary unity via

$$i^2 = -1. \tag{12.1}$$

We define a **complex number** z as an expression

$$z = a + bi,$$

given a and b real numbers, for instance $2+3i$. We call this the **algebraic representation** of z. We call $a = \text{Re } z$ the **real** part and $b = \text{Im } z$ the **imaginary part** of the number z. For instance, in $2+3i$ the real part is 2 and the imaginary part is 3. We typeset the set of all complex numbers as $\mathbb{C}$.

The former definition $z = a + bi$ is based on two real numbers a and b. This allows us to interpret a plane point labeled (a,b) as a complex number z. And the other way around, to each complex number z corresponds geometrically a unique point (a,b) in the plane. We call this plane the complex **number plane** or **Gauss plane**, named after the famous German mathematician.

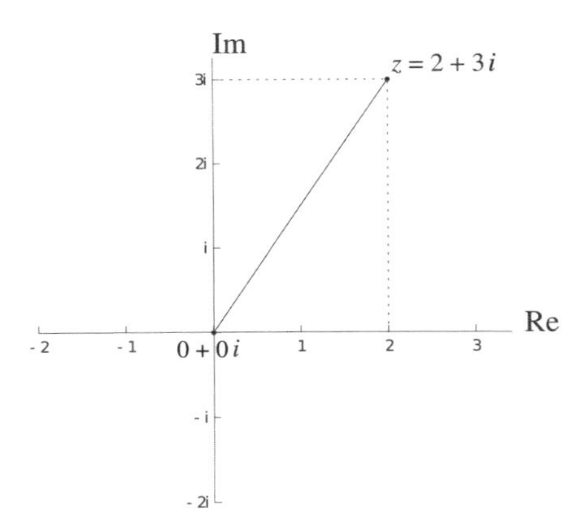

Figure 12.2: The complex number plane

Based on the correspondence of complex numbers to plane points, we can represent them alternatively by polar coordinates (see paragraph 6.3).

We define the **modulus** $|z|$ of a complex number $z = a + bi$ as the distance from the origin to the point labeled (a,b) in the complex plane. The Pythagorean theorem therefore yields

$$|z| = \sqrt{a^2 + b^2}.$$

We define the **argument** $\arg z$ of a complex number as the counterclockwise angle θ, made by the segment from the origin to the point labeled (a,b) and the positive real number line. Since θ corresponds to the polar angle, we determine it as

$$\arg z = \theta = \arctan\left(\frac{b}{a}\right) + k\pi \text{ given } k \in \{0,1\}.$$

We realize that the complex number $z = 0 + 0i$ lacks a unique trigonometrical representation.

Analogously to polar coordinates, we determine the real part a and the imaginary part b, given the known values of the modulus $|z|$ and the argument θ, straightforwardly as

$$a = Re(z) = |z| \cos\theta$$
$$b = Im(z) = |z| \sin\theta.$$

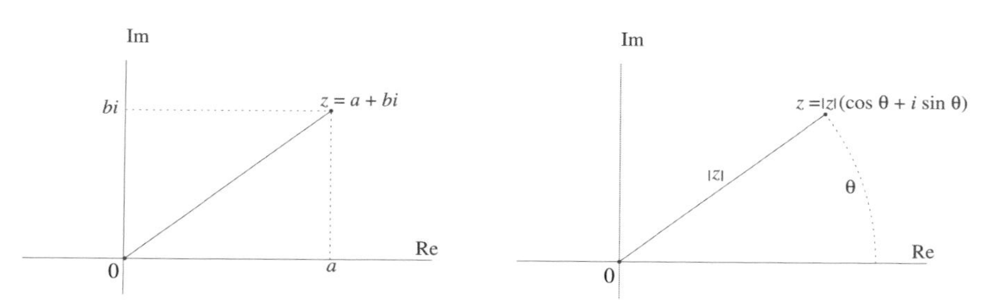

Figure 12.3: Algebraic versus trigonometric representation

Substituting the above parts in $z = a + bi$ yields $z = |z|\cos\theta + i|z|\sin\theta$, leading to the **trigonometrical representation** of a complex number as

$$z = |z|(\cos\theta + i\sin\theta).$$

Example: We represent the complex number $z = 2 + 3i$ trigonometrically. We calculate its modulus as $|z| = \sqrt{2^2 + 3^2} = \sqrt{13}$. We retrieve its argument as $\arg z = \arctan\left(\frac{3}{2}\right) = 0.98\,\text{rad} = 56.31°$. We summarize the trigonometrical representation of this complex number as

$$z = \sqrt{13}(\cos 56.31° + i\sin 56.31°).$$

12.2 Complex number arithmetics

Complex conjugate

We define the **complex conjugate** number of $z = a + bi$ as $z^* = a - bi$. For instance, the complex conjugate number of $2 + 3i$ is $(2 + 3i)^* = 2 - 3i$. Consequently, successively complex conjugating the complex conjugate z^* yields $(z^*)^* = z$ again.

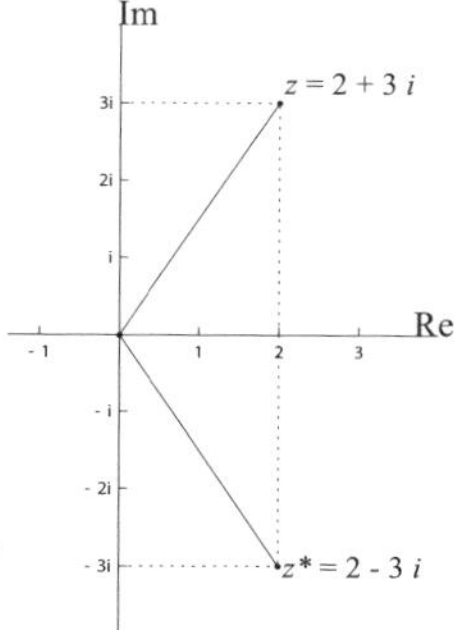

Figure 12.4: Complex conjugate number

Addition and subtraction

Each complex number corresponds geometrically to a point in the plane. We recall that each point in the plane can be interpreted as the head of a location vector. Addition of complex numbers hence corresponds to the addition of vectors (see paragraph 7.2).

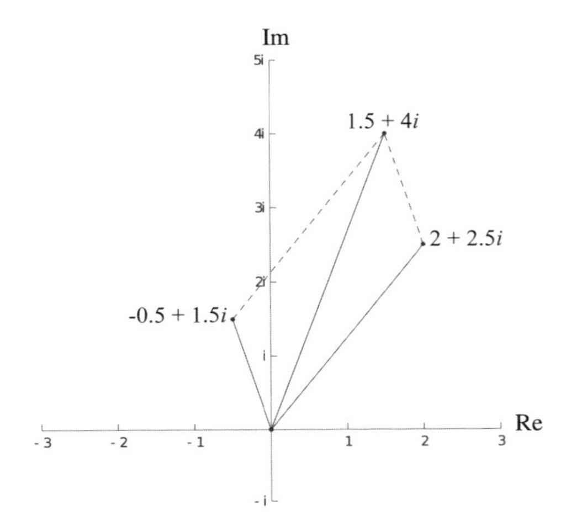

Figure 12.5: Addition of complex numbers

Example

$$(2+2.5i)+(-0.5+1.5i) = (2+(-0.5))+(2.5+1.5)i = 1.5+4i$$

learns us that we have to add the real parts and the imaginary parts, similarly to the component-wise addition of vectors. The complex addition summarizes as

$$(a+bi)+(c+di) = (a+c)+(b+d)i. \tag{12.2}$$

Consequently, the **opposite** of a complex number $z = a+bi$ is $-z = -a-bi$ because $(a+bi)+(-a-bi) = 0+0i$. We define the complex subtraction as the addition with the opposite number:

$$(a+bi)-(c+di) = (a+bi)+(-c-di) = (a-c)+(b-d)i. \tag{12.3}$$

MULTIPLICATION

We explore the complex multiplication, applying the real arithmetics combined with the extra rule $i^2 = -1$. Applying the distributive property, leads to

$$\begin{aligned}(a+bi)\cdot(c+di) &= a(c+di)+bi(c+di)\\ &= ac+adi+bci+bidi\\ &= ac+(ad+bc)i+bd\,i^2\\ &= ac+(ad+bc)i-bd\\ &= (ac-bd)+(ad+bc)i. \end{aligned} \tag{12.4}$$

Example:
$(2+3i)\cdot(-1+2i) = 2(-1+2i)+3i(-1+2i) = -2+4i-3i+6\,i^2 = -8+i$

Note that multiplying a complex number with its conjugate number returns a real number.

$$\begin{aligned} z\cdot z^* &= (a+bi)\cdot(a-bi) = a^2-abi+abi-b^2\,i^2\\ &= a^2+b^2 \end{aligned}$$

The complex multiplication can be performed more efficiently in the trigonometrical representation. Given two complex numbers $z = |z|(\cos\alpha + i\sin\alpha)$ and $w = |w|(\cos\beta + i\sin\beta)$, we simplify the complex multiplication, applying the trigonometrical Sum Identities (see paragraph 3.7).

$$\begin{aligned} z\cdot w &= |z|(\cos\alpha+i\sin\alpha)|w|(\cos\beta+i\sin\beta)\\ &= |z||w|(\cos\alpha+i\sin\alpha)(\cos\beta+i\sin\beta)\\ &= |z||w|(\cos\alpha\cos\beta+i\cos\alpha\sin\beta+i\sin\alpha\cos\beta+i^2\sin\alpha\sin\beta)\\ &= |z||w|\left((\cos\alpha\cos\beta-\sin\alpha\sin\beta)+i(\cos\alpha\sin\beta+\sin\alpha\cos\beta)\right)\\ &= \underbrace{|z||w|}_{\text{product}}(\cos\underbrace{(\alpha+\beta)}_{\text{sum}}+i\sin\underbrace{(\alpha+\beta)}_{\text{sum}}) \end{aligned}$$

The complex multiplication in trigonometric representation summarizes as:

- ▷ the modulus of a complex product $z\cdot w$ equals the product of the moduli of z and w,

$$|z\cdot w| = |z|\cdot|w|$$

- ▷ the argument of a complex product $z\cdot w$ equals the sum of the arguments of z and w.

$$\arg(z\cdot w) = \arg z + \arg w$$

Example: We recalculate the previous complex product of $z = 2 + 3i$ and $w = -1 + 2i$. Firstly, we convert both complex numbers z and w into their trigonometrical representation. We determine their moduli as $|z| = \sqrt{13}$ and $|w| = \sqrt{5}$. We retrieve their arguments as $\arg z = \arctan\left(\frac{3}{2}\right)$ and $\arg w = \arctan\left(\frac{2}{-1}\right) + \pi$.

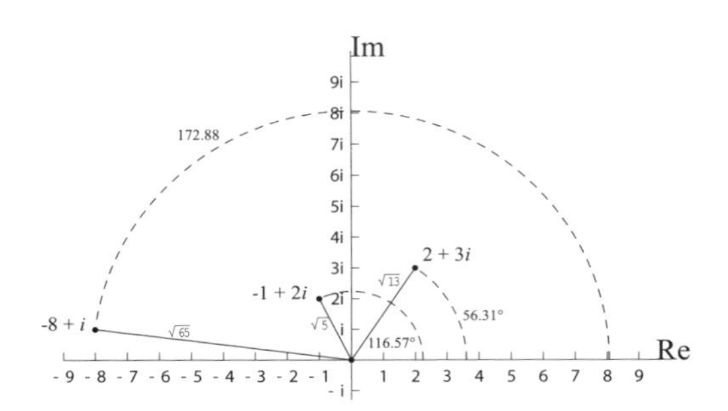

The modulus of the product $z \cdot w$ equals

$$|z \cdot w| = |z| \cdot |w| = \sqrt{13} \cdot \sqrt{5} = \sqrt{65}.$$

The argument of the product $z \cdot w$ equals

$$\arg(z \cdot w) = \arg z + \arg w = \arctan\left(\frac{3}{2}\right) + \arctan\left(\frac{2}{-1}\right) + \pi \approx 172.88^\circ.$$

We approximate the complex product of z and w as

$$z \cdot w \approx \sqrt{65}(\cos 172.88^\circ + i \sin 172.88^\circ) \approx \sqrt{65}(-0.99 + 0.12i) \approx -8 + i.$$

We extend the complex multiplication logically to complex exponentiation by successively multiplying a complex number z by itself. The modulus of the exponentiation calculates as $|z^n| = |z|^n$. The argument of the exponentiation simplifies as $\arg(z^n) = n \arg z$. The complex exponentiation of a number $z = |z|(\cos\alpha + i \sin\alpha)$ in trigonometric representation simplifies to

$$z^n = |z|^n (\cos n\alpha + i \sin n\alpha). \tag{12.5}$$

Example: We calculate z^4, given $z = 1 + i\sqrt{3}$. The modulus of z equals 2 and its argument equals $\frac{\pi}{3}$. We summarize z trigonometrically as $z = 2\left(\cos\frac{\pi}{3} + i \sin\frac{\pi}{3}\right)$. The complex exponentiation of z to the power 4, exponentiates its modulus to 2^4 and multiplies its argument to $4\frac{\pi}{3}$. We summarize this exponentiation trigonometrically as

$$z^4 = 16\left(\cos\frac{4\pi}{3} + i \sin\frac{4\pi}{3}\right).$$

Again representing this result algebraically via $\cos\frac{4\pi}{3} = \frac{-1}{2}$ and $\sin\frac{4\pi}{3} = \frac{-\sqrt{3}}{2}$, results finally into $z^4 = -8 - 8i\sqrt{3}$.

DIVISION

At a first glance, the complex division looks troublesome. For instance dividing 1 by i results in the fraction $\frac{1}{i}$, which does not fit the definition $a+bi$ of complex numbers. This looks similar to for instance dividing the integers -3 by 5 which results into the fraction $\frac{-3}{5}$, being no longer an integer. But at a second glance, we can simplify a complex fraction using the complex conjugate of its denominator. To simplify complex fractions we need to multiply both their numerator and denominator with the conjugate of the denominator. As a try-out, our above example simplifies to

$$\frac{1}{i} = \frac{1}{i} \cdot \left(\frac{-i}{-i}\right) = \frac{-i}{-i^2} = \frac{-i}{1} = -i.$$

Example: To simplify the complex fraction $\frac{1}{2+2.5i}$ we need to multiply both its numerator and denominator with the conjugate of the denominator $(2+2.5i)^* = 2-2.5i$.

$$\begin{aligned} \frac{1}{2+2.5i} &= \left(\frac{1}{2+2.5i}\right) \cdot \left(\frac{2-2.5i}{2-2.5i}\right) = \frac{2-2.5\,i}{2^2+2.5^2} \\ &= \frac{2-2.5i}{10.25} = \frac{2}{10.25} - \frac{2.5}{10.25}i \\ &\approx 0.20 - 0.24i. \end{aligned}$$

Generalizing the above simplification method proves it can never fail:

$$\begin{aligned} \frac{a+bi}{c+di} &= \left(\frac{a+bi}{c+di}\right) \cdot \left(\frac{c-di}{c-di}\right) = \frac{(a+bi)(c-di)}{c^2+d^2} \\ &= \frac{ac+bd}{c^2+d^2} + \left(\frac{-ad+bc}{c^2+d^2}\right) i. \end{aligned} \tag{12.6}$$

We advise to forget this algebraic formula, and instead to remember its procedure practically: 'multiply both numerator and denominator of complex fractions with the conjugate of its denominator'.

Similarly to the complex multiplication, the division can be performed more efficiently in the trigonometrical representation. Dividing the complex number $z = |z|(\cos\alpha + i\sin\alpha)$ by $w = |w|(\cos\beta + i\sin\beta)$, we simplify the complex division, applying the trigonometrical Sum Identities (see paragraph 3.7).

$$\begin{aligned}
\frac{z}{w} &= \frac{|z|(\cos\alpha + i\sin\alpha)}{|w|(\cos\beta + i\sin\beta)} \\
&= \frac{|z|}{|w|}\,\frac{(\cos\alpha + i\sin\alpha)}{(\cos\beta + i\sin\beta)}\,\frac{(\cos\beta + i\sin\beta)^*}{(\cos\beta + i\sin\beta)^*} \\
&= \frac{|z|}{|w|}\,\frac{(\cos\alpha\cos\beta - i\cos\alpha\sin\beta + i\sin\alpha\cos\beta + \sin\alpha\sin\beta)}{\cos^2\beta + \sin^2\beta} \\
&= \underbrace{\frac{|z|}{|w|}}_{\text{division}}\ (\cos\ \underbrace{(\alpha - \beta)}_{\text{subtraction}}\ + i\sin\ \underbrace{(\alpha - \beta)}_{\text{subtraction}}\ )
\end{aligned}$$

The complex division in trigonometric representation summarizes as:

- ▷ the modulus of a complex quotient $\frac{z}{w}$ equals the quotient of the moduli of z and w.

$$\left|\frac{z}{w}\right| = \frac{|z|}{|w|}$$

- ▷ the complex division subtracts the argument of w from the argument of z.

$$\arg\left(\frac{z}{w}\right) = \arg z - \arg w$$

Example: We divide the complex number $z = 2 + 3i$ by $w = -1 + 2i$. Firstly, we convert both complex numbers z and w into their trigonometrical representation. We determine their moduli as $|z| = \sqrt{13}$ and $|w| = \sqrt{5}$. We retrieve their arguments as $\arg z = \arctan\left(\frac{3}{2}\right)$ and $\arg w = \arctan\left(\frac{2}{-1}\right) + \pi$.

The modulus of the quotient $\frac{z}{w}$ equals

$$\left|\frac{z}{w}\right| = \frac{|z|}{|w|} = \frac{\sqrt{13}}{\sqrt{5}} = \sqrt{\frac{13}{5}}.$$

The argument of the quotient $\frac{z}{w}$ equals

$$\arg\left(\frac{z}{w}\right) = \arg z - \arg w = \arctan\left(\frac{3}{2}\right) - \left(\arctan\left(\frac{2}{-1}\right) + \pi\right) \approx 299.74^\circ.$$

We approximate the complex quotient of z and w as

$$\frac{z}{w} \approx \sqrt{\frac{13}{5}}(\cos 299.74^\circ + i\sin 299.74^\circ).$$

12.3 Complex numbers and transformations

We recall that multiplication with the factor i geometrically corresponds to a basic rotation over 90°. This allows us to realize transformations by the use of complex numbers.

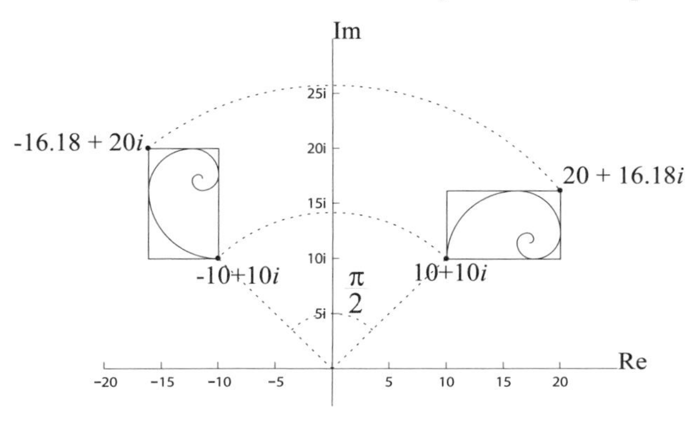

Figure 12.6: Rotation by the use of complex numbers

Figure 12.6 recalls the effect of a basic 2D rotation over 90° on the golden rectangle (see page 201). To rotate this golden rectangle by the use of complex numbers requires the multiplication of its vertices in the complex plane by the factor i. The argument of the imaginary unity i is $\frac{\pi}{2}$ and its modulus is 1. Due to the trigonometrical representation of the complex multiplication, the modulus of the vertices remains invariant and $\frac{\pi}{2}$ is added to their arguments:

$$|z \cdot i| = |z| \cdot |i| = |z| \cdot 1 = |z| \quad \text{and } \arg(z \cdot i) = \arg(z) + \arg(i) = \arg(z) + \frac{\pi}{2}.$$

We calculate the image points of the golden rectangle's defining vertices as:

$$(10 + 10i)i = -10 + 10i \text{ and } (20 + 16.18i)i = -16.18 + 20i.$$

We realize that the multiplication with a factor featuring modulus 1 and argument θ, geometrically corresponds to a basic rotation over θ. For instance the multiplication with $\frac{1}{2} + i\frac{\sqrt{3}}{2}$ corresponds to the basic rotation over 60°. The modulus of $\frac{1}{2} + i\frac{\sqrt{3}}{2}$ is 1 and its argument is 60°.

Multiplying each point of an object with a positive real factor, rescales our object according to the factor's value. Positive scale factors smaller than 1 shrink the original object and positive scale factors larger than 1 enlarge the original object. For instance, the complex multiplication by the factor $\frac{1}{2}$ geometrically corresponds to a basic scaling that halves our golden rectangle uniformly, featuring scale factors $s_x = s_y = \frac{1}{2}$.

Multiplication with an arbitrary complex number $z = |z|(\cos\theta + i\sin\theta)$ corresponds to a transformation composed of a basic scaling by the modulus $|z| > 0$ and a basic rotation

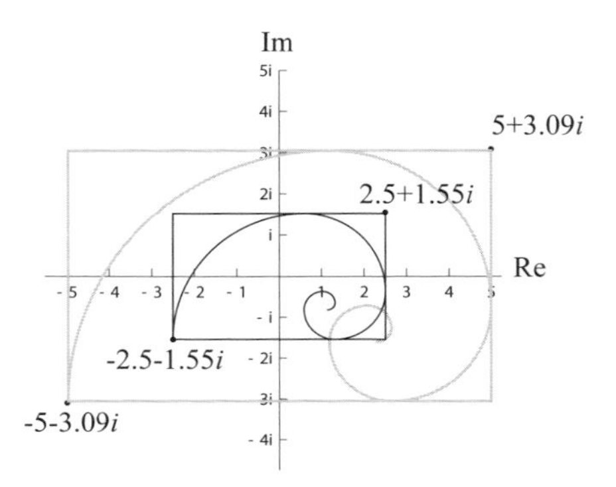

Figure 12.7: Scaling by the use of complex numbers

by the argument θ. For instance, the complex multiplication by the factor $\frac{i}{2}$ basically rescales the object to half its size and rotates it over the right angle around the origin. We reveal this effect by representing $\frac{i}{2}$ trigonometrically:

$$\frac{i}{2} = 0 + \frac{1}{2}i = \frac{1}{2}(0+i) = \frac{1}{2}(\cos 90^\circ + i \sin 90^\circ).$$

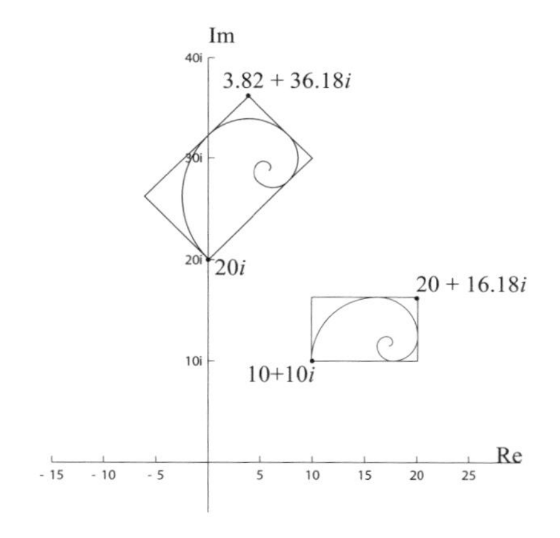

Figure 12.8: Composed transformations by the use of complex numbers

As a final example, multiplication with the complex number $z = 1 + i$ corresponds to a transformation composed of an enlargement by the scale factor $|z| = \sqrt{2}$ and a basic rotation by the argument $\theta = 45^\circ$. We apply this composed transformation to the golden rectangle by multiplying its vertices $10 + 10i$ and $20 + 16.18i$ by the factor $1 + i$.

$$(10 + 10i) \cdot (1 + i) = 20i \text{ and } (20 + 16.18i) \cdot (1 + i) = 3.82 + 36.18i$$

12.4 Complex continuation of the Fibonacci numbers

Based upon the complex plane, we extend Binet's formula to generate Fibonacci numbers (see formula (10.4)) beyond its conventional range of Fibonacci numbers in $\mathbb{N}$. We achieve this in two steps by interpreting Binet's formula as a function $f : k \mapsto f(k)$ dictated by its recipe

$$f(k) = \frac{(\Phi)^k - (\Phi')^k}{\sqrt{5}}.$$

In the first step we evaluate this function f as a mapping from the domain $\mathbb{Z}$ to the range $\mathbb{Z}$. In the second step we extend this function as a mapping from the domain $\mathbb{Q}$ to the range $\mathbb{C}$.

Integer Fibonacci numbers

We firstly consider the integer function $f : \mathbb{Z} \to \mathbb{Z} : k \mapsto f(k)$, dictated by the above recipe. Evaluating its first negative argument $k = -1$, returns the Fibonacci number $f(-1)$, which we calculate as

$$\begin{aligned} f(-1) &= \frac{1}{\sqrt{5}} \left(\left(\frac{1+\sqrt{5}}{2} \right)^{-1} - \left(\frac{1-\sqrt{5}}{2} \right)^{-1} \right) \\ &= \frac{1}{\sqrt{5}} \left(\frac{2}{1+\sqrt{5}} - \frac{2}{1-\sqrt{5}} \right) \\ &= \frac{1}{\sqrt{5}} \left(\frac{2(1-\sqrt{5}) - 2(1+\sqrt{5})}{(1+\sqrt{5})(1-\sqrt{5})} \right) \\ &= \frac{1}{\sqrt{5}} \left(\frac{-4\sqrt{5}}{-4} \right) = 1. \end{aligned}$$

Proceeding with its return values $f(-2) = \frac{1}{\sqrt{5}} \cdot \left(\left(\frac{1+\sqrt{5}}{2} \right)^{-2} - \left(\frac{1-\sqrt{5}}{2} \right)^{-2} \right) = -1$ and $f(-3) = 2$ and so on, we extrapolate the Fibonacci sequence below zero. We note that this extended function f remains being governed by the defining property of a Lucas sequence: $f(n+1) = f(n) + f(n-1)$. This allows us for each decremented argument $n-1$ to calculate f_{n-1} faster via $f_{n-1} = f_{n+1} - f_n$. Therefore, the **countable** continuation of the Fibonacci sequence to $\mathbb{Z}$ equals $\ldots, -8, 5, -3, 2, -1, 1, 0, 1, 1, 2, 3, 5, 8, 13, \ldots$.

Complex Fibonacci numbers

We now extend this function to rational arguments $f:\mathbb{Q}\to\mathbb{C}:k\mapsto f(k)$ by applying the complex exponentiation (see formula (12.5)) on the trigonometric representations of Φ and Φ'. Therefore, we can consequently calculate complex return values such as

$$\begin{aligned}
f\left(\frac{1}{2}\right) &= \frac{1}{\sqrt{5}}\left(\left(\frac{1+\sqrt{5}}{2}\right)^{\frac{1}{2}}-\left(\frac{1-\sqrt{5}}{2}\right)^{\frac{1}{2}}\right)\\
&= \frac{1}{\sqrt{5}}\left(\left(\left|\frac{1+\sqrt{5}}{2}\right|(\cos 0+i\sin 0)\right)^{\frac{1}{2}}-\left(\left|\frac{1-\sqrt{5}}{2}\right|(\cos\pi+i\sin\pi)\right)^{\frac{1}{2}}\right)\\
&= \frac{1}{\sqrt{5}}\left(\left(\frac{1+\sqrt{5}}{2}\right)^{\frac{1}{2}}\left(\cos\frac{1}{2}0+i\sin\frac{1}{2}0\right)-\left(\frac{-1+\sqrt{5}}{2}\right)^{\frac{1}{2}}\left(\cos\frac{1}{2}\pi+i\sin\frac{1}{2}\pi\right)\right)\\
&= \frac{1}{\sqrt{5}}\left(\sqrt{\frac{1+\sqrt{5}}{2}}(1+0i)-\sqrt{\frac{-1+\sqrt{5}}{2}}(0+1i)\right)\\
&= \sqrt{\frac{1}{10}(1+\sqrt{5})}-i\sqrt{\frac{1}{10}(-1+\sqrt{5})}.
\end{aligned}$$

Or we calculate return values such as $f(\frac{5}{2})=1.48931-0.134291i$, which also resides in the complex plane. This complex range of Fibonacci numbers remains countable, because the rational domain $\mathbb{Q}$ of the function f is countable.

We finally extend Binet's function to real arguments $f:\mathbb{R}\to\mathbb{C}:k\mapsto f(k)$, sustaining the above approach. In this case f returns an **uncountable** complex continuation of the Fibonacci sequence. Plotting for each real argument $k\in\mathbb{R}$ its return value $f(k)\in\mathbb{C}$, visualizes the range of the function f graphically in the complex plane.

The complex graph of this Fibonacci continuation illustrates perfectly the way f contains all integer Fibonacci numbers, governed by $f(n+1)=f(n)+f(n-1)$ as a Lucas sequence. We observe graphically that return value 1 occurs three times, which was indeed confirmed by $f(1)=f(2)=f(-1)=1$. The numbers $2,5,13,\dots$ on the odd positions in the ordinary Fibonacci sequence occur twice.

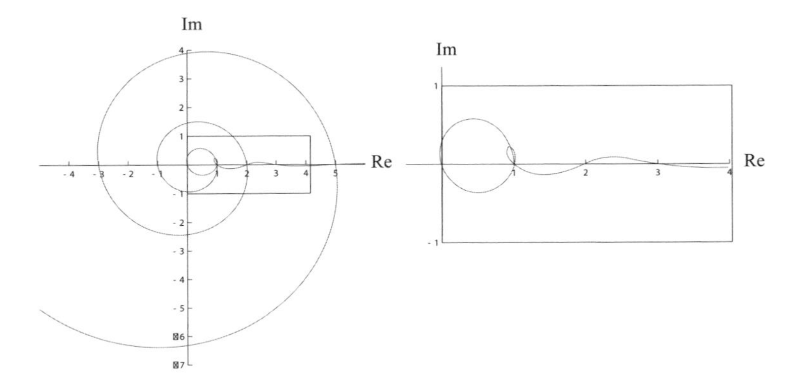

Figure 12.9: Complex continuation of the Fibonacci sequence

12.5 Quaternions

In the former paragraph we interpreted complex numbers as points in the 2D complex plane, which allowed us to multiply these points. During the 19th century, mathematicians have been seeking to multiply points similarly in a 3D complex space, concluding it required 4D.

It was the Irish mathematician **Wiliam Rowan Hamilton** (1805–1865) to discover these 4D complex numbers also known as **quaternions**. Initially Hamilton tried to extend complex arithmetics by adopting only one extra imaginary unity j. For years he investigated complex arithmetics using one real part and two imaginary parts, completely in vain. It was not until 1843 before Hamilton tried his idea of complex arithmetics based on one real part and three imaginary unities i, j and k. Meanwhile it is a famous narrative in mathematical history how Hamilton walked the Brougham Bridge in Dublin when this genius idea urged him to jag the basic rules for i, j and k in its stone. Today the Dubliners have a plaque on the Brougham Bridge, in memory of the quaternions and their inventor.

To express a quaternion takes, apart from the symbol i, two extra symbols j and k to typeset its three imaginary components. We define a **quaternion** as an expression $a+bi+cj+dk$, given a, b, c and d are real numbers and i, j and k are governed by Hamilton's basic rules

$$i^2 = j^2 = k^2 = ijk = -1. \tag{12.7}$$

The imaginary unities i, j and k are arithmetically interrelated by

$$ij = k, \quad jk = i, \quad ki = j, \quad ji = -k, \quad kj = -i, \text{ and } ik = -j. \tag{12.8}$$

For instance the numbers $7, 3+i, 1+j, 2-3i+5k$ and $9-i+2j+7k$ are quaternions. The quaternion set is a subset of the **hypercomplex number** set. We typeset the quaternion set as $\mathbb{H}$, in honour of Hamilton.

Note that $ij \neq ji, ik \neq ki$ and $jk \neq kj$ which implies the quaternion multiplication is non-commutative. Applying Hamilton's basic rules and the imaginary unities' interrelationships, allows for quaternion arithmetics.

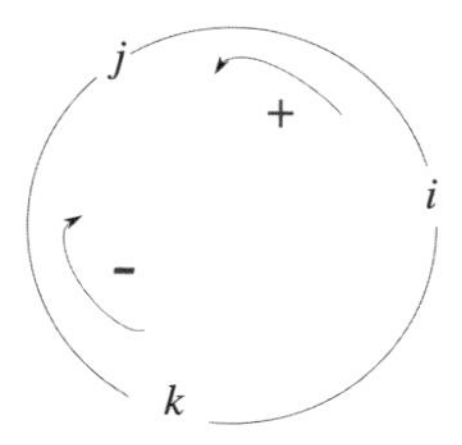

Figure 12.10: Mnemonic device for quaternion arithmetics

For instance multiplying the number $1+k$ with $1+i$ illustrates its non-commutativity:

$$(1+k)(1+i) = 1+i+k+ki = 1+i+k+j,$$
$$(1+i)(1+k) = 1+k+i+ik = 1+i+k-j.$$

Given their one real and three imaginary components, we can typeset quaternions component wise as vectors. We express a quaternion $q = q_0 + q_1 i + q_2 j + q_3 k$ as the 4D vector $[q_0, (q_1, q_2, q_3)] = [q_0, \vec{q}\,]$, separating its scalar component from its imaginary 3D vector $\vec{q} = (q_1, q_2, q_3)$. As an example the other way around, the 4D vector $[1, (1, 1, -1)]$ corresponds to the quaternion $1+i+j-k$.

Identically to vectors and complex numbers, we define the **norm** of a quaternion as

$$||q|| = \sqrt{q_0^2 + q_1^2 + q_2^2 + q_3^2}.$$

We define the **unit quaternions** as quaternions with a norm $||q||$ equal to 1. We define the **pure quaternions** as quaternions with the scalar component q_0 equal to 0. Consequently, we define the pure unit quaternions as quaternions $u = [0, (u_x, u_y, u_z)]$ featuring the real component $u_0 = 0$ and a norm $\sqrt{u_x^2 + u_y^2 + u_z^2} = 1$ and typeset them with the letter u referring to 'unit'.

12.6 Quaternion arithmetics

Applying Hamilton's basic rules and interrelationships to ordinary algebra on i, j and k, yields quaternion arithmetics.

Addition and subtraction

We add two quaternions $p = [p_0, (p_1, p_2, p_3)]$ and $q = [q_0, (q_1, q_2, q_3)]$ by adding their corresponding components, similarly to vector addition.

$$\begin{aligned} p + q &= [p_0 + q_0, (p_1 + q_1, p_2 + q_2, p_3 + q_3)] \\ &= (p_0 + q_0) + (p_1 + q_1)i + (p_2 + q_2)j + (p_3 + q_3)k \end{aligned}$$

Consequently, the opposite quaternion $-q$ of the quaternion $q = [q_0, (q_1, q_2, q_3)]$ equals $-q = [-q_0, (-q_1, -q_2, -q_3)]$. We define quaternion subtraction as addition with the opposite quaternion.

$$\begin{aligned} p - q = p + (-q) &= [p_0 - q_0, (p_1 - q_1, p_2 - q_2, p_3 - q_3)] \\ &= (p_0 - q_0) + (p_1 - q_1)i + (p_2 - q_2)j + (p_3 - q_3)k \end{aligned}$$

We base the scalar multiplication of quaternions on the repeated addition of the quaternion to itself. Extending this idea, we scalar multiply a quaternion $q = [q_0, (q_1, q_2, q_3)]$ with a number $\lambda \in \mathbb{R}$ by multiplying each component of q with λ, identically to scalar multiplication of vectors.

$$\lambda q = [\lambda q_0, (\lambda q_1, \lambda q_2, \lambda q_3)] = \lambda q_0 + (\lambda q_1)i + (\lambda q_2)j + (\lambda q_3)k$$

Multiplication

We multiply two quaternions $p = [p_0, (p_1, p_2, p_3)]$ and $q = [q_0, (q_1, q_2, q_3)]$ applying Hamilton's basic rules and interrelationships.

$$\begin{aligned} p \cdot q &= (p_0 + p_1 i + p_2 j + p_3 k) \cdot (q_0 + q_1 i + q_2 j + q_3 k) \\ &= p_0 q_0 + p_0 q_1 i + p_0 q_2 j + p_0 q_3 k + p_1 q_0 i + p_1 q_1 i^2 + p_1 q_2 ij + p_1 q_3 ik \\ &\quad + p_2 q_0 j + p_2 q_1 ji + p_2 q_2 j^2 + p_2 q_3 jk + p_3 q_0 k + p_3 q_1 ki + p_3 q_2 kj + p_3 q_3 k^2 \\ &= p_0 q_0 + p_0 q_1 i + p_0 q_2 j + p_0 q_3 k + p_1 q_0 i - p_1 q_1 + p_1 q_2 k - p_1 q_3 j \\ &\quad + p_2 q_0 j - p_2 q_1 k - p_2 q_2 + p_2 q_3 i + p_3 q_0 k + p_3 q_1 j - p_3 q_2 i - p_3 q_3 \\ &= (p_0 q_0 - p_1 q_1 - p_2 q_2 - p_3 q_3) + (p_0 q_1 + p_1 q_0 + p_2 q_3 - p_3 q_2)i \\ &\quad + (p_0 q_2 - p_1 q_3 + p_2 q_0 + p_3 q_1)j + (p_0 q_3 + p_1 q_2 - p_2 q_1 + p_3 q_0)k \end{aligned}$$

We recall the dot product of the imaginary vectors $\vec{p} = (p_1, p_2, p_3)$ and $\vec{q} = (q_1, q_2, q_3)$ as the number $\vec{p} \cdot \vec{q} = p_1 q_1 + p_2 q_2 + p_3 q_3 \in \mathbb{R}$. We recall the cross product of $\vec{p} = (p_1, p_2, p_3)$ and $\vec{q} = (q_1, q_2, q_3)$ as the vector $\vec{p} \times \vec{q} = (p_2 q_3 - p_3 q_2, p_3 q_1 - p_1 q_3, p_1 q_2 - p_2 q_1)$. Hence, we summarize the quaternion product as

$$p \cdot q = [p_0 q_0 - \vec{p} \cdot \vec{q},\ (p_0 \vec{q} + q_0 \vec{p} + \vec{p} \times \vec{q})]. \tag{12.9}$$

We note that the non-commutativity of the quaternion product is due to the non-commutativity of the cross product of the imaginary vectors.

Example: We multiply the quaternions $p = 1+2i+3j+4k = [1,(2,3,4)]$ and $q = 5-2i+6j-7k = [5,(-2,6,-7)]$. We perform their quaternion product $p \cdot q$ stepwise.

- ▷ $p_0 q_0 = 1 \cdot 5 = 5$
- ▷ $\vec{p} \cdot \vec{q} = \begin{pmatrix} 2 \\ 3 \\ 4 \end{pmatrix} \cdot \begin{pmatrix} -2 \\ 6 \\ -7 \end{pmatrix} = -14$

We calculate its scalar component as $p_0 q_0 - \vec{p} \cdot \vec{q} = 5 - (-14) = 19$.

- ▷ $p_0 \vec{q} = 1 \begin{pmatrix} -2 \\ 6 \\ -7 \end{pmatrix} = \begin{pmatrix} -2 \\ 6 \\ -7 \end{pmatrix}$
- ▷ $q_0 \vec{p} = 5 \begin{pmatrix} 2 \\ 3 \\ 4 \end{pmatrix} = \begin{pmatrix} 10 \\ 15 \\ 20 \end{pmatrix}$
- ▷ $\vec{p} \times \vec{q} = \begin{pmatrix} 2 \\ 3 \\ 4 \end{pmatrix} \times \begin{pmatrix} -2 \\ 6 \\ -7 \end{pmatrix} = \begin{pmatrix} -45 \\ 6 \\ 18 \end{pmatrix}$

We calculate its imaginary vector as $p_0 \vec{q} + q_0 \vec{p} + \vec{p} \times \vec{q} = \begin{pmatrix} -37 \\ 27 \\ 31 \end{pmatrix}$.

We summarize the product quaternion as

$$p \cdot q = [19, (-37, 27, 31)] = 19 - 37i + 27j + 31k.$$

We illustrate the non-commutativity of the quaternion product by also calculating $q \cdot p$.

- ▷ $q_0 p_0 = 5 \cdot 1 = 5$
- ▷ $\vec{q} \cdot \vec{p} = \begin{pmatrix} -2 \\ 6 \\ -7 \end{pmatrix} \cdot \begin{pmatrix} 2 \\ 3 \\ 4 \end{pmatrix} = -14$

We calculate its scalar component as $q_0 p_0 - \vec{q} \cdot \vec{p} = 5 - (-14) = 19$.

- ▷ $q_0 \vec{p} = 5 \begin{pmatrix} 2 \\ 3 \\ 4 \end{pmatrix} = \begin{pmatrix} 10 \\ 15 \\ 20 \end{pmatrix}$

▷ $p_0\vec{q} = 1\begin{pmatrix} -2 \\ 6 \\ -7 \end{pmatrix} = \begin{pmatrix} -2 \\ 6 \\ -7 \end{pmatrix}$

▷ $\vec{q} \times \vec{p} = \begin{pmatrix} -2 \\ 6 \\ -7 \end{pmatrix} \times \begin{pmatrix} 2 \\ 3 \\ 4 \end{pmatrix} = \begin{pmatrix} 45 \\ -6 \\ -18 \end{pmatrix}$

We calculate its imaginary vector as $q_0\vec{p} + p_0\vec{q} + \vec{q} \times \vec{p} = \begin{pmatrix} 53 \\ 15 \\ -5 \end{pmatrix}$.

We summarize the product quaternion as

$$q \cdot p = [19, (53, 15, -5)] = 19 + 53i + 15j - 5k.$$

Quaternion conjugate

We define the **conjugate quaternion** q^* of a quaternion q analogously to the conjugation of complex numbers. The conjugation of a quaternion keeps its scalar component and takes the opposite of its imaginary vector. Given $q = [q_0, (q_1, q_2, q_3)]$, we define its conjugate quaternion as

$$q^* = [q_0, -\vec{q}] = [q_0, -(q_1, q_2, q_3)] = q_0 - (q_1 i + q_2 j + q_3 k).$$

Example: We multiply a general quaternion $q = [q_0, (q_1, q_2, q_3)]$ with its conjugate quaternion $q^* = [q_0, -(q_1, q_2, q_3)]$.

▷ $q_0 q_0 = q_0^2$

▷ $\vec{q} \cdot (-\vec{q}) = -(q_1^2 + q_2^2 + q_3^2)$

We calculate its scalar component as

$$q_0 q_0 - \vec{q} \cdot (-\vec{q}) = q_0^2 - (-(q_1^2 + q_2^2 + q_3^2)) = q_0^2 + q_1^2 + q_2^2 + q_3^2 = \|q\|^2$$

▷ $q_0(-\vec{q}) = -q_0\vec{q}$

▷ $q_0\vec{q}$

▷ $\vec{q} \times (-\vec{q}) = \begin{pmatrix} q_1 \\ q_2 \\ q_3 \end{pmatrix} \times \begin{pmatrix} -q_1 \\ -q_2 \\ -q_3 \end{pmatrix} = \begin{pmatrix} 0 \\ 0 \\ 0 \end{pmatrix}$

We calculate its imaginary vector as $-q_0\vec{q}+q_0\vec{q}+\vec{q}\times(-\vec{q}) = \begin{pmatrix} 0 \\ 0 \\ 0 \end{pmatrix}$.

We summarize the product quaternion as

$$q \cdot q^* = [\|q\|^2, (0,0,0)] = \|q\|^2 \in \mathbb{R}.$$

INVERSE QUATERNION

We define the inverse quaternion q^{-1} of the nonzero quaternion $q \neq [0,(0,0,0)]$ as the unique factor for which $q \cdot q^{-1} = q^{-1} \cdot q = [1,(0,0,0)]$, given this product $[1,(0,0,0)] = 1$ the **identity quaternion**. We determine the inverse of a quaternion via its conjugate quaternion

$$\underbrace{q^{-1} \cdot q}_{1} \cdot q^* = q^* \quad \text{as}$$

$$\begin{aligned} q^{-1} &= \frac{q^*}{q \cdot q^*} \\ &= \frac{q^*}{\|q\|^2}. \end{aligned}$$

In case of a unit quaternion u for which $\|u\| = 1$, the above calculation simplifies to $u^{-1} = u^*$, meaning its inverse is equal to its conjugate (which is way easier to determine).

12.7 Quaternions and rotation

Contemporary interactive 3D computer graphics thrive on quaternions to handle fast changing location vectors. Multiplying with unit quaternions u features exactly the same properties as applying 3D rotators (see page 218). Therefore, we may use unit quaternions u to rotate location vectors around an axis of rotation in 3D space.

We define the **rotation quaternion** r for rotating around a unit direction vector $\vec{n}$ over an angle θ as

$$r = [r_0, \vec{r}] = \left[\cos\frac{\theta}{2}, \vec{n}\sin\frac{\theta}{2}\right].$$

We prove that rotation quaternions r are unit quaternions, as $||r||$ equals 1.

$$||r|| = \sqrt{\left(\cos\frac{\theta}{2}\right)^2 + ||\vec{n}||^2\left(\sin\frac{\theta}{2}\right)^2} = \sqrt{\left(\cos\frac{\theta}{2}\right)^2 + 1\cdot\left(\sin\frac{\theta}{2}\right)^2}$$
$$= \sqrt{1} = 1$$

We augment the location vector $\vec{p}$ which heads the point P by one extra dimension to the pure quaternion $p = [0, \vec{p}\,]$. Without a proof we express the quaternion rotation of a point P around a unit direction vector $\vec{n}$ over an angle θ, yielding the image point P', as

$$\begin{aligned} p' &= r\cdot p\cdot r^{-1} \\ &= r\cdot[0,\vec{p}]\cdot r^{-1}. \end{aligned}$$

This rotation expression is not valid for arbitrary unit quaternions u! It only suits the formerly defined rotation quaternions $r = \left[\cos\frac{\theta}{2}, \vec{n}\sin\frac{\theta}{2}\right]$ given $||\vec{n}|| = 1$.

Since our defined rotation quaternion r is a unit quaternion, we simplify the above by $r^{-1} = r^*$. Conclusively, we realize the rotation of the point P by the quaternion r to the image point P' as

$$\begin{aligned} [0,\vec{p'}] &= r\cdot[0,\vec{p}]\cdot r^* \\ &= \left[\cos\frac{\theta}{2}, \vec{n}\sin\frac{\theta}{2}\right]\cdot[0,\vec{p}]\cdot\left[\cos\frac{\theta}{2}, -\vec{n}\sin\frac{\theta}{2}\right]. \end{aligned} \tag{12.10}$$

Example 1: We rotate the point $P(6,2,4)$ around an axis of rotation along the direction vector $\begin{pmatrix}1\\1\\1\end{pmatrix}$ over an angle of 120°, applying quaternions. For this purpose, we need to normalize $\begin{pmatrix}1\\1\\1\end{pmatrix}$ to $\vec{n} = \frac{1}{\sqrt{3}}\begin{pmatrix}1\\1\\1\end{pmatrix}$. Consequently, we find the rotation quaternion r as

$$\begin{aligned} r &= \cos 60^\circ + i\,\frac{\sin 60^\circ}{\sqrt{3}} + j\,\frac{\sin 60^\circ}{\sqrt{3}} + k\,\frac{\sin 60^\circ}{\sqrt{3}} \\ &= \frac{1}{2} + \frac{1}{2}\,i + \frac{1}{2}\,j + \frac{1}{2}\,k \\ &= \left[\frac{1}{2}, \left(\frac{1}{2},\frac{1}{2},\frac{1}{2}\right)\right]. \end{aligned}$$

We verify that r is a unit quaternion by calculating its norm $||r|| = 1$. The conjugate quaternion of r equals $r^* = \left[\frac{1}{2}, -\left(\frac{1}{2},\frac{1}{2},\frac{1}{2}\right)\right]$. We augment the point $P(6,2,4)$ to its pure quaternion $p = [0,(6,2,4)]$. We perform the specified rotation by the quaternion expression

$$\begin{aligned}
p' &= r \cdot p \cdot r^* \\
&= \left[\frac{1}{2}, \left(\frac{1}{2}, \frac{1}{2}, \frac{1}{2}\right)\right] \cdot [0,(6,2,4)] \cdot \left[\frac{1}{2}, \left(-\frac{1}{2}, -\frac{1}{2}, -\frac{1}{2}\right)\right] \\
&= [-6,(4,2,0)] \cdot \left[\frac{1}{2}, \left(-\frac{1}{2}, -\frac{1}{2}, -\frac{1}{2}\right)\right] \\
&= [0,(4,6,2)] .
\end{aligned}$$

Conclusively, the rotation of the point $P(6,2,4)$ around the direction vector $\begin{pmatrix} 1 \\ 1 \\ 1 \end{pmatrix}$ over an angle of 120°, results in the image point $P'(4,6,2)$.

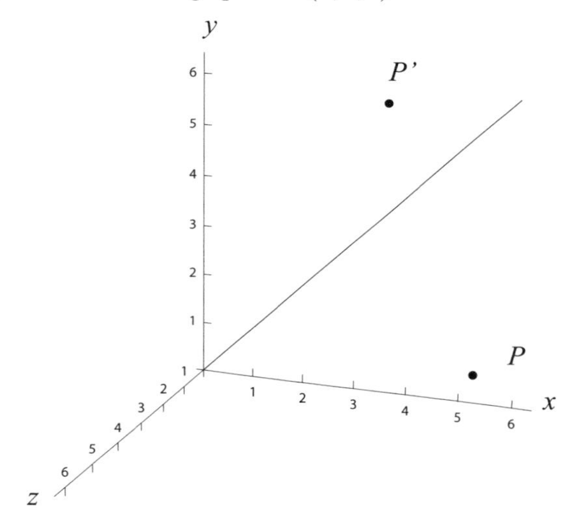

Figure 12.11: Rotation by quaternions

Example 2: We rotate the general point $P(x,y,z)$ around the z-axis over a general angle θ, applying quaternions. For this purpose, we do not need to normalize $\vec{e}_z = \begin{pmatrix} 0 \\ 0 \\ 1 \end{pmatrix}$ since it is already a unit vector. Consequently, we find the rotation quaternion r as

$$\begin{aligned} r &= \left[\cos\frac{\theta}{2}, \vec{e}_z \sin\frac{\theta}{2}\right] \\ &= \left[\cos\frac{\theta}{2}, (0,0,1)\sin\frac{\theta}{2}\right] \\ &= \left[\cos\frac{\theta}{2}, \left(0,0,\sin\frac{\theta}{2}\right)\right]. \end{aligned}$$

We verify that r is a unit quaternion by calculating its norm $\|r\| = 1$. The conjugate quaternion of r equals $r^* = \left[\cos\frac{\theta}{2}, -\left(0,0,\sin\frac{\theta}{2}\right)\right]$. We augment the point $P(x,y,z)$ to its pure quaternion $p = [0,(x,y,z)]$. We perform the specified rotation by the quaternion expression (see exercise 113)

$$\begin{aligned} p' &= r \cdot p \cdot r^* \\ &= \left[\cos\frac{\theta}{2}, \left(0,0,\sin\frac{\theta}{2}\right)\right] \cdot [0,(x,y,z)] \cdot \left[\cos\frac{\theta}{2}, \left(0,0,-\sin\frac{\theta}{2}\right)\right] \\ &= \left[-z\sin\frac{\theta}{2}, \left(x\cos\frac{\theta}{2} - y\sin\frac{\theta}{2}, x\sin\frac{\theta}{2} + y\cos\frac{\theta}{2}, z\cos\frac{\theta}{2}\right)\right] \cdot \left[\cos\frac{\theta}{2}, \left(0,0,-\sin\frac{\theta}{2}\right)\right] \\ &= [0,(x\cos\theta - y\sin\theta, x\sin\theta + y\cos\theta, z)]. \end{aligned}$$

The resulting pure quaternion p' contains the image point

$$P'(x\cos\theta - y\sin\theta, x\sin\theta + y\cos\theta, z).$$

We found the same result after applying the matrix rotator for a roll in 3D of point P around the z-axis (see formula (11.10)).

$$\begin{pmatrix} \cos\theta & -\sin\theta & 0 & 0 \\ \sin\theta & \cos\theta & 0 & 0 \\ 0 & 0 & 1 & 0 \\ 0 & 0 & 0 & 1 \end{pmatrix} \begin{pmatrix} x \\ y \\ z \\ 1 \end{pmatrix} = \begin{pmatrix} x\cos\theta - y\sin\theta \\ x\sin\theta + y\cos\theta \\ z \\ 1 \end{pmatrix}$$

Quaternion rotation is popular in computer applications because it needs only one move for the rotation around a direction vector $\vec{n}$ over an angle θ. Whereas the same effect takes three moves (around the major axes) in matrix rotation, which can even cause complications.

In this table we list some basic 3D rotations around the major axes and their corresponding rotation quaternion $r \in \mathbb{H}$. Again note that all rotation quaternions are unit quaternions.

rotation around the x-axis	rotation quaternion	rotation around the y-axis	rotation quaternion	rotation around the z-axis	rotation quaternion
over 180°	$[0,(1,0,0)]$	over 180°	$[0,(0,1,0)]$	over 180°	$[0,(0,0,1)]$
over 90°	$\left[\frac{\sqrt{2}}{2},\left(\frac{\sqrt{2}}{2},0,0\right)\right]$	over 90°	$\left[\frac{\sqrt{2}}{2},\left(0,\frac{\sqrt{2}}{2},0\right)\right]$	over 90°	$\left[\frac{\sqrt{2}}{2},\left(0,0,\frac{\sqrt{2}}{2}\right)\right]$
over −90°	$\left[\frac{\sqrt{2}}{2},\left(-\frac{\sqrt{2}}{2},0,0\right)\right]$	over −90°	$\left[\frac{\sqrt{2}}{2},\left(0,-\frac{\sqrt{2}}{2},0\right)\right]$	over −90°	$\left[\frac{\sqrt{2}}{2},\left(0,0,-\frac{\sqrt{2}}{2}\right)\right]$
over 60°	$\left[\frac{\sqrt{3}}{2},\left(\frac{1}{2},0,0\right)\right]$	over 60°	$\left[\frac{\sqrt{3}}{2},\left(0,\frac{1}{2},0\right)\right]$	over 60°	$\left[\frac{\sqrt{3}}{2},\left(0,0,\frac{1}{2}\right)\right]$

12.8 Exercises

Exercise 108 Represent the following four complex numbers algebraically and trigonometrically.

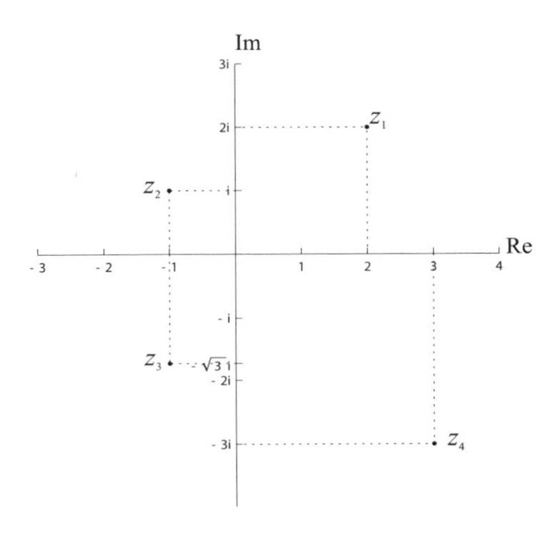

Exercise 109 Calculate the following expressions, given $z_1 = -4i$, $z_2 = 3 - 2i$ and $z_3 = -1 + i$.

1) $z_1 - 2z_2 + 3z_3$
2) $2z_1z_2^*$
3) z_2z_3
4) $\frac{z_1^*z_2}{z_3}$
5) $z_1(2z_2^* - z_1) + z_3^*$
6) $\frac{z_1 - z_2^*}{3z_3^*}$

Exercise 110 Consider the rectangle spanned by the vertices $A(1,1)$, $B(\sqrt{3},1)$, $C(\sqrt{3},-1)$ and $D(1,-1)$.

1) Give the complex number for a basic rotation around the origin over an angle of 120°. Apply this complex number to calculate the rectangle's image vertices after rotation.

2) Give the complex number for a basic rotation around the origin over an angle of 80° and a basic scaling with scale factor 3.

Exercise 111 Apply the function $f : \mathbb{Z} \to \mathbb{Z} : k \mapsto f(k) = \frac{(\Phi)^k - (\Phi')^k}{\sqrt{5}}$ to calculate the elements $f(-1), f(-2)$ and $f(-3)$ of the continued Fibonacci sequence. Both $\Phi = \frac{1+\sqrt{5}}{2}$ and $\Phi' = \frac{1-\sqrt{5}}{2}$ are known, respectively as the golden number and its variant.

Exercise 112 Calculate the following expressions, for $q=2-4i+2k$, $w=1-2i+j-2k$, $s=-1+i+j+k$ and $t=i+2j-k$.

1) $q+w^*$
2) $\|s\|$
3) $s\cdot t^*$
4) $q-2s+t^*$
5) $q\cdot w$
6) t^{-1}

Exercise 113 Calculate the quaternion product

$$\left[\cos\frac{\theta}{2},\left(0,0,\sin\frac{\theta}{2}\right)\right]\cdot[0,(x,y,z)]\cdot\left[\cos\frac{\theta}{2},\left(0,0,-\sin\frac{\theta}{2}\right)\right].$$

Exercise 114

1) Construct the rotation matrix for a roll over 90° and apply this roll to the point $P(1,0,0)$.
2) Repeat this rotation applying a rotation quaternion.

Exercise 115

1) Construct the rotation matrix for a pitch over 240° and apply this pitch to the point $P(2,3,1)$.
2) Repeat this rotation applying a rotation quaternion.

Exercise 116 Calculate the image coordinates of the point $P(1,1,0)$ after the rotation over 180° around the direction vector $\begin{pmatrix}1\\2\\1\end{pmatrix}$, applying quaternion arithmetics.

Exercise 117 The given matrix expression Z behaves identically to the complex number $z=a+bi$. It even adopts a 'complex' conjugate expression Z^*. We define

$$Z=a\begin{pmatrix}1&0\\0&1\end{pmatrix}+b\begin{pmatrix}0&-1\\1&0\end{pmatrix},$$

given $a,b\in\mathbb{R}$ and its conjugate expression as

$$Z^*=a\begin{pmatrix}1&0\\0&1\end{pmatrix}-b\begin{pmatrix}0&-1\\1&0\end{pmatrix}.$$

Calculate $Z+Z^*, Z-Z^*$ and $Z\cdot Z^*$.

Chapter 13 · Fractals

As we know, straight lines, smooth curves and surfaces are insufficient to model nor to understand nature's rich and complicated patterns. This shortage can be bridged by fractal geometry, which in its short existence already influences our world through a wide variety of applications. Fractals are used to design nature-like textures and arty screensavers, but also to compress graphical images or build the internal antennas of hand-held devices. The digital movie production as well as the computer game industry sometimes choose for fractal landscapes as the setting for their stories.

13.1 The concept of a fractal

Study the structure of a cauliflower, the leaf of a fern or the shape of a snowflake. It are perfect examples of fractals and we find many of them in nature: clouds, lightning, coastlines, mountains, Elm trees, cauliflowers, broccoli, river deltas, frost on glass, dendrites, ... Understanding their growth mathematically, enables us to create them.

Figure 13.1: Fractal examples

The term 'fractal' originates from the Latin word '*fractus*' which means '*broken*', being introduced by the French mathematician **Benoit Mandelbrot** (1924 - 2010). Mandelbrot distinguishes between fractals and smooth euclidean objects. No matter how we scale them, fractal objects keep ridged, flocky or dusty.

We define a **fractal** as a geometrical object featuring

- ▷ self-similarity (see paragraph 13.2) and
- ▷ a broken dimension (see paragraph 13.3).

Fractals are created by applying a procedure (a motif or a function) iteratively or recursively. Let us first of all define these major concepts and then meet some well-known fractals, to learn from these examples.

We define **iteration** as applying a procedure repeatedly, using the output from one iteration as the input to the next. Iteration of apparently simple motifs can lead to intriguing results.

We define **recursion** as a function calling itself, using the return from one call as the argument to the next call. Recursion of apparently simple functions can lead to surprisingly complicated behaviours.

THE SIERPINSKI GASKET

In 1915 the Polish mathematician **Waclaw Sierpinski** (1882 - 1969) designed a spectacular fractal named after him (see the chapter 3 image on page 42). We construct the **Sierpinski gasket** starting from an equilateral triangle in which we connect its midpoints by line segments. Erasing the inner triangle, we repeat connecting the midpoints in each of the three remaining subtriangles. Again erasing the hearts of all three of them, we remain with nine smaller triangles. We iterate this process of erasing midpoint based triangles endlessly, which yields the Sierpinski gasket.

Figure 13.2: Sierpinski gasket

THE KOCH SNOWFLAKE

In 1904 the Swedish mathematician **Helge von Koch** (1870 - 1924) designed a fascinating curve named after him. The Koch curve has no tangents and each of its parts, no matter how restricted, has an infinite length. We construct the **Koch curve** starting from a line segment of which we cut out its center third part, replacing it by an isosceles wedge with legs of length equal to the missing third part (see figure 13.3). In other words, each of the four edges of this disturbed segment equals $\frac{1}{3}$. We repeat cutting out the center third part of each new edge, replacing it by an isosceles wedge and leaving us with sixteen edges of length $\frac{1}{9}$. We iterate this process of cutting out and replacing center thirds endlessly, which yields the Koch curve.

Figure 13.3: Koch curve

Replacing the edges of equilateral polygons by their Koch curve leads to the socalled **Koch snowflakes**. However their perimeter is infinite, Koch snowflakes enclose a finite area.

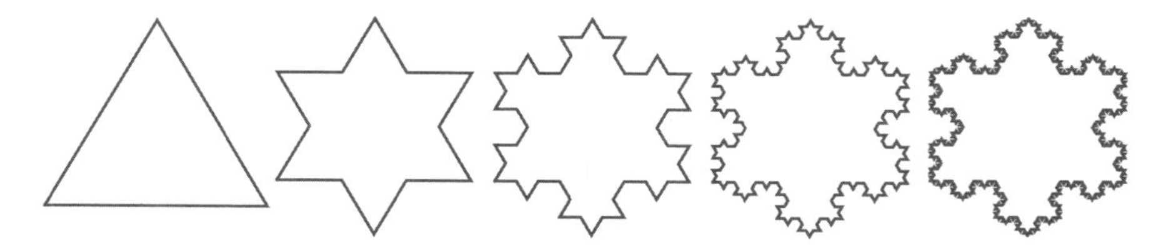

Figure 13.4: Koch snowflake

THE MINKOWSKI ISLAND

It was possibly the Russian mathematician **Hermann Minkowski** (1864 - 1909) who designed another fascinating curve. We construct the **Minkowski sausage** starting from a line segment which we replace by an 8-side generator shape (see figure 13.5). We repeat replacing each of the eight new segments by the same 8-side generator motif. Already the first few iterations modify the appearance of this curve heavily. We iterate this process of replacing segments by the 8-side generator shape endlessly, which yields the Minkowski sausage.

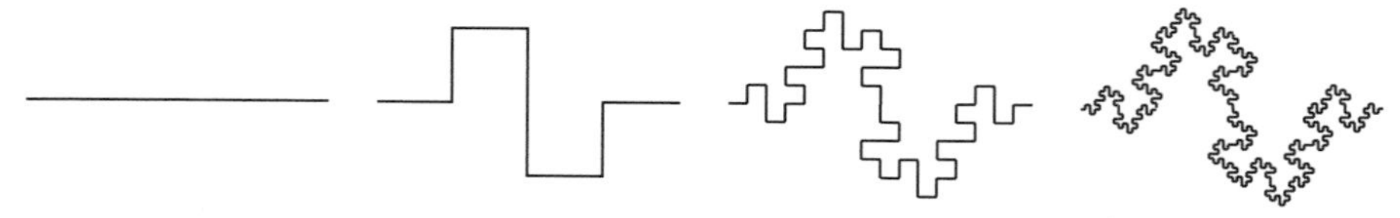

Figure 13.5: Minkowski sausage

The length of the Minkowski sausage doubles by each iteration step: the length of its segments is divided by four while the number of them is multiplied by eight. Various Minkowski sausages of iteration depth 2 are in use as miniature antennas built-in in mobile phones.

Similar to the Koch snowflakes, replacing the edges of equilateral polygons by their Minkowski sausage leads to the interesting **Minkowski islands**. Although their perimeter is infinite, Minkowski islands enclose a finite area.

Figure 13.6: Minkowski island

The Cantor set

In 1870 the German mathematician **Georg Cantor** (1845 - 1918) designed the oldest artificial fractal, called the Cantor set. We construct the **Cantor set** starting from the closed **unity interval** $[0, 1] \subset \mathbb{R}$, minus its center *open* interval $]\frac{1}{3}, \frac{2}{3}[$, thus leaving its boundary elements $\frac{1}{3}$ and $\frac{2}{3}$ in the subintervals. We remain with two separated subintervals of length one third of the original unity interval. We repeat set subtracting the center intervals from each subinterval: the *open* interval $]\frac{1}{9}, \frac{2}{9}[$ from the former and $]\frac{7}{9}, \frac{8}{9}[$ from the latter. After this step we remain with four subintervals of length one ninth of the original unity interval. We iterate this process of set subtracting endlessly, which yields the Cantor set.

Figure 13.7: Cantor set

The Pythagoras Tree

In 1942 the Dutch engineer **Albert E. Bosman** (1891 - 1961) sketched an arty fractal, named after the Greek mathematician Pythagoras of Samos. We construct the **Pythagoras tree** by picturing the Pythagorean theorem in a right isosceles triangle. In step 1 we adopt both of its squared sides as the squared hypotenuses of two child right isosceles triangles. In step 2 we repeat picturing the Pythagorean theorem upon each new squared side, growing four new branches to the tree. Generally, step k adds 2^k new branches to this impressive tree. Iterating this process endlessly, yields the Pythagoras tree.

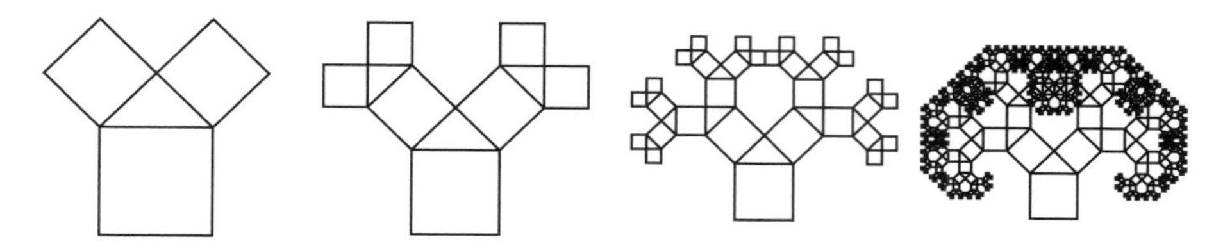

Figure 13.8: Pythagoras tree

13.2 Self-similarity

Geometrical objects are **similar** if they share the same shape. Similar objects can be obtained from each other by uniformly scaling them, their scale factor s given by the constant ratio of their corresponding measures (see paragraph 11.2). We illustrate this similarity by a counterexample (the top couple of rectangles) and an example (the bottom couple of rectangles, see figure 13.9).

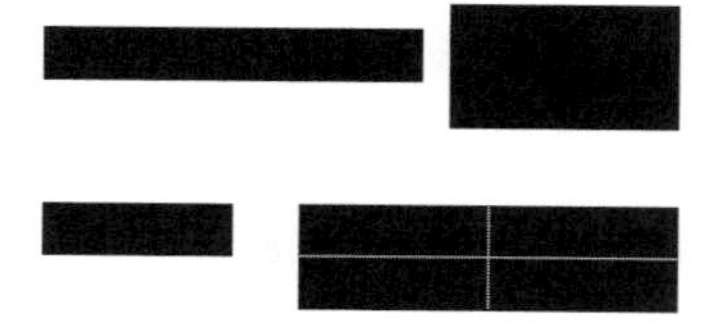

Figure 13.9: Counterexample versus example of similarity

The bottom couple of rectangles are similar with scale factor $s = 2$, which uniformly doubles the original into the second rectangle. This means that as well its length as its width grew twice as large through the same scale factor. The constant ratio of the corresponding lengths equals the ratio of their widths, which we typeset as 2:1. For this reason we call the bottom couple of rectangles similar.

We define a **self-similar** object as similar to a part of itself. Note that this definition does not necessarily refer to fractal shapes, as shown by a self-similar trapezium which is not fractal (see figure 13.10).

All small trapezia are scaled copies of the big trapezium they compose. Applying scale factor $s = 2$ onto a small trapezium, returns the big trapezium. For a profound insight, we

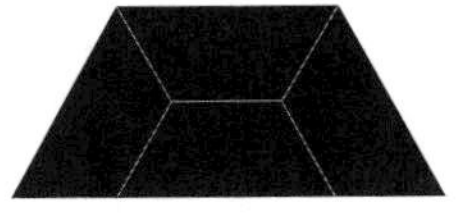

Figure 13.10: Self-similar trapezium

illustrate this property by a gallery of self-similar planar objects (see figure 13.11).

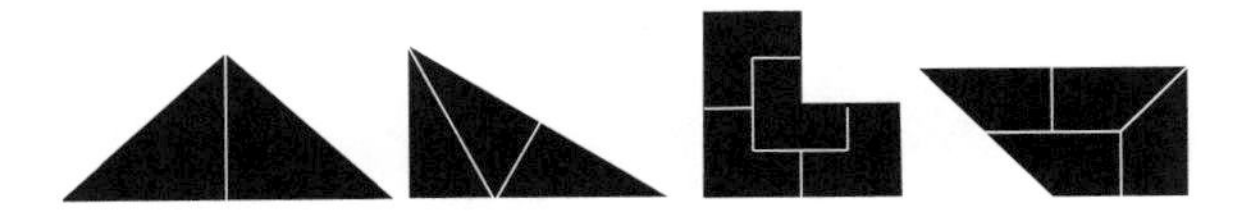

Figure 13.11: Various self-similar objects

Let us now examine this self-similarity of all fractal examples we have met so far.

Example 1: Applying scale factor $s = 2$ onto the bottom right subtriangle of the Sierpinski gasket, returns the original Sierpinski gasket. Applying scale factor $s = 4$ onto one of the nine subtriangles of the second iteration $k = 2$, returns the Sierpinski gasket itself. In general, applying the scale factor $s = 2^k$ onto a subtriangle of iteration step k, returns the total Sierpinski gasket. For this reason we conclude the Sierpinski gasket to be self-similar.

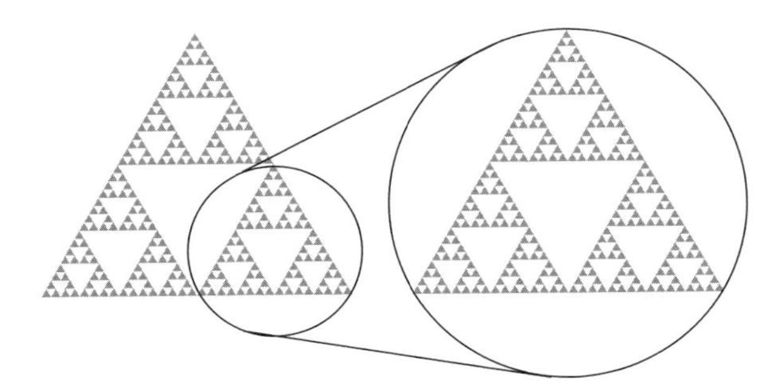

Figure 13.12: Self-similarity of the Sierpinski gasket

Example 2: Applying scale factor $s = 3$ onto the first out of four edges of the Koch curve, returns the original Koch curve. Each iteration step generates four shrinked copies of the original. In general, applying the scale factor $s = 3^k$ onto one of the 4^k edges of iteration step k, returns the total Koch curve. It is obvious that the Koch curve is self-similar.

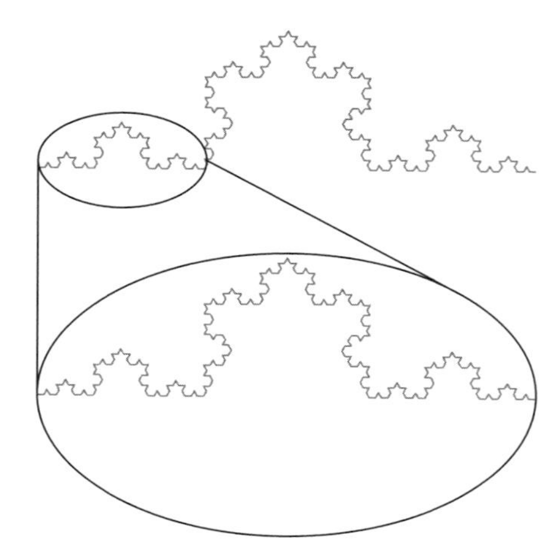

Figure 13.13: Self-similarity of the Koch curve

Example 3: Applying scale factor $s = 4$ onto the 8th side of the Minkowski sausage, returns the original Minkowski sausage. We can as well explain this the other way around. Each iteration step divides a side by 4, replacing it by eight smaller ones. Therefore, the Minkowski sausage is clearly self-similar.

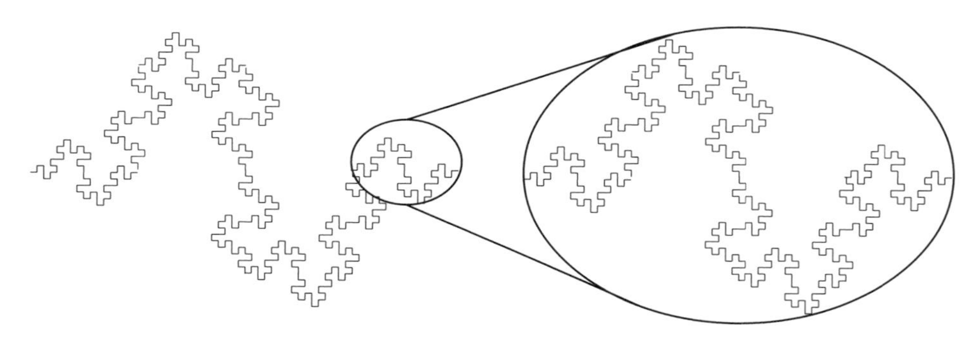

Figure 13.14: Self-similarity of the Minkowski sausage

Example 4: Applying scale factor $s = 3$ onto the second out of two branches of the Cantor set, returns the original Cantor set. Each iteration step of the Cantor set divides the interval by 3 and yields two new copies of the original. We conclude that also the Cantor set is self-similar.

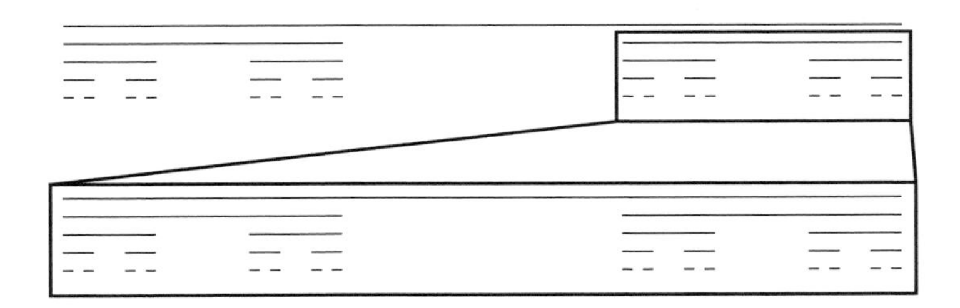

Figure 13.15: Self-similarity of the Cantor set

Example 5: Applying scale factor $s = \sqrt{2}$ onto the second of the two branches of the Pythagoras tree, returns the original tree. Each iteration step of the Pythagoras tree shrinks it by divisor $\sqrt{2}$ and yields two new copies of the original. The Pythagoras tree grows self-similar on each iteration level.

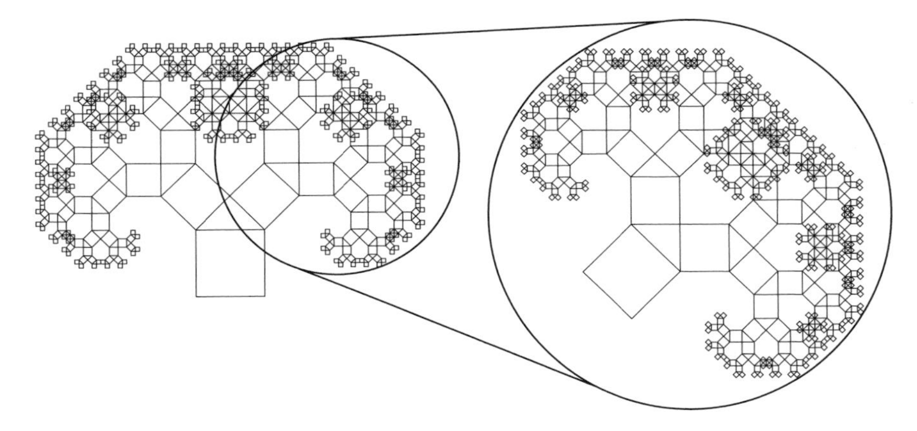

Figure 13.16: Self-similarity of the Pythagoras tree

13.3 Fractal dimension

Euclidean dimension

We intuitively count mutual perpendicular directions to determine the **euclidean dimension** of objects. For instance, we consider points as zero-dimensional, lines as one-dimensional, planes as two-dimensional (given their perpendicular length and width) and cubes as three-dimensional (because of their mutual perpendicular length, width and depth). We notice how all euclidean dimensions are natural numbers: $0, 1, 2, 3$. We can even pursue euclidean dimensions beyond 3D (for which we refer to the specialized literature).

Talking about dimensions, fractal objects cause trouble since they have broken dimensions. Fractals can be of intermediate dimensions like 1.6 or 2.4. In other words, we are challenged to extend the former definition of dimension to fractal objects, whilst conserving the euclidean dimensions. The renewed 'fractal' dimension of a line segment, a square and a cube should still yield $1, 2$ and 3, respectively.

Doubling the length of a line segment yields two copies of the original. In general, applying scale factor s onto a line segment yields s copies of it. Doubling the length of the side of a square yields four copies of the original. In general, applying scale factor s onto the side of a square yields s^2 copies of it. Doubling the length of the edge of a cube yields eight copies of the original. In general, applying scale factor s onto the edge of a cube yields s^3 copies of it. We hereby seem to rediscover the euclidean dimension as the corresponding exponent of the number of similar copies, given the copying factor s.

Hausdorff dimension

The above approach inspired the German mathematician **Felix Hausdorff** (1868–1942) in 1918 to successfully extend the concept of dimension, applicable to fractal objects.

We define the **fractal dimension** or **hausdorff dimension** of an object as the exponent h of its number of self-similar copies by uniformly scaling it with factor s. Alternatively, given the fractal dimension h and the uniform scaling factor s of an object, the power

$$s^h = n \tag{13.1}$$

calculates its number of self-similar copies, for each iteration.

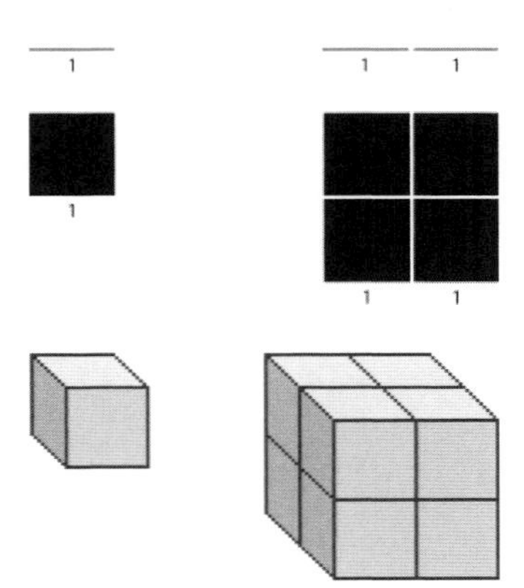

Figure 13.17: Extending the euclidean dimension

The concept of a logarithm

We define the **logarithm** in base g of a number a as the exponent $x \in \mathbb{R}$ to which we need to raise this base g in order to yield the number a. In symbols:

$$g^x = a \Leftrightarrow x = \log_g(a), \tag{13.2}$$

given the double condition

$$\text{base } g \in \mathbb{R}^+ \backslash \{0, 1\}$$

and

$$\text{argument } a \in \mathbb{R}^+ \backslash \{0\}.$$

In other words, we can easily memorize this concept by an overstatement: '*the logarithm is the corresponding exponent*'.

Illustrations

Therefore, we redefine the fractal dimension of an object as the logarithm

$$h = \log_s(n), \tag{13.3}$$

for a given uniform scale factor $s \neq 1$ producing n self-similar copies of the original.

Examples: We hereby list some fractal dimensions h for a variety of geometrical and natural objects.

object	s	n	h
line segment	2	2	$\log_2(2) = 1$
square	2	4	$\log_2(4) = 2$
cube	2	8	$\log_2(8) = 3$
Sierpinski gasket	2	3	$\log_2(3) \approx 1.585$
	4	9	$\log_4(9) \approx 1.585$
Koch curve	3	4	$\log_3(4) \approx 1.262$
Minkowski sausage	4	8	$\log_4(8) = 1.5$
Cantor set	3	2	$\log_3(2) \approx 0.631$
Pythagoras tree	$\sqrt{2}$	2	$\log_{\sqrt{2}}(2) = 2$
cauliflower	3	13	$\log_3(13) \approx 2.335$

13.4 The Mandelbrot and Julia Sets

The Mandelbrot set and the various Julia sets are fractals which reside in the complex plane (see paragraph 12.1). During his computer experiments at the IBM company, Benoit Mandelbrot was the very first to poorly visualize on a screen the complex fractal named after him. This first screen image just gave an impression, as Mandelbrot was able to see the kidney shape attached to the disk shape, but apart from the antenna and some blurred isles no more details were visible. Thanks to the increasing computing power and screen resolution, recent images show this fractal's frayed continent body surrounded by many crisp isles, some of them self-similar copies of the continent. Nowadays we are aware that especially the edge of the Mandelbrot set is of a huge complexity. Scaling parts of the edge reveals surprisingly new complex patterns.

Dynamical systems

The complex function $q : \mathbb{C} \to \mathbb{C} : z \mapsto z^2$ maps an argument $z = a + bi$ onto its square

$$(a+bi) \mapsto (a+bi)^2 = (a^2 - b^2) + 2abi.$$

We iterate this function q by recursively applying it, which we typeset as

$$z_{k+1} = z_k^2.$$

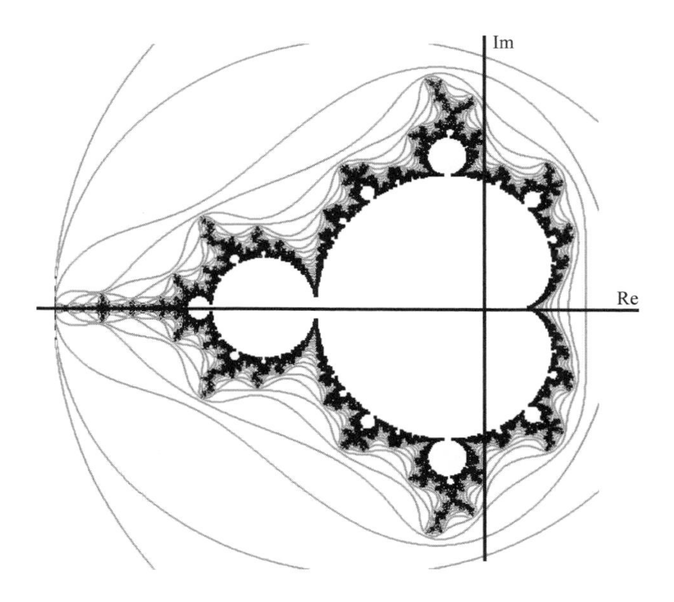

Figure 13.18: Mandelbrot set

In other words, starting with an initial value z_0, its return value z_1 becomes the successive argument to the function q. We define such a recursion generally as a **dynamical system**. The graph of its returned sequence of points $z_0, z_1, z_2, \ldots$ in the complex plane, is called the **trajectory** of the initial value $z_0 \in \mathbb{C}$.

Example: Calculating the trajectory of the initial value $z_0 = 0.12 + 0.78i$, yields

$$\begin{aligned} z_0 &= 0.12 + 0.78i \\ z_1 &= -0.59 + 0.18i \\ z_2 &= 0.31 - 0.22i \\ &\vdots \end{aligned}$$

We discover different behaviours according to the chosen initial value. Some initial values, like $z_0 = 1 + 2i$, cause their trajectory to run away from the complex origin. Alternatively, initial values like $z_0 = \frac{1}{2} + \frac{1}{2}i$ perform a trajectory tending towards $0 + 0i$. In the latter case we define $O(0,0)$ as an attracting point or **attractor** for such a converging trajectory.

The Mandelbrot Set

What we call the Mandelbrot set, is a collection of points in the complex plane. In other words, the Mandelbrot set is a subset of the set of complex numbers. For each complex number we can tell whether or not it is an element of the Mandelbrot set. One way to define the Mandelbrot set, is by testing each complex number separately for it. The test requires calculating for various $c \in \mathbb{C}$ its trajectory recursively like this: $c, c^2+c, (c^2+c)^2+c, \ldots$ via subsequent squaring and adding to itself.

For this reason, we extend the former complex squaring q. Adding a complex term $c \in \mathbb{C}$ to the recipe of q, yields

$$f : \mathbb{C} \to \mathbb{C} : z \mapsto z^2 + c, \qquad \text{given} \quad c \in \mathbb{C}. \tag{13.4}$$

We iterate this adapted function f by recursively applying it, which we typeset as

$$z_{k+1} = z_k^2 + c. \tag{13.5}$$

We can hereby base the Mandelbrot set on recursively applying f for the constant initial value $z_0 = 0 + 0i$ and various numbers $c = a + bi \in \mathbb{C}$. In other words, the dynamical system to determine the Mandelbrot set runs like this:

$$\begin{aligned} z_0 &= 0 + 0i \\ z_1 &= (0+0i)^2 + c = c \\ z_2 &= c^2 + c \\ z_3 &= (c^2+c)^2 + c \\ &\vdots \end{aligned}$$

For this dynamical system, there are only two behaviours according to the chosen $c \in \mathbb{C}$. Either the trajectory of c runs away from the complex origin, or either c performs a trajectory tending towards $0 + 0i$. We test these behaviours based on the moduli of the return values (see page 226). For increasing recursion steps $k \in \mathbb{N}$ applied on the complex function f the moduli $|z_k|$ either

- ▷ grow without bound, or
- ▷ converge to a constant value.

We define the **Mandelbrot set** $\mathbb{M} \in \mathbb{C}$ as the set of all complex numbers c for which their recursive moduli converge, given the dynamical system (13.5) for the initial value $z_0 = 0 + 0i$. We note that the unique Mandelbrot set is defined by one constant initial value z_0 and varying values of the complex term c.

We recall that to each complex number $c = a + bi$ corresponds its point $C(a, b)$ in the complex plane. Therefore, we visualize the Mandelbrot set $\mathbb{M}$ in the complex plane by

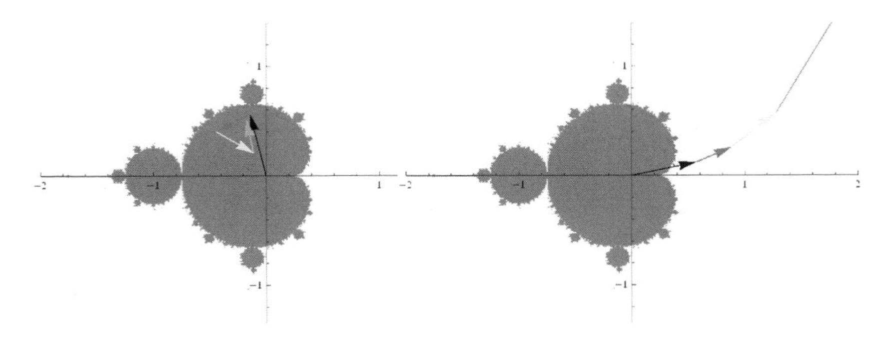

Figure 13.19: Trajectories of two different complex values c

colouring the points C which correspond to converging trajectories. We define the **convergence rate** of a point C as the number of iterations required for c to acquire a constant modulus. This definition enables us to even colour the points C of the Mandelbrot set according to their different convergence rates, revealing much more detail. The self-similarity which occurs through this colouring is of a non-linear kind, unlike the former examples (see paragraph 13.2). But subsequent scalings of the Mandelbrot set $\mathbb{M}$ clearly reveal self-similar copies of $\mathbb{M}$, some of them transformed.

We illustrate this by five successive enlargements of an initially visualized Mandelbrot set in the complex plane (see figure 13.20).

Its complicated self-similarity confirms that $\mathbb{M}$ is a complex fractal.

The Julia Sets

What we call the Julia sets, are also extraordinary collections of points in the complex plane. We define them by the same dynamical system (13.5), but with the variability the other way around: for a selected constant term $c \in \mathbb{C}$ we vary the initial values z_0 for it. In other words, we determine the Julia set based on the term c by recursively applying f for various initial values $z_0 \in \mathbb{C}$. Consequently, the latter dynamical system runs like this:

$$\begin{aligned}
z_0 &= a+bi \\
z_1 &= z_0^2+c \\
z_2 &= (z_0^2+c)^2+c \\
z_3 &= \left((z_0^2+c)^2+c\right)^2+c \\
&\vdots
\end{aligned}$$

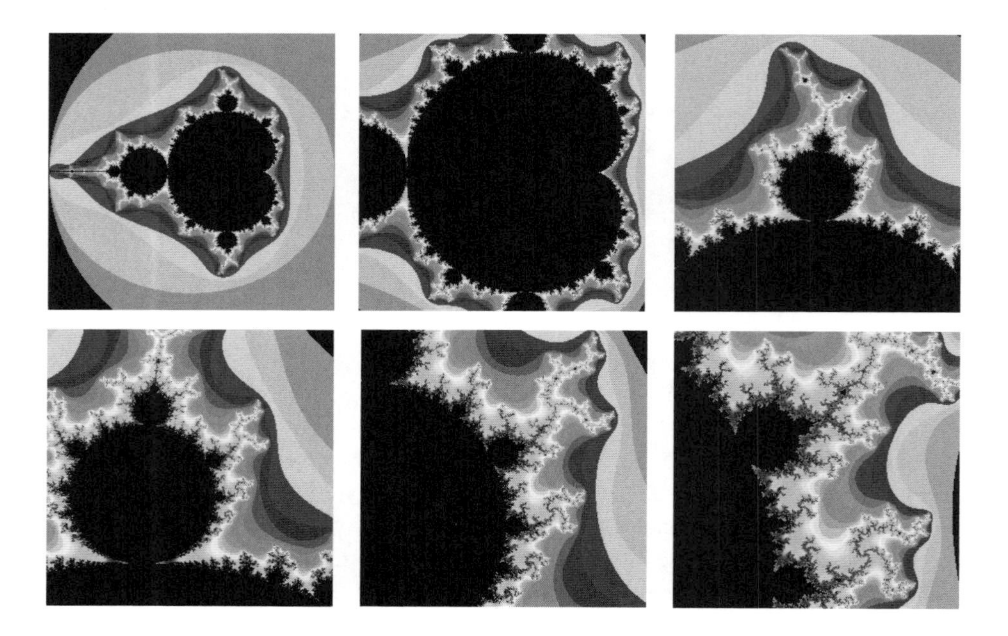

Figure 13.20: Successively magnifying the Mandelbrot set $\mathbb{M}$

It goes without saying that since we need to let vary $z_0 = a + bi$ for Julia sets, these initial values are not restricted to only $0 + 0i$ at all. Again, there are only two behaviours according to the tested initial value $z_0 \in \mathbb{C}$. Either the trajectory of z_0 runs away from the complex origin, or either z_0 performs a trajectory tending towards $0 + 0i$. We test these behaviours based on the moduli of the return values z_k. For increasing recursion steps $k \in \mathbb{N}$ applied on the complex function f the moduli $|z_k|$ either

- ▷ grow without bound, or
- ▷ converge to a constant value.

The French mathematician **Gaston Julia** (1893 - 1978) defined a **Julia set** $\mathbb{J}_c \in \mathbb{C}$ based on the term c, as the set of all complex values z_0 for which their recursive moduli converge, given the dynamical system (13.5). We note that each Julia set $\mathbb{J}_c$, as based on its term c, is defined by testing a wide variety of initial values z_0 for convergence. Unlike the uniqueness of the Mandelbrot set $\mathbb{M}$, we understand there are an uncountable number of Julia sets $\mathbb{J}_c$ as their complex constants c are uncountable. Each complex number $c = a + bi \in \mathbb{C}$, equivalently each point $C(a,b)$ in the complex plane, determines a corresponding Julia set $\mathbb{J}_c$.

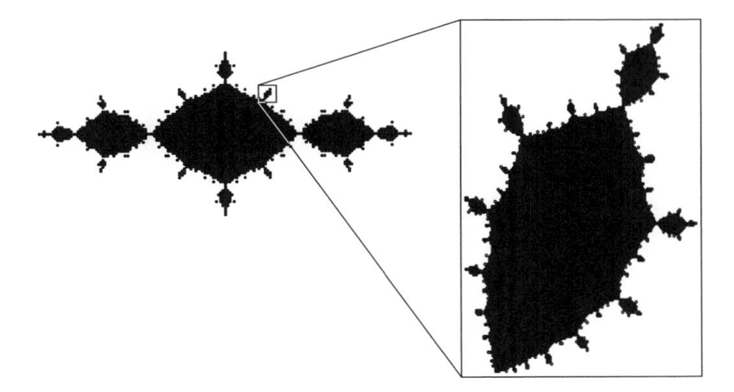

Figure 13.21: The Julia set $\mathbb{J}_{-1.1+0i}$ plus enlarged detail

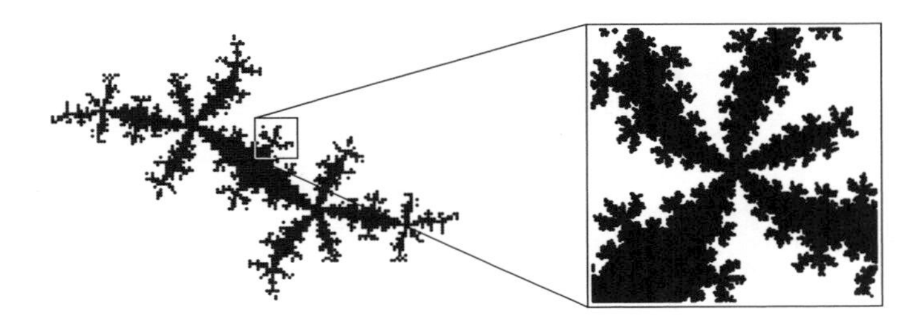

Figure 13.22: The Julia set $\mathbb{J}_{-0.53+0.6i}$ plus enlarged detail

We analogously visualize a Julia set $\mathbb{J}_c$ in the complex plane by colouring the points Z_0 which correspond to converging trajectories. Colouring its points Z_0 according to their different convergence rates, reveals much more detail. The self-similarity which occurs through this colouring is again of the non-linear kind. Successive scalings of a Julia set $\mathbb{J}_c$ clearly display details which are self-similar to the original. Again, this complicated but obvious self-similarity confirms that all Julia sets $\mathbb{J}_c$ are complex fractals.

We define a set of points as **connected** if, for any two points in the set, there is at least one path consisting entirely of points in the set, which leads from one point to the other. A given Julia set $\mathbb{J}_c$ is connected or disconnected, depending on its selected term $c = a + bi$. Each Julia set is either a connected set or a Cantor set (see page 253). A connected set consists of one piece, whereas a Cantor set consists of an uncountably infinite set of disjoint points. If we colour the corresponding selected points $C(a,b)$ in the complex plane, for which their corresponding Julia sets $\mathbb{J}_c$ are connected, we rediscover the unique Mandelbrot set M. Points $C(a,b)$ that are located outside of the Mandelbrot set correspond to complex terms $c \in \mathbb{C} \setminus \mathbb{M}$ which label all disconnected Julia sets $\mathbb{J}_c$ (see figure 13.23).

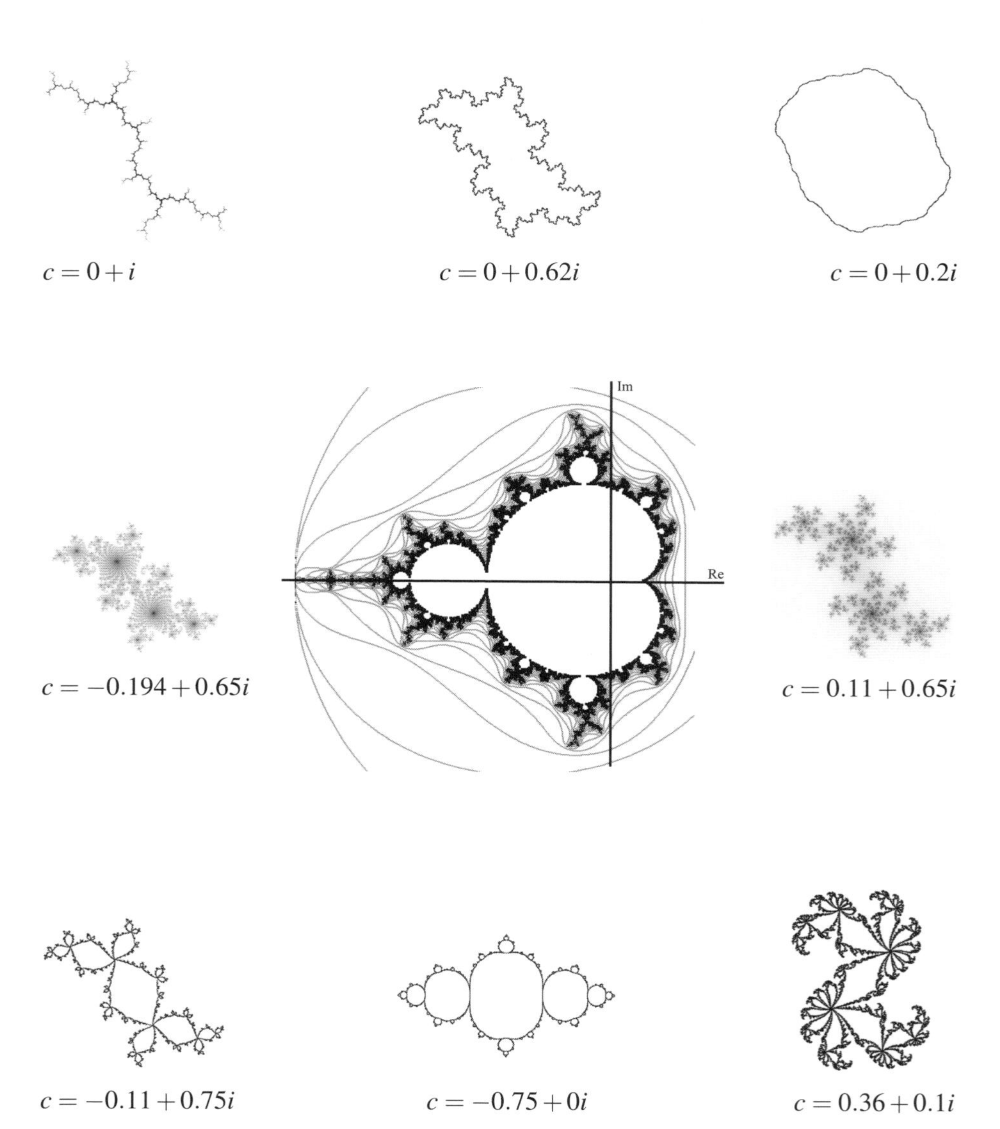

Figure 13.23: The Mandelbrot set binds all Julia sets $\mathbb{J}_c$ which are connected

13.5 Exercises

Exercise 118 Try this fractal origami.

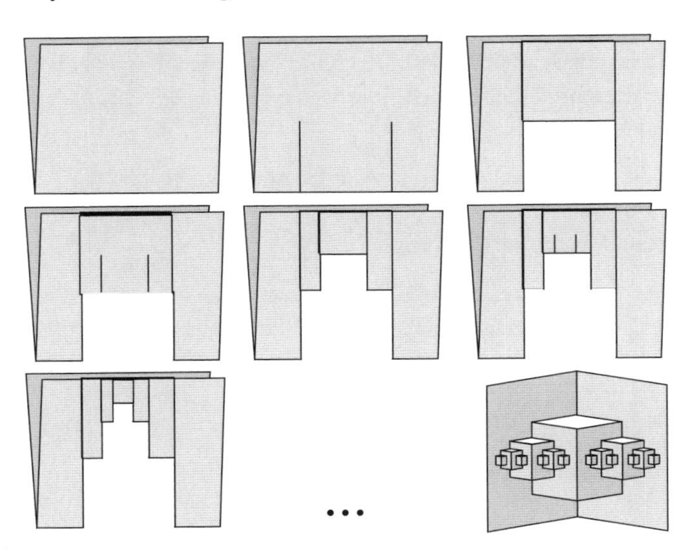

Exercise 119 The Austrian mathematician **Karl Menger** (1902 - 1985) designed a fractal named after him.

- ▷ We construct the **Menger sponge** starting from a square, giving its two division points on each of its sides, to divide the side into three equal segments. Connecting the corresponding division points of opposite sides, yields a smaller inner square. Erasing the inner square, we repeat connecting the division points in each of the eight remaining subsquares. Again erasing the hearts of all eight of them, we remain with sixty-four smaller squares. We iterate this process of erasing division point based squares endlessly, which yields the Menger sponge.
- ▷ We started with a square having side 1, therefore we typeset the initial area $a_0 = 1$. Find the remaining area after iteration step 1 by expressing a_1 in terms of the initial area a_0. Extend this sequence of areas up to a_2, a_3, a_4 expressing all of them in terms of a_0.
- ▷ Generalizing this sequence, also express the kth area a_k in terms of a_0.
- ▷ Calculate the fractal dimension of this planar Menger sponge.

Exercise 120 Try to answer some questions about the Koch snowflake (see both figures 13.3 and 13.4).

- ▷ For k iteration steps, find the number of edges which we typeset as $e_0, e_1, e_2, e_3, \ldots, e_k$ including the general number of edges e_k.
- ▷ Its equilateral edges are initially of length 1, therefore we typeset length $l_0 = 1$. For k iteration steps, find the length of its edges typeset as $l_0, l_1, l_2, l_3, \ldots, l_k$ and express the general length l_k.
- ▷ For k iteration steps, find the circumference of the Koch snowflake stages as p_0, p_1, $p_2, p_3, \ldots, p_k$ up to the general perimeter p_k. Prove that its circumference increases without bound, based on the formula for p_k.
- ▷ Calculating the final area enclosed by the fractal Koch snowflake is hard. Alternatively, prove that the Koch snowflake is boxed in a square with sides $\frac{2\sqrt{3}}{3}$. Therefore, we can conclude that the area of the Koch snowflake has an upper bound. Find it.

Exercise 121 Determine the fractal dimension of

- ▷ the Sierpinski pyramid (left),
- ▷ the 3D Koch snowflake as portrayed by its first three iteration steps (right).

Exercise 122 Determine whether the following complex numbers belong to the Mandelbrot set.

1) $c = 0.25 + 0.25i$

2) $c = 1 + i$

3) $c = 2i$

Exercise 123 Calculate and draw the trajectories of the following initial values, using the dynamical system of $\mathbb{J}_{0+0i}$.

1) $z_0 = 1 + 2i$

2) $z_0 = \frac{1}{2} + \frac{1}{2}i$

3) $z_0 = i$

Chapter 14 · Bezier curves

Programming multimedia applications often involves plotting lines or curves from point to point. Amongst the variety of methods to achieve this, the most popular and practical algorithm is based on Bezier or B-spline segments. Those segments are named after the French engineer **Pierre Bézier** (1910–1999) who discovered a mathematical approach to design smooth curves for the company Renault, which ran on their first industrial computers. Simultaneously, the French mathematician **Paul de Casteljau** (1930) developed a similar plot algorithm for the competiting company Citroën, to digitally plot smooth curves.

Ever since, apart from plotting smooth lines, Bezier segments are in use for 2D and 3D shapes, to model letter fonts and to design timelines for animation. All professional design, graphical and animation software makes use of Bezier segments or B-splines.

14.1 Vector equation of segments

Renault-engineer Pierre Bézier developed all underneath parametrical segments for the company's computer aided design during the sixties.

Linear Bezier segment

We recall the vector equation of a line as $\vec{x} = \vec{b} + \lambda(\vec{a} - \vec{b})$, given parameter λ or often $t \in \mathbb{R}$ (see figure 8.1). We also recall that in euclidean space every two different points span one straight line.

If we limit the values for the parameter t to the **unity interval** $[0,1] \subset \mathbb{R}$, we limit the head point X to the segment $[BA]$. Let us explore this by rearranging the vector equation to

$$\vec{x} = (1-t)\vec{b} + t\vec{a},$$

which for the parameter sequence $t = 0, \frac{1}{2}, 1$, heads its location vector $\vec{x}$ respectively to the points $B, \dfrac{B+A}{2}, A$. We define such a line segment as a **linear** or **2-point Bezier segment**. We express the initial and the terminal point of the line segment $[P_0P_1]$ by their respective location vectors $\vec{p}_0$ and $\vec{p}_1$ and typeset this Bezier segment, a function of t, as $\vec{b}_{01}(t)$. Its indices 0 and 1 refer to its initial and terminal point P_0 and P_1. We summarize the above into the parametric equation of the linear Bezier segment from P_0 to P_1:

$$\vec{b}_{01}(t) = (1-t)\vec{p}_0 + t\vec{p}_1 \quad \text{given} \quad t \in [0,1] \subset \mathbb{R}. \tag{14.1}$$

The parametric equation of $\vec{b}_{01}(t)$ confirms it indeed is linear in the parameter t and it runs through its initial point $B_{01}(0) = P_0$ and through its terminal point $B_{01}(1) = P_1$.

Quadratic Bezier segment

Plotting a Bezier segment running from P_0 to P_2 and curved by one control point P_1, is somehow more involving. For a better insight we sample the continuous parameter $t \in [0,1] \subset \mathbb{R}$ by a long sequence of discrete parameter values $t_0 = 0, t_1, t_2, t_3, \ldots, t_{n-1}, t_n = 1$. Starting at the parameter value $t_0 = 0$, we plot

- ▷ for every subsequent parameter value t_1 as well the point $B_{01}(t_1)$ as the point $B_{12}(t_1)$,
- ▷ and on their fresh Bezier segment $[B_{01}B_{12}]$ again the point $B_{012}(t_1)$ corresponding to the parameter value t_1

and we iterate both steps for the incrementing parameter values $t_2, t_3, \ldots, t_{n-1}$ up to $t_n = 1$.

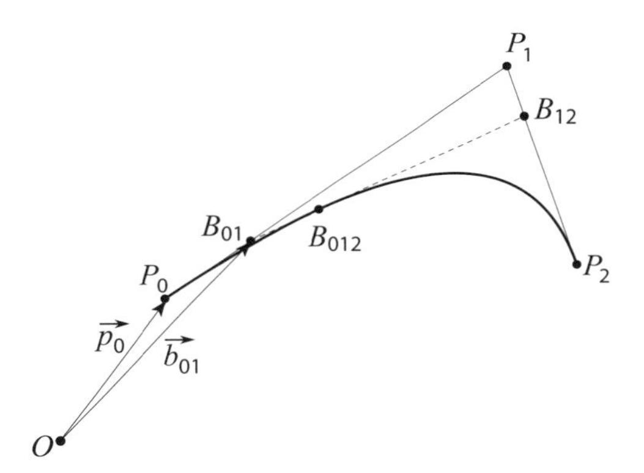

Figure 14.1: Quadratic or 3-point Bezier segment

We define the obtained (and theoretical continuous) curve $\vec{b}(t)$ as the **quadratic** or **3-point Bezier segment** $\vec{b}_{012}(t)$ defined by its **control points** P_0, P_1 and P_2. These control points P_0, P_1 and P_2 are indexed by the Bezier's indices $0, 1$ and 2.

Based on the above plot algorithm, we elaborate its parametric equation as:

$$\begin{aligned}\vec{b}_{012}(t) &= (1-t)\cdot\vec{b}_{01}(t) + t\cdot\vec{b}_{12}(t) \\ &= (1-t)\left((1-t)\vec{p}_0 + t\vec{p}_1\right) + t\left((1-t)\vec{p}_1 + t\vec{p}_2\right) \\ &= (1-t)^2\vec{p}_0 + 2(1-t)t\vec{p}_1 + t^2\vec{p}_2. \qquad (14.2)\end{aligned}$$

The parametric equation of $\vec{b}_{012}(t)$ confirms it indeed is quadratic in the parameter t and it runs through its initial point $B_{012}(0) = P_0$ and through its terminal point $B_{012}(1) = P_2$, whilst curved by the attracting control point P_1.

We recall that three noncollinear points define exactly one plane (see figure 14.1). In general the control points P_0, P_1 and P_2 are noncollinear, which implies that their corresponding location vectors $\vec{p}_0, \vec{p}_1$ and $\vec{p}_2$ are at least plane vectors:

$$\vec{b}_{012}(t) = (1-t)^2 \begin{pmatrix} x_0 \\ y_0 \end{pmatrix} + 2(1-t)t \begin{pmatrix} x_1 \\ y_1 \end{pmatrix} + t^2 \begin{pmatrix} x_2 \\ y_2 \end{pmatrix}.$$

The plane linear and quadratic Bezier segments are default built-in in the 2D software Adobe Flash®.

Cubic Bezier segment

Let us plot a Bezier segment running from P_0 to P_3 and curved by two control points P_1 and P_2. Again starting at the parameter value $t_0 = 0$, we plot

- ▷ for every subsequent parameter value t_1 the points $B_{01}(t_1)$, $B_{12}(t_1)$ and $B_{23}(t_1)$,
- ▷ and on their fresh Bezier segments $[B_{01}B_{12}]$ and $[B_{12}B_{23}]$ the point $B(t_1)$ corresponding to the parameter value t_1, which we typeset as B_{012} and B_{123} respectively,
- ▷ and finally on the line segment $[B_{012}B_{123}]$ the point $B_{0123}(t_1)$ corresponding to the parameter value t_1,

and we iterate all three steps for the incrementing parameter values $t_2, t_3, \ldots, t_{n-1}$ up to $t_n = 1$.

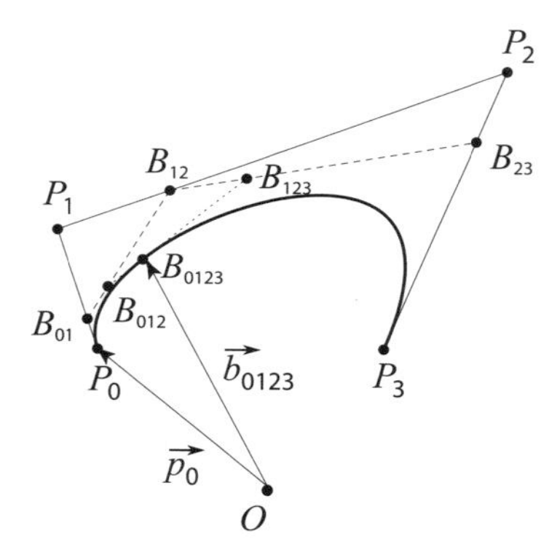

Figure 14.2: Cubic or 4-point Bezier segment

We define the obtained path $\vec{b}_{0123}(t)$ as the **cubic** or **4-point Bezier segment** defined by its control points P_0, P_1, P_2 and P_3. The control points are indexed by the Bezier indices $0, 1, 2$ and 3.

Based on the above plot algorithm, we elaborate its parametric equation as:

$$\begin{aligned}\vec{b}_{0123}(t) &= (1-t)\cdot\vec{b}_{012}(t)+t\cdot\vec{b}_{123}(t) \qquad \text{and via formula (14.2)}\\ &= (1-t)\left((1-t)^2\vec{p}_0+2(1-t)t\vec{p}_1+t^2\vec{p}_2\right)+t\left((1-t)^2\vec{p}_1+2(1-t)t\vec{p}_2+t^2\vec{p}_3\right)\\ &= (1-t)^3\vec{p}_0+3(1-t)^2t\vec{p}_1+3(1-t)t^2\vec{p}_2+t^3\vec{p}_3. \qquad (14.3)\end{aligned}$$

The parametric equation of $\vec{b}_{0123}(t)$ confirms it indeed is a cubic polynomial in t, running from its initial point $B_{0123}(0) = P_0$ to its terminal point $B_{0123}(1) = P_3$, whilst curved by the attracting control points P_1 and P_2.

We define four points P, Q, U and V as **coplanar** if they lie in a common plane. In general the four control points P_0, P_1, P_2 and P_3 are noncollinear and non coplanar, which implies that their corresponding location vectors $\vec{p}_0, \vec{p}_1, \vec{p}_2$ and $\vec{p}_3$ are at least 3D vectors:

$$\vec{b}_{0123}(t) = (1-t)^3\begin{pmatrix} x_0 \\ y_0 \\ z_0 \end{pmatrix} + 3(1-t)^2t\begin{pmatrix} x_1 \\ y_1 \\ z_1 \end{pmatrix} + 3(1-t)t^2\begin{pmatrix} x_2 \\ y_2 \\ z_2 \end{pmatrix} + t^3\begin{pmatrix} x_3 \\ y_3 \\ z_3 \end{pmatrix}.$$

The 3D Bezier segments are naturally built-in in 3D software such as CAD applications and 3ds Max®. Such segments offer - even in case of coplanarity - a much wider variety of curves, up to looped profiles. We note the order of the four control points indeed matters (see figure 14.3). The cubic plot routine features reusability for quadratic and linear Bezier segments, by selecting their control points backwards compatibly.

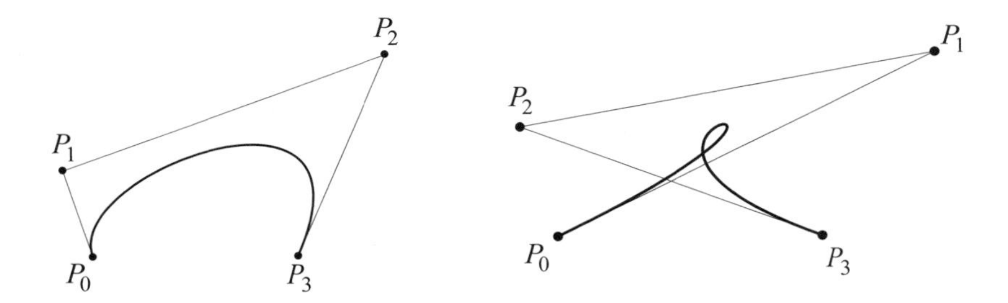

Figure 14.3: Cubic plane Bezier segment, regular versus looped

Example: The parametric equation of the cubic Bezier segment in 3D defined by the location vectors $\vec{p}_0 = \begin{pmatrix} 1 \\ 0 \\ 0 \end{pmatrix}$, $\vec{p}_1 = \begin{pmatrix} 0 \\ 1 \\ 0 \end{pmatrix}$, $\vec{p}_2 = \begin{pmatrix} 0 \\ 0 \\ 1 \end{pmatrix}$ and $\vec{p}_3 = \begin{pmatrix} 1 \\ 1 \\ 1 \end{pmatrix}$ simplifies to

$$\begin{aligned}
\vec{b}_{0123}(t) &= (1-t)^3\begin{pmatrix}1\\0\\0\end{pmatrix}+3(1-t)^2t\begin{pmatrix}0\\1\\0\end{pmatrix}+3(1-t)t^2\begin{pmatrix}0\\0\\1\end{pmatrix}+t^3\begin{pmatrix}1\\1\\1\end{pmatrix}\\
&= \begin{cases} x(t)=(1-t)^3+t^3\\ y(t)=3(1-t)^2t+t^3\\ z(t)=3(1-t)t^2+t^3\end{cases}\\
&= \begin{cases} x(t)=1-3t+3t^2\\ y(t)=3t-6t^2+4t^3\\ z(t)=3t^2-2t^3\end{cases}
\end{aligned}\tag{14.4}$$

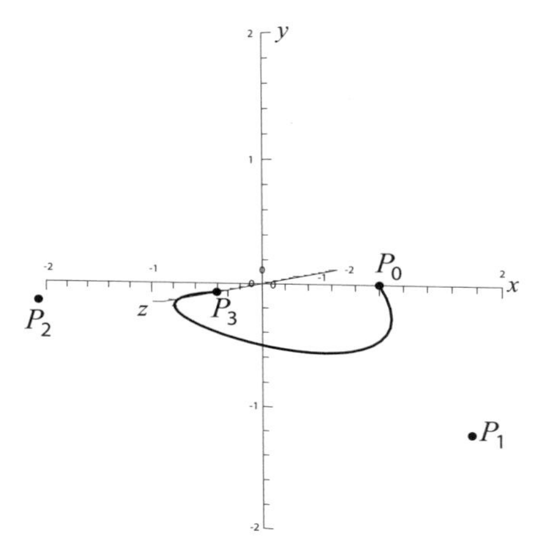

Figure 14.4: Cubic Bezier segment in 3D

Bezier segments of higher degree

We can extend the above routine to generally plot $(n+1)$-point Bezier segments running from P_0 to P_n and curved by intermediate control points $P_1, P_2, P_3, \ldots, P_{n-1}$ by applying Pierre Bézier's method. Skipping its details, we note that the resulting polynomial will keep fitting Pascal's Triangle (see page 24):

$$\vec{b}_{01234}(t)=\mathbf{1}(1-t)^4\vec{p}_0+\mathbf{4}(1-t)^3t\vec{p}_1+\mathbf{6}(1-t)^2t^2\vec{p}_2+\mathbf{4}(1-t)t^3\vec{p}_3+\mathbf{1}t^4\vec{p}_4.$$

14.2 De Casteljau algorithm

Simultaneously to Pierre Bézier at Renault, Paul de Casteljau developed in 1959 at Citroën his own construction algorithm for Bezier segments. We outline **de Casteljau construction** for a cubic Bezier segment defined by the four control points P_0, P_1, P_2 and P_3.

Construction: We plot the Bezier segment from its initial point P_0 to its terminal point P_3, curved by its control points P_1 and P_2. Both the points P_0 and P_3 lie by design on the Bezier segment.

1) We construct six midpoints (see page 45) like this:

   A as the midpoint of the line segment $[P_0P_1]$,

   B as the midpoint of the line segment $[P_2P_3]$,

   C as the midpoint of the line segment $[P_1P_2]$,

   A' as the midpoint of the line segment $[AC]$,

   B' as the midpoint of the line segment $[BC]$,

   C' as the midpoint of the line segment $[A'B']$.

2) This first step corresponds to Pierre Bézier's plot method for parameter value $t = \frac{1}{2}$. Its final midpoint C' equals the point $B_{0123}\left(\frac{1}{2}\right)$ and is it the only midpoint lying on the Bezier segment. Meanwhile the original set of control points P_0, P_1, P_2, P_3 splits into two new sets P_0, A, A', C' and C', B', B, P_3. For each of them we repeat the first step:

   ▷ plot the Bezier segment from the initial point P_0 to the terminal point C', curved by the control points A and A',

   ▷ plot the Bezier segment from the initial point C' to the terminal point P_3, curved by the control points B' and B.

   The above outlined de Casteljau construction is clearly recursive (see page 251).

3) Iterate the above steps until the distance between the initial and the terminal control points becomes smaller than a fixed minimum $\varepsilon > 0$, and then connect them both by a line segment.

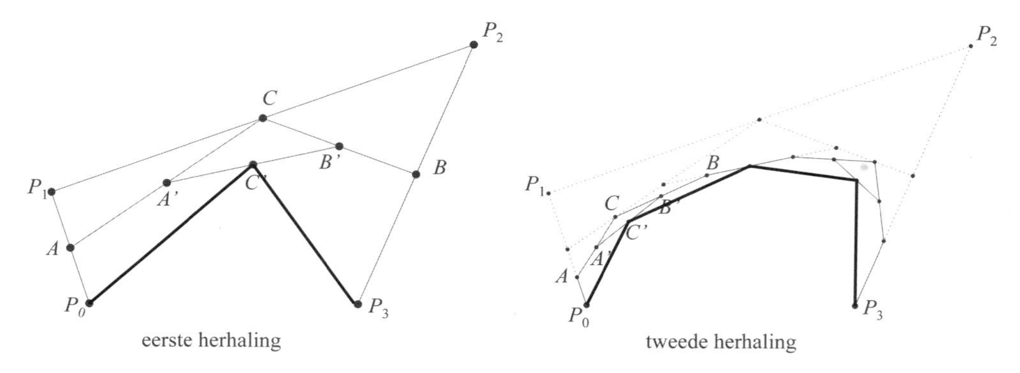

Figure 14.5: First two iterations of $\vec{b}_{0123}(t)$ via de Casteljau construction

Via de Casteljau construction we acquire the Bezier segment as a sequence of hundreds of connected points, starting at P_0 and ending at P_3. It can be proved that this recursively generated sequence of hundreds of connected midpoints equals the Bezier segment $\vec{b}_{0123}(t)$ as defined by the corresponding location vectors $\vec{p}_0, \vec{p}_1, \vec{p}_2$ and $\vec{p}_3$ (see formula (14.3)). We visualize the first two iterations of $\vec{b}_{0123}(t)$ by de Casteljau construction in figure 14.5. For the complete proof of its equivalence with figure 14.2, we refer to the expert literature. We note that the de Casteljau construction is **numerically stable** and has a fractal property, since it can be used to split a Bezier segment into two smaller Bezier segments.

14.3 Bezier curves

Concatenation

In case we smoothly concatenate two or more Bezier segments, for instance a cubic segment defined by P_0, P_1, P_2 and P_3, and a quadratic segment defined by Q_0, Q_1 and Q_2, by coinciding the terminal point P_3 and the initial point Q_0 as the contact point, we obtain a **Bezier curve**.

We define the concatenation having **positional** or **zeroth order continuity** in case the contact point $P_3 = Q_0$ is not lying on the line segment $[P_2Q_1]$. In this situation both segments meet continuously, however each segment has a different tangent line at the contact point $P_3 = Q_0$. Alternatively, we may call such a positional continuous concatenation **non smooth**. For a correct understanding of all continuity concepts, we emphasize that P_2P_3 is the tangent line to the first segment at P_3 and Q_0Q_1 is the tangent line to the second seg-

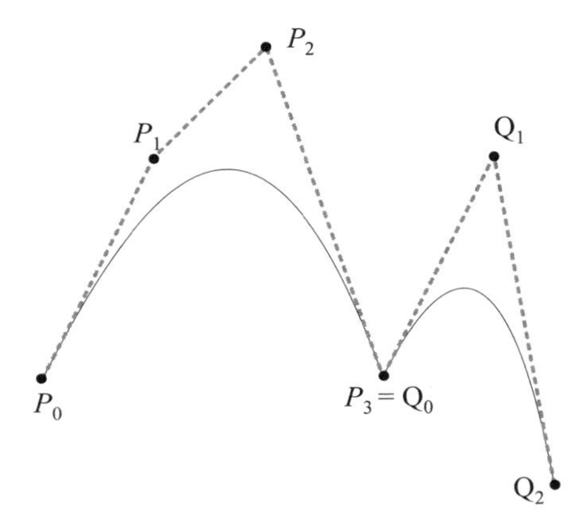

Figure 14.6: Positional or zeroth order continuous join of two Bezier segments

ment at Q_0. Knowing this immediately offers a deeper insight in the geometrical meaning of the intermediate control points such as P_2 and Q_1. Each Bezier segment is at its initial point P_0 tangent to the line P_0P_1 and at its terminal point P_n tangent to the line $P_{n-1}P_n$.

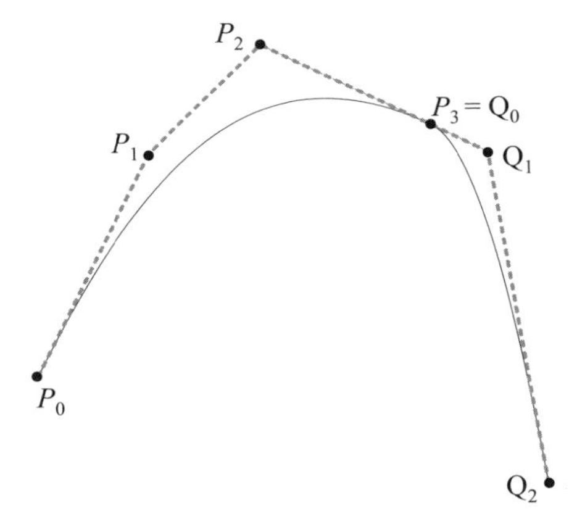

Figure 14.7: Tangential or first order continuous join of two Bezier segments

We define the concatenation having **tangential** or **first order continuity** in case the junction $P_3 = Q_0$ is lying on the line segment $[P_2Q_1]$. In this situation both segments meet continuously and each segment shares the same tangent line $P_2P_3 = Q_0Q_1$ at the junction $P_3 = Q_0$. Alternatively, we may call such a tangential continuous concatenation **smooth**.

For an outline of higher order continuities such as **curvature** or **second order continuity**, we refer to the expert literature on this topic.

Linear transformations

Linear transforming (translating, scaling, rotating or shearing) an object made out of Bezier segments, only requires transforming all of the control points of its Bezier segments.

Proof:

We prove this property for instance for a quadratic Bezier segment $\vec{b}_{012}(t)$, based on the linearity of the transformation operator L.

$$\begin{aligned}\vec{b}_{012}(t)' &= L(\vec{b}_{012}(t))\\ &= L\left((1-t)^2\vec{p}_0+2(1-t)t\vec{p}_1+t^2\vec{p}_2\right)\\ &= (1-t)^2L(\vec{p}_0)+2(1-t)tL(\vec{p}_1)+t^2L(\vec{p}_2)\\ &= (1-t)^2\vec{p}_0{}'+2(1-t)t\vec{p}_1{}'+t^2\vec{p}_2{}' \qquad \blacksquare\end{aligned}$$

Amongst all linear transformations L, especially the scaling S realizes a powerful graphical advantage through this property (see page 197). Due to their internal use of location vectors, applications and file types which visualize via Bezier segments, are called respectively vector graphics software (e.g. InkScape, Adobe Illustrator® and CorelDRAW®) and vector image formats (for instance .svg or scalable vector graphics). Scalable vector formats do not suffer **aliasing** by zooming, given the redrawing of their Bezier segments does not suffer any quality losses (see figure 14.8).

Figure 14.8: Scaling with and without loss of quality

Illustrations

Example 1: Continuous function graphs consist of an uncountable infinite amount of points $(x, f(x))$, impossible to display on a discrete screen. Therefore we need to sample a function f by a finite sequence of its points $(x_0, f(x_0)), (x_1, f(x_1)), (x_2, f(x_2)), \ldots, (x_n, f(x_n))$, which we display assembled by linear Bezier segments. In figure 14.9 we approximate in this way the graph of $\sin x$, initially by a sequence of only ten sample points $(x_0, \sin x_0), (x_1, \sin x_1), (x_2, \sin x_2), \ldots, (x_9, \sin x_9)$, which we subsequent double up to a Bezier curve of fourty segments.

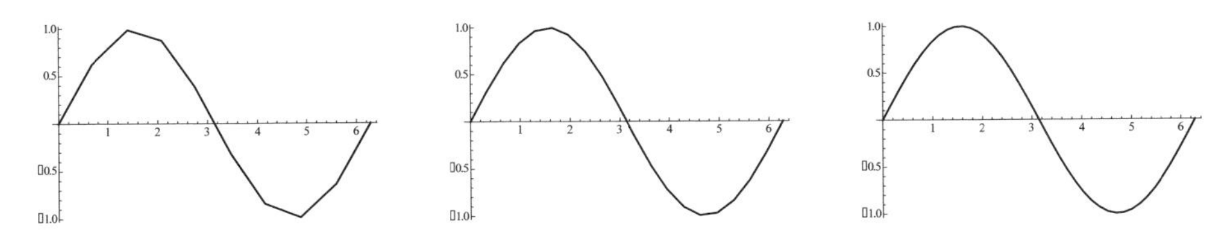

Figure 14.9: Approximating a continuous function graph by Bezier curves

Example 2: We illustrate **vector** or **outline fonts** such as TrueType® and OpenType® by designing a capital letter ‘A’ as a concatenation of ten linear and four quadratic Bezier segments. We recall that a linear transformation L of this vector font only requires the linear transformation L of all of its control points. This powerful property allows for reflecting, rotating or scaling such fonts without any loss of graphical quality. For instance italizing this vector font only requires the shearing of all of its control points along the x-axis.

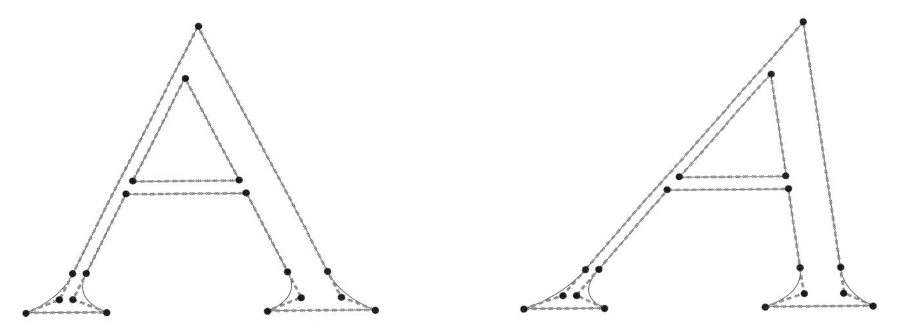

Figure 14.10: An outline or vector font is a Bezier curve

Example 3: We illustrate Bezier based shapes by designing a ‘circle’ approximated by a smoothly closed concatenation of four cubic Bezier segments (see figure 14.11). Of course, a mathematical circle is far from a Bezier curve, but a good approximation might convince and may offer a better plot performance. Moreover, Bezier based default shapes guarantee efficient transformation without loss of graphical quality.

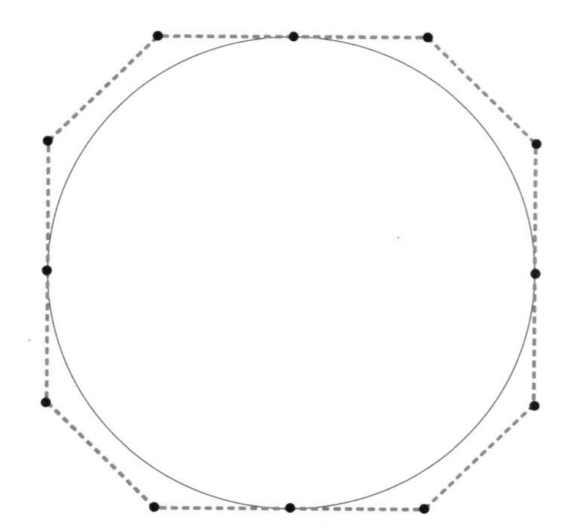

Figure 14.11: Approximating a circle by a closed Bezier curve

14.4 Matrix representation

LINEAR BEZIER SEGMENT

We present the matrix representation of the linear Bezier segment defined by two control points P_0 and P_1 from its parametric equation.

$$\begin{aligned}\vec{b}_{01}(t) &= (1-t)\vec{p}_0 + t\vec{p}_1 \quad \text{with} \quad t \in [0,1] \subset \mathbb{R} \\ &= \begin{pmatrix} (1-t) & t \end{pmatrix} \cdot \begin{pmatrix} \vec{p}_0 \\ \vec{p}_1 \end{pmatrix} \\ &= \begin{pmatrix} 1-t & t \end{pmatrix} \cdot \begin{pmatrix} \vec{p}_0 \\ \vec{p}_1 \end{pmatrix} \\ &= \begin{pmatrix} 1 & t \end{pmatrix} \cdot \begin{pmatrix} 1 & 0 \\ -1 & 1 \end{pmatrix} \cdot \begin{pmatrix} \vec{p}_0 \\ \vec{p}_1 \end{pmatrix}\end{aligned}$$

We express the linear Bezier segment as the dot product of three matrices $\vec{b}_{01}(t) = T \cdot B \cdot P$, given

$$T = \begin{pmatrix} 1 & t \end{pmatrix}$$

containing Bezier's **parameter base** and

$$B = \begin{pmatrix} 1 & 0 \\ -1 & 1 \end{pmatrix}$$

which we define as its **characteristic coefficient matrix**. Expanding this matrix product via $\vec{b}_{01}(t) = T \cdot (B \cdot P)$, we encounter

$$
\begin{aligned}
\vec{b}_{01}(t) &= \begin{pmatrix} 1 & t \end{pmatrix} \cdot \left(\begin{pmatrix} 1 & 0 \\ -1 & 1 \end{pmatrix} \cdot \begin{pmatrix} \vec{p}_0 \\ \vec{p}_1 \end{pmatrix} \right) \\
&= \begin{pmatrix} 1 & t \end{pmatrix} \cdot \begin{pmatrix} \vec{p}_0 \\ -\vec{p}_0 + \vec{p}_1 \end{pmatrix} \\
&= \vec{p}_0 + t(\vec{p}_1 - \vec{p}_0) \\
&= \vec{p}_0 + t\overrightarrow{P_0P_1}
\end{aligned}
$$

as the parametric equation of the line segment $[P_0P_1]$ (see page 272).

Quadratic Bezier segment

We present the matrix representation of the quadratic Bezier segment defined by three control points P_0, P_1 and P_2 from its parametric equation.

$$
\begin{aligned}
\vec{b}_{012}(t) &= (1-t)^2\vec{p}_0 + 2(1-t)t\vec{p}_1 + t^2\vec{p}_2 \quad \text{with} \quad t \in [0,1] \subset \mathbb{R} \\
&= \begin{pmatrix} (1-t)^2 & 2(1-t)t & t^2 \end{pmatrix} \cdot \begin{pmatrix} \vec{p}_0 \\ \vec{p}_1 \\ \vec{p}_2 \end{pmatrix} \\
&= \begin{pmatrix} 1-2t+t^2 & 2t-2t^2 & t^2 \end{pmatrix} \cdot \begin{pmatrix} \vec{p}_0 \\ \vec{p}_1 \\ \vec{p}_2 \end{pmatrix} \\
&= \begin{pmatrix} 1 & t & t^2 \end{pmatrix} \cdot \begin{pmatrix} 1 & 0 & 0 \\ -2 & 2 & 0 \\ 1 & -2 & 1 \end{pmatrix} \cdot \begin{pmatrix} \vec{p}_0 \\ \vec{p}_1 \\ \vec{p}_2 \end{pmatrix}
\end{aligned}
$$

We express the quadratic Bezier segment as the dot product of three matrices $\vec{b}_{012}(t) = T \cdot B \cdot P$, given its parameter base

$$
T = \begin{pmatrix} 1 & t & t^2 \end{pmatrix} \quad \text{and} \quad B = \begin{pmatrix} 1 & 0 & 0 \\ -2 & 2 & 0 \\ 1 & -2 & 1 \end{pmatrix}
$$

known as its characteristic coefficient matrix.

Expanding the matrix product in a reversed order as $\vec{b}_{012}(t) = T \cdot (B \cdot P)$, leads to

$$\begin{aligned}
\vec{b}_{012}(t) &= \begin{pmatrix} 1 & t & t^2 \end{pmatrix} \cdot \left(\begin{pmatrix} 1 & 0 & 0 \\ -2 & 2 & 0 \\ 1 & -2 & 1 \end{pmatrix} \cdot \begin{pmatrix} \vec{p}_0 \\ \vec{p}_1 \\ \vec{p}_2 \end{pmatrix} \right) \\
&= \begin{pmatrix} 1 & t & t^2 \end{pmatrix} \cdot \begin{pmatrix} \vec{p}_0 \\ -2\vec{p}_0 + 2\vec{p}_1 \\ \vec{p}_0 - 2\vec{p}_1 + \vec{p}_2 \end{pmatrix} \\
&= \vec{p}_0 + 2(-\vec{p}_0 + \vec{p}_1)t + (\vec{p}_0 - 2\vec{p}_1 + \vec{p}_2)t^2.
\end{aligned}$$

We define such a reversed expansion of the matrix product as the **performant equivalent** of the quadratic Bezier segment,

$$\vec{b}_{012}(t) = \vec{p}_0 + 2(-\vec{p}_0 + \vec{p}_1)t + (\vec{p}_0 - 2\vec{p}_1 + \vec{p}_2)t^2.$$

Cubic Bezier segment

To present the matrix representation of the cubic Bezier segment defined by four control points P_0, P_1, P_2 and P_3, we rewrite its parametric equation

$$\vec{b}_{0123}(t) = (1-t)^3\vec{p}_0 + 3(1-t)^2 t\vec{p}_1 + 3(1-t)t^2\vec{p}_2 + t^3\vec{p}_3$$

to the matrix product

$$\vec{b}_{0123}(t) = \begin{pmatrix} (1-t)^3 & 3t(1-t)^2 & 3t^2(1-t) & t^3 \end{pmatrix} \cdot \begin{pmatrix} \vec{p}_0 \\ \vec{p}_1 \\ \vec{p}_2 \\ \vec{p}_3 \end{pmatrix}.$$

We expand the elements of the row matrix to

$$\vec{b}_{0123}(t) = \begin{pmatrix} 1 - 3t + 3t^2 - t^3 & 3t - 6t^2 + 3t^3 & 3t^2 - 3t^3 & t^3 \end{pmatrix} \cdot \begin{pmatrix} \vec{p}_0 \\ \vec{p}_1 \\ \vec{p}_2 \\ \vec{p}_3 \end{pmatrix}.$$

Consequently, we again rewrite this row matrix into a dot product.

$$\vec{b}_{0123}(t) = \left(\begin{pmatrix} 1 & t & t^2 & t^3 \end{pmatrix} \cdot \begin{pmatrix} 1 & 0 & 0 & 0 \\ -3 & 3 & 0 & 0 \\ 3 & -6 & 3 & 0 \\ -1 & 3 & -3 & 1 \end{pmatrix} \right) \cdot \begin{pmatrix} \vec{p}_0 \\ \vec{p}_1 \\ \vec{p}_2 \\ \vec{p}_3 \end{pmatrix}. \tag{14.5}$$

The first matrix factor is the parameter base T. The second matrix factor is the characteristic coefficient matrix of a cubic Bezier segment. Subsequently we apply the associative property of the dot product $(T \cdot B) \cdot P = T \cdot (B \cdot P)$, to further rewrite this matrix representation.

$$\vec{b}_{0123}(t) = \begin{pmatrix} 1 & t & t^2 & t^3 \end{pmatrix} \cdot \left(\begin{pmatrix} 1 & 0 & 0 & 0 \\ -3 & 3 & 0 & 0 \\ 3 & -6 & 3 & 0 \\ -1 & 3 & -3 & 1 \end{pmatrix} \cdot \begin{pmatrix} \vec{p}_0 \\ \vec{p}_1 \\ \vec{p}_2 \\ \vec{p}_3 \end{pmatrix} \right)$$

$$= \begin{pmatrix} 1 & t & t^2 & t^3 \end{pmatrix} \cdot \begin{pmatrix} \vec{p}_0 \\ -3(\vec{p}_0 - \vec{p}_1) \\ 3(\vec{p}_0 - 2\vec{p}_1 + \vec{p}_2) \\ -\vec{p}_0 + 3\vec{p}_1 - 3\vec{p}_2 + \vec{p}_3 \end{pmatrix}$$

This reordered expansion of the matrix representation eventually yields the performant equivalent of the cubic Bezier segment,

$$\vec{b}_{0123}(t) = \vec{p}_0 - 3(\vec{p}_0 - \vec{p}_1)t + 3(\vec{p}_0 - 2\vec{p}_1 + \vec{p}_2)t^2 + (-\vec{p}_0 + 3\vec{p}_1 - 3\vec{p}_2 + \vec{p}_3)t^3.$$

At a first glance the performant equivalent equation does not simplify the initial parametric equation, but it may offer some plot performance compared to the previous methods (see pages 275 and 277).

For each parameter value $t_0 = 0, t_1, t_2, t_3, \ldots, t_{n-1}, t_n = 1$ during the plot session, a smaller amount of arithmetic operations is required. This is due to the stored constant coefficients $-3(\vec{p}_0 - \vec{p}_1)$ and $3(\vec{p}_0 - 2\vec{p}_1 + \vec{p}_2)$ and $(-\vec{p}_0 + 3\vec{p}_1 - 3\vec{p}_2 + \vec{p}_3)$, which only need to be multiplied with powers of t. We recall that the parametric equation (14.3) on the contrary requires a complete recalculation for each new parameter value t. Even the pen and paper practicing of Bezier segments will benefit from the performant equivalent equation, because of the easy evaluation of its control point location vectors $\vec{p}_0, \vec{p}_1, \vec{p}_2$ and $\vec{p}_3$.

Example: We evaluate the performant equivalent equation of the cubic Bezier segment in 3D for the location vectors $\vec{p}_0 = \begin{pmatrix} 1 \\ 0 \\ 0 \end{pmatrix}$, $\vec{p}_1 = \begin{pmatrix} 0 \\ 1 \\ 0 \end{pmatrix}$, $\vec{p}_2 = \begin{pmatrix} 0 \\ 0 \\ 1 \end{pmatrix}$ and $\vec{p}_3 = \begin{pmatrix} 1 \\ 1 \\ 1 \end{pmatrix}$.

$$\begin{aligned}
\vec{b}_{0123}(t) &= \begin{pmatrix}1\\0\\0\end{pmatrix} - 3\left(\begin{pmatrix}1\\0\\0\end{pmatrix} - \begin{pmatrix}0\\1\\0\end{pmatrix}\right)t \\
&\quad +3\left(\begin{pmatrix}1\\0\\0\end{pmatrix} - 2\begin{pmatrix}0\\1\\0\end{pmatrix} + \begin{pmatrix}0\\0\\1\end{pmatrix}\right)t^2 \\
&\quad + \left(-\begin{pmatrix}1\\0\\0\end{pmatrix} + 3\begin{pmatrix}0\\1\\0\end{pmatrix} - 3\begin{pmatrix}0\\0\\1\end{pmatrix} + \begin{pmatrix}1\\1\\1\end{pmatrix}\right)t^3 \\
&= \begin{pmatrix}1\\0\\0\end{pmatrix} - 3\begin{pmatrix}1\\-1\\0\end{pmatrix}t + 3\begin{pmatrix}1\\-2\\1\end{pmatrix}t^2 + \begin{pmatrix}0\\4\\-2\end{pmatrix}t^3 \\
&= \begin{cases} x(t) = 1 - 3t + 3t^2 \\ y(t) = 3t - 6t^2 + 4t^3 \\ z(t) = 3t^2 - 2t^3 \end{cases}
\end{aligned}$$

We note that the evaluated result of the performant equivalent equation indeed corresponds to the parametric equation (14.4) of the same Bezier segment $\vec{b}_{0123}(t)$.

14.5 B-splines

Bezier segments make a subset of the so-called **B-splines**, referring to Bezier and a spline, meaning flexible strip (of wood or rubber). In other words, there are many more techniques to assemble a finite sequence of points into a smooth curve.

Cubic B-splines

Cubic B-splines generally need to contain not even one of their four control points P_0, P_1, P_2 or P_3. Constructing a B-spline $\vec{s}_{0123}(t)$ is analogous to the former construction of a 4-point Bezier segment $\vec{b}_{0123}(t)$, for which we refer to further expert literature. Defining B-splines exactly is technically involving, for which we once more refer to the specific literature. In this book we illustrate B-splines by constructing only one cubic B-spline through its matrix representation.

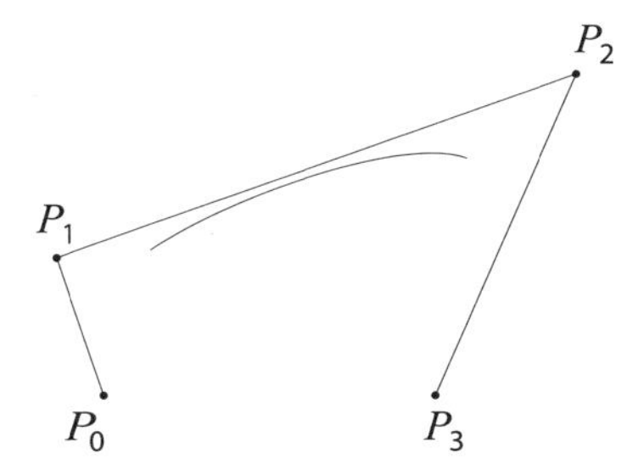

Figure 14.12: Cubic B-spline defined by four control points

Matrix representation

Skipping its exact definition, we are able to express a cubic B-spline $\vec{s}_{0123}(t)$ defined by the control points P_0, P_1, P_2 and P_3, analogously to the former matrix product for ordinary Bezier segments (14.5),

$$\vec{s}_{0123}(t) = \begin{pmatrix} 1 & t & t^2 & t^3 \end{pmatrix} \cdot \frac{1}{6} \begin{pmatrix} 1 & 4 & 1 & 0 \\ -3 & 0 & 3 & 0 \\ 3 & -6 & 3 & 0 \\ -1 & 3 & -3 & 1 \end{pmatrix} \cdot \begin{pmatrix} \vec{p}_0 \\ \vec{p}_1 \\ \vec{p}_2 \\ \vec{p}_3 \end{pmatrix}. \tag{14.6}$$

Completely similar to the Bezier segment, the parameter t runs through the real unity interval $t \in [0,1] \subset \mathbb{R}$, while the characteristic coefficient matrix

$$B = \frac{1}{6} \begin{pmatrix} 1 & 4 & 1 & 0 \\ -3 & 0 & 3 & 0 \\ 3 & -6 & 3 & 0 \\ -1 & 3 & -3 & 1 \end{pmatrix}, \tag{14.7}$$

determines a specified B-spline $\vec{s}_{0123}(t)$. Similarly, the row matrix

$$T = \begin{pmatrix} 1 & t & t^2 & t^3 \end{pmatrix}$$

contains the parameter base for a 4-point B-spline.

By expanding the above dot product of matrices as $T \cdot (B \cdot P)$, we obtain the performant equivalent equation of the B-spline $\vec{s}_{0123}(t)$ defined by $\vec{p}_0, \vec{p}_1, \vec{p}_2$ and $\vec{p}_3$.

$$\vec{s}_{0123}(t) = \begin{pmatrix} 1 & t & t^2 & t^3 \end{pmatrix} \cdot \left(\frac{1}{6} \begin{pmatrix} 1 & 4 & 1 & 0 \\ -3 & 0 & 3 & 0 \\ 3 & -6 & 3 & 0 \\ -1 & 3 & -3 & 1 \end{pmatrix} \cdot \begin{pmatrix} \vec{p}_0 \\ \vec{p}_1 \\ \vec{p}_2 \\ \vec{p}_3 \end{pmatrix} \right)$$

$$= \frac{1}{6} \begin{pmatrix} 1 & t & t^2 & t^3 \end{pmatrix} \cdot \begin{pmatrix} \vec{p}_0 + 4\vec{p}_1 + \vec{p}_2 \\ -3\vec{p}_0 + 3\vec{p}_2 \\ 3\vec{p}_0 - 6\vec{p}_1 + 3\vec{p}_2 \\ -\vec{p}_0 + 3\vec{p}_1 - 3\vec{p}_2 + \vec{p}_3 \end{pmatrix}$$

We finalize the matrix expansion which results in the performant equivalent equation,

$$\begin{aligned} \vec{s}_{0123}(t) &= \frac{1}{6}(\vec{p}_0 + 4\vec{p}_1 + \vec{p}_2) + \frac{1}{2}(-\vec{p}_0 + \vec{p}_2)t \\ &+ \frac{1}{2}(\vec{p}_0 - 2\vec{p}_1 + \vec{p}_2)t^2 + \frac{1}{6}(-\vec{p}_0 + 3\vec{p}_1 - 3\vec{p}_2 + \vec{p}_3)t^3. \end{aligned} \tag{14.8}$$

The performant equivalent expression reveals this B-spline $\vec{s}_{0123}(t)$ does not contain either its initial control point $P_0 \neq S_{0123}(0)$ nor its terminal control point $P_3 \neq S_{0123}(1)$. Its stored constant coefficients $\frac{1}{6}(\vec{p}_0 + 4\vec{p}_1 + \vec{p}_2)$, $\frac{1}{2}(-\vec{p}_0 + \vec{p}_2)$, $\frac{1}{2}(\vec{p}_0 - 2\vec{p}_1 + \vec{p}_2)$ and $\frac{1}{6}(-\vec{p}_0 + 3\vec{p}_1 - 3\vec{p}_2 + \vec{p}_3)$ are calculated independently from the running parameter t.

Example: We evaluate the performant equivalent equation of the cubic B-spline in 3D for the location vectors $\vec{p}_0 = \begin{pmatrix} 1 \\ 0 \\ 0 \end{pmatrix}$, $\vec{p}_1 = \begin{pmatrix} 0 \\ 1 \\ 0 \end{pmatrix}$, $\vec{p}_2 = \begin{pmatrix} 0 \\ 0 \\ 1 \end{pmatrix}$ and $\vec{p}_3 = \begin{pmatrix} 1 \\ 1 \\ 1 \end{pmatrix}$.

$$\begin{aligned} \vec{s}_{0123}(t) &= \frac{1}{6}\left(\begin{pmatrix} 1 \\ 0 \\ 0 \end{pmatrix} + 4 \begin{pmatrix} 0 \\ 1 \\ 0 \end{pmatrix} + \begin{pmatrix} 0 \\ 0 \\ 1 \end{pmatrix} \right) + \frac{1}{2}\left(- \begin{pmatrix} 1 \\ 0 \\ 0 \end{pmatrix} + \begin{pmatrix} 0 \\ 0 \\ 1 \end{pmatrix} \right) t \\ &+ \frac{1}{2}\left(\begin{pmatrix} 1 \\ 0 \\ 0 \end{pmatrix} - 2 \begin{pmatrix} 0 \\ 1 \\ 0 \end{pmatrix} + \begin{pmatrix} 0 \\ 0 \\ 1 \end{pmatrix} \right) t^2 \\ &+ \frac{1}{6}\left(- \begin{pmatrix} 1 \\ 0 \\ 0 \end{pmatrix} + 3 \begin{pmatrix} 0 \\ 1 \\ 0 \end{pmatrix} - 3 \begin{pmatrix} 0 \\ 0 \\ 1 \end{pmatrix} + \begin{pmatrix} 1 \\ 1 \\ 1 \end{pmatrix} \right) t^3 \end{aligned}$$

$$\vec{s}_{0123}(t) = \frac{1}{6}\begin{pmatrix} 1 \\ 4 \\ 1 \end{pmatrix} + \frac{1}{2}\begin{pmatrix} -1 \\ 0 \\ 1 \end{pmatrix} t + \frac{1}{2}\begin{pmatrix} 1 \\ -2 \\ 1 \end{pmatrix} t^2 + \frac{1}{6}\begin{pmatrix} 0 \\ 4 \\ -2 \end{pmatrix} t^3$$

$$= \begin{cases} x(t) = \frac{1}{6} - \frac{1}{2}t + \frac{1}{2}t^2 \\ y(t) = \frac{2}{3} - t^2 + \frac{2}{3}t^3 \\ z(t) = \frac{1}{6} + \frac{1}{2}t + \frac{1}{2}t^2 - \frac{1}{3}t^3 \end{cases}$$

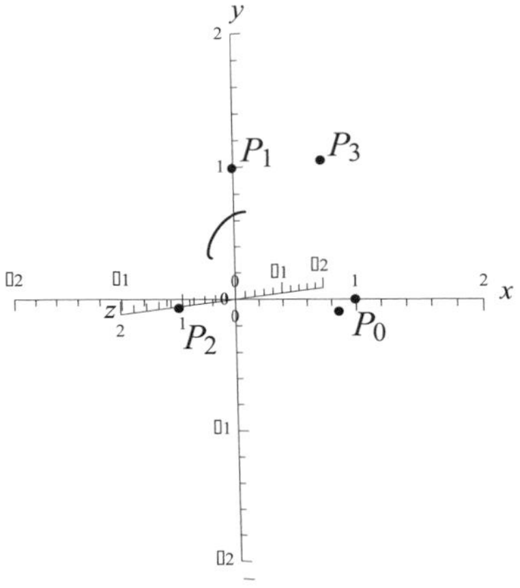

Figure 14.13: Cubic B-spline in 3D space

We note that our exemplary B-spline $\vec{s}_{0123}(t)$ differs from the ordinary Bezier segment (14.4) defined by the same control points as headed by their location vectors

$$\vec{p}_0 = \begin{pmatrix} 1 \\ 0 \\ 0 \end{pmatrix}, \vec{p}_1 = \begin{pmatrix} 0 \\ 1 \\ 0 \end{pmatrix}, \vec{p}_2 = \begin{pmatrix} 0 \\ 0 \\ 1 \end{pmatrix} \text{ and } \vec{p}_3 = \begin{pmatrix} 1 \\ 1 \\ 1 \end{pmatrix}.$$

De Boor's algorithm

The German-American mathematician **Carl de Boor** (1937) generalized the popular de Casteljau construction to his **de Boor construction** as the numerically stable alternative to draw or to split B-splines. We skip a complete outline of the de Boor construction, for which we refer to the expert literature. B-splines and B-spline surfaces still reign our contemporary computer aided design (see figures 14.14 and 14.15), ending this chapter where it historically started.

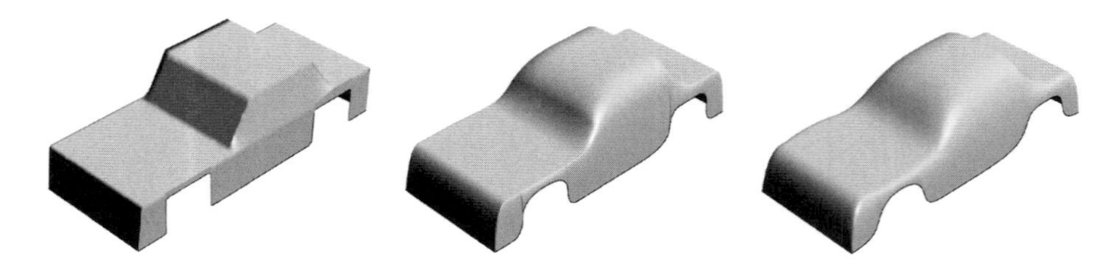

Figure 14.14: CAD using linear, quadratic and cubic B-spline surfaces

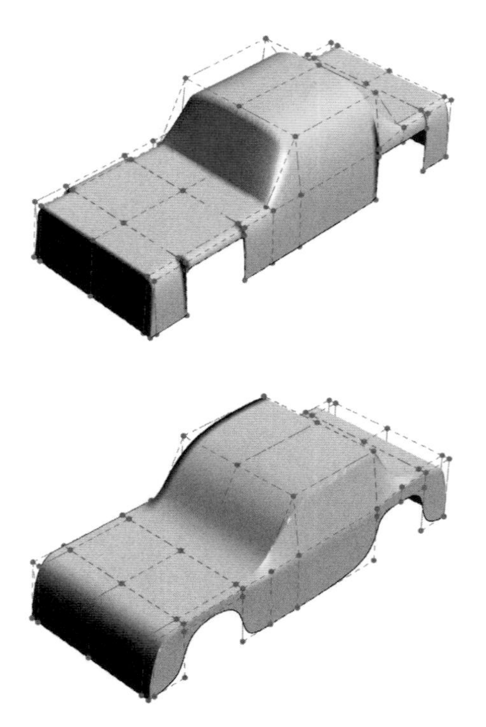

Figure 14.15: CAD applying linear and cubic B-splines orthogonally

14.6 Exercises

Exercise 124 Verify that we can express a linear Bezier segment $\vec{b}_{01}(t)$ as a quadratic Bezier segment $\vec{b}_{021}(t)$ by locating its *intermediate* location vector $\vec{p}_2 = \dfrac{\vec{p}_0 + \vec{p}_1}{2}$ at the midpoint of $[P_0P_1]$.

Exercise 125 Calculate the parametric components $x(t), y(t)$ and $z(t)$ of the cubic Bezier segment described by the parametric equation

$$\vec{b}_{0123}(t) = (1-t)^3\vec{p}_0 + 3(1-t)^2 t\vec{p}_1 + 3(1-t)t^2\vec{p}_2 + t^3\vec{p}_3$$

and the location vectors $\vec{p}_0 = \begin{pmatrix} 0 \\ 3 \\ 0 \end{pmatrix}$, $\vec{p}_1 = \begin{pmatrix} 1 \\ -1 \\ 1 \end{pmatrix}$, $\vec{p}_2 = \begin{pmatrix} -2 \\ 4 \\ 0 \end{pmatrix}$ and $\vec{p}_3 = \begin{pmatrix} 3 \\ 0 \\ 2 \end{pmatrix}$.

Use a computer to plot this Bezier segment in 3D space.

Exercise 126 For the parametric equations of Bezier segments, the sum of their coefficients in t equals 1. This is easily verified in case of linear Bezier segments via $(1-t)+t = 1$. Verify this property in case of quadratic and cubic Bezier segments.

Exercise 127 Draw the control points $P_0(-2,-1)$, $P_1(0,3)$, $P_2(4,6)$ and $P_3(5,0)$ in three different (x,y)-frames. Dedicate one by one an (x,y)-frame to draw

1) the linear Bezier segment $\vec{b}_{03}(t)$ from P_0 to P_3,
2) the third iteration from the de Casteljau construction of the quadratic Bezier segment from P_0 to P_3 curved by the control point P_2,
3) the second iteration from the de Casteljau construction of the cubic Bezier segment $\vec{b}_{0123}(t)$ defined by the control points P_0, P_1, P_2 and P_3.

Exercise 128 Calculate the 2D components $x(t)$ and $y(t)$ based upon the performant equivalent equations of:

1) the linear Bezier segment $\vec{b}_{03}(t)$ from P_0 to P_3,
2) the quadratic Bezier segment from P_0 to P_3 curved by the control point P_2,
3) the cubic Bezier segment $\vec{b}_{0123}(t)$ defined by P_0, P_1, P_2 and P_3,

and the location vectors $\vec{p}_0 = \begin{pmatrix} -2 \\ -1 \end{pmatrix}$, $\vec{p}_1 = \begin{pmatrix} 0 \\ 3 \end{pmatrix}$, $\vec{p}_2 = \begin{pmatrix} 4 \\ 6 \end{pmatrix}$ and $\vec{p}_3 = \begin{pmatrix} 5 \\ 0 \end{pmatrix}$.

Use a computer to plot each of these Bezier segments in 2D space.

Exercise 129 Calculate the components $x(t)$ and $y(t)$ via the performant equivalents of:

1) $\vec{b}_{0123}(t)$ given $\vec{p}_0 = \begin{pmatrix} -2 \\ -1 \end{pmatrix}$, $\vec{p}_1 = \begin{pmatrix} -3 \\ 3 \end{pmatrix}$, $\vec{p}_2 = \begin{pmatrix} 4 \\ 6 \end{pmatrix}$ and $\vec{p}_3 = \begin{pmatrix} 0 \\ 0 \end{pmatrix}$.

2) $\vec{b}_{0123}(t)$ given $\vec{p}_0 = \begin{pmatrix} -2 \\ -1 \end{pmatrix}$, $\vec{p}_1 = \begin{pmatrix} 4 \\ 6 \end{pmatrix}$, $\vec{p}_2 = \begin{pmatrix} -3 \\ 3 \end{pmatrix}$ and $\vec{p}_3 = \begin{pmatrix} 0 \\ 0 \end{pmatrix}$.

Use a computer to plot each of these Bezier segments in an (x,y)-frame. Anything remarkable to notice?

Exercise 130 Calculate the components $x(t)$ and $y(t)$ of both the cubic Bezier segment $\vec{b}_{0123}(t)$ and of the B-spline $\vec{s}_{0123}(t)$, based upon their respective matrix representations (14.5) and (14.6), and given the location vectors
$\vec{p}_0 = \begin{pmatrix} -2 \\ -1 \end{pmatrix}$, $\vec{p}_1 = \begin{pmatrix} -3 \\ 3 \end{pmatrix}$, $\vec{p}_2 = \begin{pmatrix} 4 \\ 6 \end{pmatrix}$ and $\vec{p}_3 = \begin{pmatrix} 0 \\ 0 \end{pmatrix}$.

Use a computer to plot both segments $\vec{b}_{0123}(t)$ and $\vec{s}_{0123}(t)$ in a different colour in the same (x,y)-frame.

Exercise 131 Verify whether this B-spline, in its performant equivalent equation,

$$\begin{aligned}\vec{s}_{0123}(t) &= \frac{1}{6}(\vec{p}_0 + 4\vec{p}_1 + \vec{p}_2) + \frac{1}{2}(-\vec{p}_0 + \vec{p}_2)t \\ &\quad + \frac{1}{2}(\vec{p}_0 - 2\vec{p}_1 + \vec{p}_2)t^2 + \frac{1}{6}(-\vec{p}_0 + 3\vec{p}_1 - 3\vec{p}_2 + \vec{p}_3)t^3,\end{aligned}$$

contains its initial and its terminal control point.

Exercise 132 Calculate the components $x(t)$ and $y(t)$ of the cubic B-spline $\vec{s}_{0123}(t)$ described by its performant equivalent equation (14.8), given the location vectors
$\vec{p}_0 = \begin{pmatrix} 0 \\ 3 \end{pmatrix}$, $\vec{p}_1 = \begin{pmatrix} 1 \\ 1 \end{pmatrix}$, $\vec{p}_2 = \begin{pmatrix} -2 \\ 0 \end{pmatrix}$ and $\vec{p}_3 = \begin{pmatrix} 1 \\ 0 \end{pmatrix}$.

Exercise 133 Given $\vec{p}_0 = \begin{pmatrix} -1 \\ 0 \end{pmatrix}$, $\vec{p}_1 = \begin{pmatrix} 1 \\ 2 \end{pmatrix}$, $\vec{p}_2 = \begin{pmatrix} -2 \\ 1 \end{pmatrix}$ and $\vec{p}_3 = \begin{pmatrix} 0 \\ 3 \end{pmatrix}$, find

1) the linear Bezier segment $\vec{b}_{01}(t)$ from V_0 to V_1,
2) the quadratic Bezier segment $\vec{b}_{012}(t)$ from V_0 to V_2,
3) and eventually the cubic Bezier segment $\vec{b}_{0123}(t)$ from V_0 to V_3.
4) Eliminate the parameter t from the parametric equation of the linear Bezier segment $\vec{b}_{01}(t)$, in order to find the linear function by its explicit recipe $y = ax + b$.
5) Find the intersection of the linear Bezier segment $\vec{b}_{01}(t)$ and the cubic Bezier segment $\vec{b}_{0123}(t)$. Hint: substitute the above step 3 in step 4.

Annex A · Real numbers in computers

Real numbers are stored identically into the computer. From the irrational numbers such as π to giant integers such as 10^{10}, radicals and negative fractions, they all fit into the machine in the same way. Let us add also -1 billion to our example list.

A.1 Scientific notation

The storage of real numbers into computers is based on their **scientific notation** which separates the sign and the precision from the order of magnitude of each exact number x, arranged into the product

$$x = (-1)^t \times M_{10} \times 10^{E_{10}}.$$

The first factor $(-1)^t$ shows the **sign** of x, the second factor M_{10} is the **decimal mantissa** lying between 1 and 10 and finally the exponent E_{10} indicates the **decimal order of magnitude** of x.

exact value x	decimally displayed	decimally scientifically displayed
$\frac{1}{10}$	0.1	$+1 \times 10^{-1}$
π	3.141592653...	$+3.141592653\ldots \times 10^0$
0.00001234	0.00001234	$+1.234 \times 10^{-5}$
-1 billion	$-1000000000.$	-1.000000000×10^9

This normalized scientific notation allows us to simulate the storage of our real examples into a decimal machine which allocates a standardized digit sequence for each of them.

A.2 The decimal computer

Let us straightforwardly consider a decimal computer which stores one digit denoting the sign, one digit indicating the order of magnitude and stores four **significant digits** of the original value x.

scientific notation	uniform machine precision	stored **machine number** x'
$+1 \times 10^{-1}$	$(-1)^0 \times 1 \times 10^{-1}$	$(-1)^0 \times 1.000 \times 10^{-1}$
$+3.141592653\ldots \times 10^0$	$(-1)^0 \times 3.142 \times 10^0$	$(-1)^0 \times 3.142 \times 10^0$
$+1.234 \times 10^{-5}$	$(-1)^0 \times 1.234 \times 10^{-5}$	$(-1)^0 \times 1.234 \times 10^{-5}$
-1.000000000×10^9	$(-1)^1 \times 1.000 \times 10^9$	$(-1)^1 \times 1.000 \times 10^9$

Our simplified decimal computer stores exact values $x \in \mathbb{R}$ systematically in a fixed digit sequence x' containing the sign (1 digit), the exponent (1 digit) and the mantissa (4 digits). This computer is limited to store only four mantissa digits and consequently standardizes its stored numbers x' with a fixed **machine precision** (of 4 significant digits). We call the finite subset of real numbers x' which are inevitably **rounded** to fit into the computer, **machine numbers**. The accompanying figure shows all positive machine numbers, from the smallest to the largest one in $\mathbb{R}^+$.

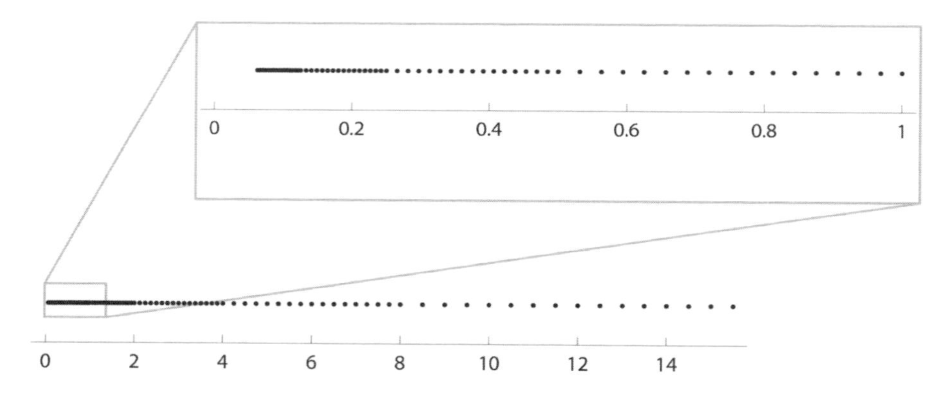

Figure A.1: The subset of machine numbers x' in $\mathbb{R}^+$

A.3 Special values

Calculations which result smaller than the smallest machine number, suffer **real underflow**. Arithmetical outputs which are larger than the largest machine number, feature **real overflow**. For instance storing the real number *zero* is an issue, since it would require the exponent $E_{10} = -\infty$. To be able to store the number zero, and similarly the *infinities* requiring exponent $E_{10} = +\infty$ and the indeterminate such as $\frac{0}{0}$ in our decimal machine, we predefine these exceptions respectively as **NULL**, **INFINITY** and **NAN** abbreviating 'Not A Number'.

Annex B · Notations and Conventions

B.1 Alphabets

Latin alphabet

meaning	symbol
constants and coefficients	$a, b, c, \ldots$
unknown quantities and variables	$x, y, z, \ldots$
points	$P, Q, R, \ldots$
lines	$r, s, t, \ldots$
planes	$\nu_R, \nu_P, \ldots$
vectors	$\vec{v}, \vec{w}, \ldots$
matrices	$A, B, C, \ldots$
angles	$\hat{A}, \hat{B}, \hat{C}, \ldots$
Bezier segment	$\vec{b}_{012\ldots n}$
B-spline	$\vec{s}_{012\ldots n}$

Greek alphabet

Traditionally, we use Greek characters to denote angles (especially in trigonometry). We also choose Greek characters for typesetting mathematical and physical constants.

name	Greek character	name	Greek character
alpha	α	nu	ν
beta	β	xi	ξ
gamma	γ	omikron	o
delta	δ, Δ	pi	π
epsilon	ε	rho	ρ
zeta	ζ	sigma	σ
eta	η	tau	τ
theta	θ	ypsilon	υ
iota	ι	phi	ϕ, Φ
kappa	κ	chi	χ
lambda	λ	psi	ψ
mu	μ	omega	ω

B.2 Mathematical symbols

Sets

number sets including zero	symbol
natural numbers (unsigned integers)	$\mathbb{N}$
integer numbers (integers)	$\mathbb{Z}$
rational numbers or fractions	$\mathbb{Q}$
real numbers (floating points)	$\mathbb{R}$
complex numbers	$\mathbb{C}$
hypercomplex numbers or quaternions	$\mathbb{H}$
Mandelbrot set	$\mathbb{M}$
Julia sets	$\mathbb{J}_c$

We chain these number sets as

$$\mathbb{N} \subset \mathbb{Z} \subset \mathbb{Q} \subset \mathbb{R} \subset \mathbb{C} \subset \mathbb{H}.$$

We chain the fractal complex subsets as

$$\mathbb{M} \subset \mathbb{C},$$

$$\mathbb{J}_c \subset \mathbb{C}.$$

Mathematical symbols

name	symbol
empty set	$\{\}$
subtract-sign for sets	$\backslash$
in-sign for elements of sets	$\in$
cardinality-sign for sets (number of elements)	#
equals-sign	$=$
equivalent-sign	$\Leftrightarrow$
implies-sign	$\Rightarrow$
distance	d
fractal or hausdorff dimension	h
difference	Δ
degrees	$^\circ$
infinity (unbound large value)	∞
sum-sign	Σ
dotproduct	$\cdot$
crossproduct	$\times$
reciprocal	$^{-1}$
transpose	^{T}
conjugated	*
imaginary unities	i, j, k
cartesian coordinates	$(\ )_{cc}$
polar coordinates	$(\ )_{pc}$

Mathematical keywords

name	symbol
radians	rad
sine	sin
cosine	cos
tangent	tan
cotangent	cot
arcsine	arcsin
arccosine	arccos
arctangent	arctan
determinant	det
absolute value	abs

NUMBERS

symbol	value
circle's constant	$\pi \approx 3.1415$
radian	$1 \text{ rad} \approx 57.30°$
silver number	$\delta = 1 + \sqrt{2} \approx 2.4142$
golden number	$\Phi = \frac{1+\sqrt{5}}{2} \approx 1.6180$
paired golden number	$\Phi' = \frac{1-\sqrt{5}}{2} \approx -0.6180$
imaginary unity	i
natural base (Euler's constant)	$e \approx 2.7182$
gravitational acceleration	$g \approx 9.8066$

Annex C · Companion website

Our companion weblink `http://www.multimediamaths.be` to this book, offers you interactivity, downloads and the answers to the exercises of each chapter. We do welcome your feedback, corrections and suggestions.

C.1 Interactivities

Especially to distant learning students, we recommend our online selection of applets to acquire a maximal mathematical insight through 'learning by doing' on the companion site. Users of the scientific software tool Mathematica® can even download and experience these supportive applets locally.

C.2 Solutions

We provide the solutions to all exercises, as presented in the final paragraph 'Exercises' of each chapter, via the companion site. To readers who lack a scientific software tool, we kindly recommend the weblink or the mobile app *Wolfram|Alpha* [24].

Bibliography

BOOKS

1) L. Ammeraal (1998). *Computer Graphics for Java Programmers*, John Wiley & Sons Ltd.
2) M. de Gee (2001). *Wiskunde in werking deel 1. Vectoren en matrices toegepast*, Epsilon Uitgaven.
3) P.A. Egerton, W.S. Hall (1999). *Computer Graphics, mathematical First Steps*, Pearson Education.
4) A.J. Hanson (2006). *Visualizing Quaternions. Series in interactive 3D technology*, Morgan Kaufmann.
5) M. Kamminga-van Hulsen, P.E.J.M. Gondrie, G.A.T.M. van Alst (1994). *Toegepaste wiskunde met computeralgebra*, Academic Service.
6) W. Kleijne, T. Konings (2000). *De Gulden Snede*, Epsilon Uitgaven.
7) P. Lothar (1993). *Wiskunde voor het hoger technisch onderwijs deel 1*, Academic Service.
8) P. Lothar (1993). *Wiskunde voor het hoger technisch onderwijs deel 2*, Academic Service.
9) W. Stahler, D. Clingman, K. Kaveh (2004). *Beginning Math and Physics for Game Programmers*, Pearson Education.
10) J. van de Craats, R. Bosch (2005). *Basisboek wiskunde*, Pearson Education.
11) I. De Pauw, B. Masselis (2012). *Wiskunde voor IT*, LannooCampus.
12) B. Langerock (2011). *Wiskunde voor ontwerpers*, LannooCampus.
13) T. Crilly (2008). *Vijftig inzichten wiskunde*, Veen Magazines.
14) J. Gielis (2001). *De uitvinding van de cirkel*, Geniaal.
15) D. Marsh (2005). *Applied Geometry for Computer Graphics and CAD*, Springer.
16) G. Farin (2002). *Curves and Surfaces for CAGD*, Morgan Kauffman.
17) J. Kaahwa, M. Quinn (2000). *School Mathematics of East Africa*, Cambridge University Press.

18) M. F. Barnsley (1993). *Fractals Everywhere*, Academic Press.

19) J. M. Van Verth, L. M. Bishop (2008). *Essential Mathematics for Games and Interactive Applications*, Morgan Kauffman.

20) E. Guyon, H. E. Stanley (1991). *Fractal Forms*, Elsevier.

WEBSITES

21) Wolfram: Corporate website (`http://www.wolfram.com/`) (access June 2013).

22) Wolfram MathWorld: The Web's Most Extensive Mathematics Resource (`http://mathworld.wolfram.com/`) (access June 2013).

23) Wolfram Demonstrations Project: Powered by CDF Technology (`http://demonstrations.wolfram.com`) (access June 2013).

24) Wolfram|Alpha: computational knowledge engine (`http://www.wolframalpha.com/`) (access June 2013).

25) D. Klingens: Gulden snede en Fibonacci (`http://www.pandd.demon.nl/sectioaurea.htm`) (access April 2009).

26) University of Surrey, Faculty of Engineering and Physical Sciences: (`http://www.mcs.surrey.ac.uk`) (access April 2009).

27) J. Wassenaar: Two dimensional curves (`http://www.2dcurves.com`) (access April 2009).

28) NVvW – Nederlandse Vereniging van Wiskundeleraren: (`http://www.nvvw.nl`) (access April 2009).

29) Dimensions: Chapters 5 and 6 (`http://www.dimensions-math.org/Dim_CH5_NL.htm`) (access February 2009).

30) Startpagina | regionaal steunpunt Eindhoven: (`http://www.win.tue.nl/wiskunded/files/public/Complexe_getallen/complex-2007-06.1s.pdf`) (access February 2009).

31) Graphics: Programma (`http://staff.science.uva.nl/~jose/graphics/matrix.pdf`) (access April 2009).

32) www.bik5.com: (`www.bik5.com/rasterizing.doc`) (access April 2009).

33) Point in triangle test: (`http://www.blackpawn.com/texts/pointinpoly/default.html`) (access April 2009).

34) senocular.com Tutorial: Understanding the Transformation Matrix in Flash 8 (http://www.senocular.com/flash/tutorials/transformmatrix/) (access April 2009).

35) Playground site: Gielis' superformula (http://www.procato.com/superformula/) (access July 2011).

36) Genicap: Corporate website (http://www.genicap.com/) (access July 2011).

37) Online Math Learning: complete from Kindergarten to University (http://www.onlinemathlearning.com/) (access June 2013).

38) Mathwords: Terms and Formulas from Beginning Algebra to Calculus (http://www.mathwords.com/) (access June 2013).

39) Wikipedia: free online Encyclopedia (http://en.wikipedia.org/) (access June 2013).

40) CLEAR at Rice: Introduction to Computer Graphics (http://www.clear.rice.edu/comp360/lectures/) (access June 2013).

41) Ultra Fractal: software application (http://www.ultrafractal.com/) (access June 2013).

42) Fractal Foundation: Science, Maths and Arts (http://fractalfoundation.org/) (access June 2013).

43) Illuminations: Online Teaching Resources (http://illuminations.nctm.org/) (access June 2013).

44) Kairos: Academic Repository (http://kairos.laetusinpraesens.org/) (access June 2013).

Index

2-point Bezier segment, 272
3-point Bezier segment, 273
3D rotation, 202, 242
4-point Bezier segment, 274

abscissa, 96
absolute value, 97
acute, 44
algebraic representation, 225
algebraic structure, 116, 172
aliasing, 280
altitude, 44
amplitude, 70
anchor point, 135, 142
angle, 42
angle bisector, 44
angular frequency, 70
anticommutative, 127
antidiagonal, 169
antiparallel, 112, 128
arccosine, 57
arcsine, 57
arctangent, 57
argument, 226
array, 109, 113, 115, 168
attractor, 261

B-spline, 286
base, 16, 44
base vector, 112
basic transformation, 198
Bezier curve, 278
Binet's formula, 188
binomial, 23
block matrix, 181

Cantor set, 253
cartesian, 96
cartesian equation, 137
center, 149
characteristic coefficient matrix, 283, 285, 287
circle, 149
coefficient, 24
coefficient matrix, 185
cofactor, 170
collinear, 156, 274, 275
column, 174, 176
column matrix, 168
column vector, 168
commutative property, 113
complementary angles, 54
complex conjugate, 227
complex number, 225
components, 111, 117, 124
composed operator matrix, 210
composed transformation, 208
conjugate quaternion, 241
connected, 265
contradiction, 19
control point, 273, 274, 286
convergence, 90
convergence rate, 263
coplanar, 275
corkscrew-rule, 110, 126
cosine, 48, 50
cotangent, 48
countable, 235
crosier curve, 102
cross product, 126, 170, 240
cubic Bezier segment, 274

curvature continuity, 279
curve, 67

de Boor construction, 289
de Casteljau construction, 277
decimal mantissa, 293
decimal order of magnitude, 293
degree, 19
degree of freedom, 33
degrees, 43
determinant, 169
dimension, 120
direction, 111
direction angles, 124
direction cosine, 125
direction vector, 135, 141, 142
discriminant, 26
distributive law, 16
division, 15
domain, 62
dot product, 120, 176
dynamical system, 261

element, 14, 175
elementary row operations, 180, 181
ellipse, 98
empty set, 14
equality, 19
equation, 18
equilateral triangle, 44
equivalent systems, 32
euclidean dimension, 258
expanding, 22
exponent, 16

factoring, 22
false italic, 205
Fibonacci numbers, 89
Fibonacci operator, 187
Fibonacci sequence, 88, 187
first order continuity, 279
foot, 44
fps, 148
fractal, 250, 278
fractal dimension, 258, 259
fraction, 16
frame rate, 148
free vector, 112, 117
full angle, 43
function, 62

Gauss plane, 226
gaussian elimination, 181
golden number, 80, 82
golden rectangle, 83
golden spiral, 85
golden triangle, 82
gravitational acceleration, 135
group, 116, 172

hausdorff dimension, 258
head, 112
head-to-tail rule, 113
homogeneous components, 194
homogeneous coordinates, 194
horizontal parabola, 67
hypercomplex number, 238
hypotenuse, 44

idempotent, 190
identity matrix, 169, 178, 180, 181
identity quaternion, 242
imaginary part, 225
implicit, 149
inconsistent, 34
indeterminate, 294
indeterminate part, 19
inequality, 18
INFINITY, 294
infinity, 180, 294
intercept, 64, 72
intersecting, 139

inverse matrix, 179
invertible, 180
irrational number, 82
isosceles triangle, 44
iteration, 251

Julia sets, 264

Koch curve, 251
Koch snowflake, 251

Law of Cosines, 51
Law of Sines, 51
left-handed frame, 110
lemniscate, 103
linear, 192
linear Bezier segment, 272
linear combination, 119
linear equation, 21
linear function, 63
linear system, 32
Linear transformations, 192
linearly dependent, 141
linearly independent, 120, 141
location vector, 112
logarithm, 259
Lucas numbers, 88
Lucas sequences, 88

machine numbers, 294
machine precision, 294
magnitude, 111
main diagonal, 169
major axis, 86
Mandelbrot set, 262
matrix, 168
matrix column, 168
matrix difference, 171
matrix element, 168
matrix equality, 169
matrix form, 186
matrix power, 178
matrix product, 176
matrix row, 168
matrix sum, 171
median, 44
Menger sponge, 268
Minkowski island, 252, 253
Minkowski sausage, 252
minor, 170
minor axis, 86
modulus, 226, 262, 264
monomial, 19
monomials of the same kind, 19

NAN, 294
nilpotent, 190
norm, 111, 112, 238
normal map, 129
normal vector, 129
normalize, 129
NULL, 294
null matrix, 169
null vector, 109
number line, 224
number plane, 226
numerically stable, 278, 289

obtuse, 44
opposite, 15, 228
opposite matrix, 173
opposite vector, 115
oppositely signed angles, 54
ordinate, 96
orientation, 111
orthogonality, 123
outline fonts, 281

pair, 14
parabola opening down, 66
parabola opening up, 66
parallel, 112, 139
parallellogram method, 113

parameter, 33
parametric curve, 96
parametric equation, 134, 137
parametric form, 96
Pascal's Triangle, 24
Pell sequence, 94
perfect product, 23
perfect square, 23
performant equivalent, 284, 285
perpendicular bisector, 44
phase, 70
polar angle, 99
polar axis, 99
polar curve, 102
polar equation, 102
pole, 99
polygon, 129, 156
polynomial, 19
polynomial function, 63
position vector, 135, 142
positional continuity, 278
power, 16
product matrix, 175, 176
pulsation, 70
pure product expression, 18
pure quaternion, 238
pure sum expression, 18
Pythagoras, 45, 50, 253
Pythagoras tree, 253

quadratic Bezier segment, 273
quadratic equation, 25
quadratic function, 65
quaternion, 237
quaternion addition, 239
quaternion conjugate, 241
quaternion inverse, 242
quaternion product, 240
quaternion rotation, 242

radial coordinate, 99
radians, 43
radius, 149
range, 62
real overflow, 294
real part, 225
real underflow, 294
reciprocal, 15
recursion, 118, 251
reflection, 204, 214, 215
reflex, 44
right, 44
right triangle, 44
right-hand rule, 110, 126
right-handed frame, 110
root, 26, 62
rose, 102
rotation, 200, 210, 218
rotation quaternion, 242
round, 294
row, 174, 176
row matrix, 168
row operation, 180
row vector, 168

scalar, 108
scalar multiplication, 114, 173
scalene triangle, 44
scaling, 197, 213
scarabaeus curve, 106
scientific notation, 293
second order continuity, 279
self-similarity, 254, 263, 265
semi-axes, 98, 103
semi-major axis, 86
semi-minor axis, 86
sense, 111
setminus, 14
shaping parameters, 103
shear strain, 206, 207
shearing, 205
side bisector, 44

Sierpinski gasket, 251
sign, 293
significant digits, 293
silver number, 94
silver rectangle, 94
silver section, 94
similarity, 45, 254
sine, 48, 50, 69
singleton, 14
singular, 180
skew, 139
slope, 63
smooth, 278, 279
solution set, 33
spanning set, 120
sphere, 150
spiral of Theodore of Cyrene, 105
square matrix, 169
square system, 33
straight, 44
subtraction, 15
superformula, 103
symmetric matrix, 174

tail, 112
tangent, 48
tangential continuity, 279
Three-Step Rule, 43
trajectory, 117, 261
translation, 192, 193
transpose of a matrix, 174
tree form, 18
triangular system, 33
trigonometrical representation, 227
trinomial, 22

uncountable, 236, 264
underdetermined, 33
unit circle, 49, 97
unit normal vector, 129
unit quaternion, 238
unit vector, 112, 158
unity interval, 253, 272

vector, 108, 111
vector components, 109
vector equation, 135, 141
vector fonts, 281
vertex of a parabola, 66

zero, 44, 294
zero divisor, 177, 178
zero matrix, 169, 172, 178
zero vector, 109, 116
zeroth order continuity, 278